Health and Disease in the Prehistoric Pacific Islands

Hallie R. Buckley

BAR International Series 2792

2016

First Published in 2016 by
British Archaeological Reports Ltd
United Kingdom

BAR International Series 2792

Health and Disease in the Prehistoric Pacific Islands

ISBN: 978 1 4073 1443 3

Cover Image:

Young children from Tanna Island, Vanuatu (photo by author).

Printed in England

All BAR titles are available from:

British Archaeological Reports Ltd
Oxford
United Kingdom
Phone +44 (0)1865 310431
Fax +44 (0)1865 316916
Email: info@barpublishing.com
www.barpublishing.com

Acknowledgements as of 2001

Many people have provided support and guidance during this undertaking.

To Dr Nancy Tayles who very generously took me under her wing in Thailand for my first experience of excavating human remains. After that, I was hooked. Nancy has also been extremely patient and supportive over the last four years, I could not have done this without her. To my great friend Dr Kate Domett who has listened to my prattling on about diseased bones and shared my love of good wine and food. Many other people in the Department have helped me along the way, even if to just stop and listen to my ideas. Always enthusiastic and supportive, fellow PhD student and friend, Trish Mahoney and Drs Susan Mercer and George Dias. Also, Mark and Andrew, the coffee was in violation of the health code I'm sure, but the conversation was always fun!

Most important is my HOD, Professor Gareth Jones, for his continued support of the 'bone people' in this department and despite my forgetting a teaching evaluation appointment, has still given me a job! This department has provided a nurturing environment for me during this research for which I will always be grateful. Also to Sally-Ann, Katherine and Kerry; thanks for all the answers to my silly questions. I have been very fortunate to have had a wonderful professional team of technicians to help me along the way; Thanks Mary Nolan for the photographic images, to Robbie McPhee for dealing with incomprehensible computer programmes, and to Andrew and Ross the coolest computer geeks in the world.

A special acknowledgement must go to Foss Leach and Janet Davidson who collectively excavated the skeletal material used in this study. They have also been very helpful with my many enquires regarding the archaeological context of the material.

My family and friends have provided the emotional support and reassurance for me to carry on with my PhD. To my mum and my brother Karl, thanks for your boundless enthusiasm for what I have been doing with the last few years of my life, even if the level of understanding has diminished with time. Also to my father, John Buckley, who has recently returned to my life and proved a valuable proof reader and friend.

And to David who when we met unwittingly took on a PhD student as well, with all that entails. Amazingly he has stuck around to keep me well looked after with wonderful food, consolation, and a happy life. While, by his own admission, David has no idea what I am talking about, he has remained unfailingly interested, supportive and loyal. You are my rock! Finally, to Oliver and Courtney, yes I have finished my PhT!

NB: A huge thank you to Dr Chris Stantis for formatting the original thesis into monograph format for publication. A great deal of work was also required to update the figures, many of which were not in digital form. Most of this work was carried out by Chris with great patience for my lack of technical savvy. Thanks Dr Stantis!

Contents

List of Figures

List of Tables

Preface

This volume is a historical document that presents my doctoral thesis on health and disease in the Pacific Islands, completed in 2001. The study was conducted using a sample from the Solomon islands in Melanesia and another sample from two burial mounds in Tonga, Polynesia. When I conducted my doctoral studies palaeopathology in this region was in its infancy and most of the population-based studies had been carried out in Micronesia only. Importantly, the study of infant and child health in the past has received a great deal of attention in the last 10 years or so (e.g. Halcrow et al., 2016, 2008; Lewis, 2007; Lewis and Gowland, 2007; Lewis, 2010). However, at the time this work was completed there was very little literature on the subject available to aid in the interpretation of the pathology observed.

Some parts of this volume have been updated while others remain essentially the same as when the thesis was first completed. I have updated the first chapter to include some of the more recent literature on palaeopathology and its role in assessing overall population health in the past. The sections on Pacific prehistory have also been updated. In chapter two the sections on the materials used in this study now include the projects my students have conducted on the samples and the publications based on these samples completed to date. The other chapters remain as they were in 2001. The data presented will hopefully serve future scholars interested in the role of disease in prehistory when compiling regional syntheses.

Certain aspects of infectious disease and non-specific stress indicators from these samples have been published already (as cited in chapter two), however there is great deal of individual and elements specific skeletal pathology data in this volume which have not been published previously.

Also at the time of this study there was also very little known about the Holocene colonisers of Near and Remote Oceania known as Lapita. As such there is only passing reference to these elusive groups in the volume. Over the last ten years there have been a series of fortuitous discoveries of Lapita cemeteries in Vanuatu (Bedford et al., 2009, 2011) and pre-Latte skeletons in Micronesia (DeFant, 2008; Nelson et al., 2016) that have greatly improved our understanding of the biological costs of colonisation of virgin island landscapes. My research group has been fortunate enough to have the opportunity to analyse the diet and disease of the Lapita people (Buckley et al., 2008; Bedford et al., 2009, 2011; Foster et al., 2013) which has begun to build population level profiles using multi-disciplinary techniques including isotope studies of diet and migration (Bentley et al., 2007; Kinaston et al., 2009, 2014a, 2014b, 2015, 2016; Shaw et al., 2009, 2011). The consequences of dietary insufficiency and genetically regulated metabolic diseases have also been reported in these samples (Buckley, 2007, 2011; Buckley et al., 2014).

The Pacific Islands are often forgotten when considering global issues of health and subsistence change. However, this region has much to offer with regards to understanding human adaptation to different environments during and after colonisation and the biosocial responses to disease. One of the main drivers for publishing this volume after all this time is an attempt to give this region more of a voice in global discussions of health and disease in prehistory.

Abstract

The main impetus of this research was to assess the role of infectious disease, particularly malaria, on the health of human populations in the prehistoric Pacific Islands. A significant difference in the disease environments of the Pacific Islands is the presence of malaria in most of Melanesia and its absence in Polynesia. This research is the first attempt to assess the impact of malaria on the health of prehistoric Pacific Islanders.

The materials used were two samples of human skeletal remains from different regions of the Pacific Islands; one from Taumako, an island in the malarial zone of the Solomon Islands, Melanesia and one from Tongatapu, the Kingdom of Tonga, Polynesia where malaria has always been absent.

The objectives of this research were to record several parameters of health and disease in these samples. Firstly, the demographic profile of each population was compiled to test whether the mortality rates of the Taumako people may have been affected by the presence of malaria. Secondly, non-specific indicators of growth disruption in dental material were recorded to assess whether levels of childhood stress were greater at Taumako. The stature of adults was also compared between Taumako and Tonga as a measure of the individual ability to achieve a genetic potential for growth. Finally, the skeletal indicators of iron-deficiency anaemia and infectious disease were recorded to test whether the prevalence of disease differed and whether these differences may be attributed to malaria.

The results of the analyses of these parameters of health and disease showed significantly higher prevalences of prenatal and childhood growth disturbance at Taumako compared to Tonga. The stature of adults was similar between the two regions but the range of heights was greater at Taumako. Similarly, a higher prevalence and more severe expression of iron-deficiency anaemia and significantly more proliferative skeletal lesions were found at Taumako. However, a significantly greater number of subadults were affected with postcranial proliferative lesions at Tonga than Taumako, although, the affected children were older at Tonga. The mortality rates of the two populations were similar, although males were more vulnerable to early death than females at Taumako.

A differential diagnosis proposed that endemic yaws was the most likely infectious disease causing many of the skeletal lesions at Taumako; while at Tonga a more non-specific pattern of disease was proposed. The possibility of multiple causes for the skeletal lesions was also proposed for some individuals at both sites.

The discussion of the results found they were consistent with the premise that the presence of malaria in Melanesia may have caused chronic growth disturbance and exacerbated the expression of anaemia and infectious disease in prehistory. However, it is also argued that differences in diet may have had an equally strong role in the observed patterns of health and disease.

In conclusion, the results of this research did not unequivocally demonstrate the role of malaria on the health of prehistoric populations. However, this study is an initial step in the investigation of the impact of malaria on human populations, while not excluding other factors such as diet.

Chapter 1

Introduction

> *We need to step back from our modern, first world perspective of relatively healthy populations,...to consider the role of parasitic and infectious diseases in prehistory. Archaeologists have paid too little attention to these matters, partly because the evidence for disease among prehistoric populations can be difficult to obtain. (Kirch, 2000a, 56)*

Through the analysis of evidence of disease in human skeletal remains, the role of infectious disease in the prehistory of Europe and the Americas, with particular reference to the adoption of agriculture, has been a matter for intensive research for decades (Cohen and Armelagos, 1984; Cohen and Crane-Kramer, 2007). However, as Kirch (2000a) suggests above the role of infectious disease in the prehistory of the Pacific Islands has received relatively less attention.

By analysing markers for disease in human skeletal remains from two islands with different disease environments, this thesis aims to begin to address the role of disease in adaptation to the environment influencing the success of human settlement in the prehistoric Pacific Islands. Central to this aim is an assessment of health in a skeletal sample from on one island with endemic malaria and of another sample from the eastern Pacific where malaria is absent. Further, it is intended to test whether the presence of malaria may have exacerbated the effects of other infectious diseases.

To achieve these aims the bone material from two skeletal samples excavated from the Pacific Islands is assessed for evidence of infectious disease and anaemia in bones and non-specific markers of growth disruption in dental material. One sample is from an archaeological site on malarious Taumako island, a Polynesian outlier in the Duff Group, south-east Solomon Islands and the other sample is from two burial mounds from Tonga, in Polynesia, where malaria has always been absent east of Buxton's Line (Figure 1.1). Within a broader context, the findings of this thesis may also shed some light on the role of malaria and other infectious disease on the success of human settlement in the Pacific Island region.

The disciplines covered in this thesis are multi-faceted and information is drawn from clinical, anthropological, palaeopathological and historical literature. The literature reviewed in this introductory chapter provides the background against which the research aims of this study are presented, while the following chapters review the literature relevant to the subject covered in each. The theoretical basis of the study of disease patterns in prehistoric skeletal material is also reviewed in this chapter.

1.1 Infectious Disease and the Prehistoric Settlement of the Pacific Islands

Anthropologists have been interested in changing patterns of health, and the impact of infectious disease during periods of technological and subsistence transition for decades. In Europe a considerable amount of research has been devoted to changes in health and disease with the adoption of a sedentary life-style associated with the intensification of agriculture, where this subsistence transition led to a deterioration in health (Cohen and Armelagos, 1984; Steckel et al., 2002; Cohen and Crane-Kramer, 2007).

Central to this research is the investigation of the role of infectious disease in the success of human adaptation to a changing environment. For example, research has shown that populations dependent on a foraging lifestyle are relatively free of crowd dependent infectious diseases, while sedentism and population growth associated with the adoption of agriculture exacerbated the effects of some zoonotic diseases (Cockburn, 1963; Armelagos et al., 2005; Harper and Armelagos, 2013).

The following discussion provides a brief outline of the current theories of human settlement in the Pacific Islands and the possible role of infectious disease on the success of this initial settlement. It should be noted that I am aware of Roger Green's disestablishment of the 19th century geographic boundaries of the Pacific Island groups of Micronesia, Melanesia and Polynesia. Green's (1991) concept of Near and Remote Oceania is reviewed in relation to island ecology and subsistence patterns in chapter two. Near Oceania includes New Guinea the Bismarck Archipelago and the western Solomon Islands as far east as San Cristobal and Santa Ana (Figure 1.1). Near Oceania is the region of greater biodiversity and longer human occupation compared to Remote Oceania (Green, 1991). The terms Melanesia and Polynesia are used consistently throughout this thesis. These terms are used only as geographic zones and

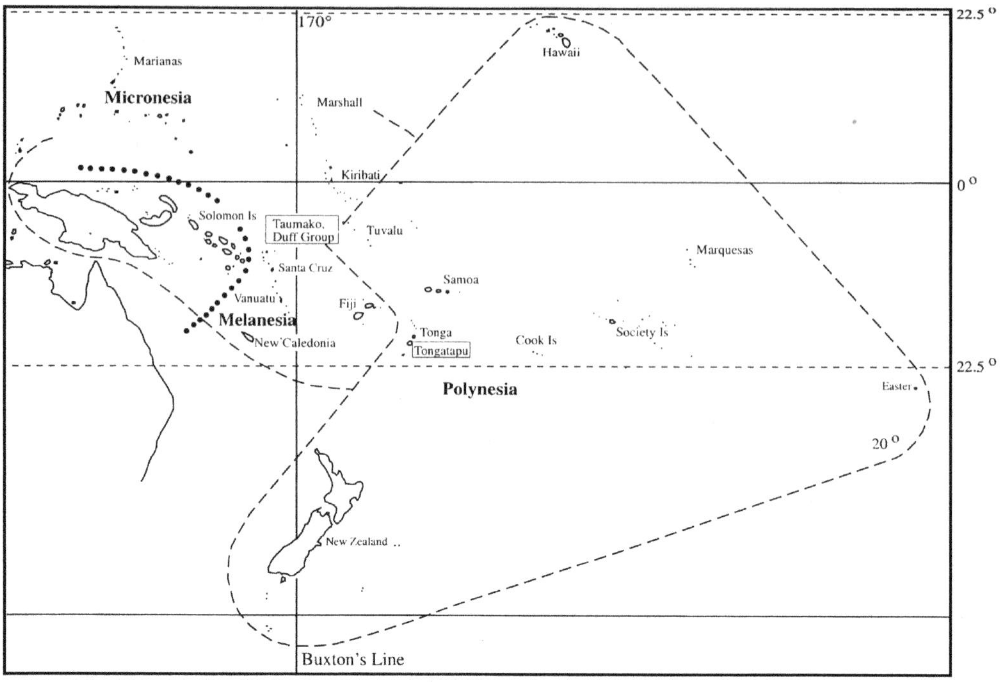

Figure 1.1. *A map of the Pacific Islands with the islands of Taumako and Tongatapu outlined. The thick dotted line represents the divisions of Near Oceania to the west and Remote Oceania to the east.*

are not meant to over-simplify the cultural, linguistic, and biological variation of the human populations of Melanesia.

1.1.1 Settlement of Near Oceania

Archaeological evidence suggests that the settlement of the western Pacific Islands began during the last Ice Age, as early as 40,000 years ago. At this time Australia and New Guinea (NG) were joined together as the large continent, Sahul. The early inhabitants of the region are thought to have migrated from the continent of Sunda, now modern South East Asia, by way of land bridges and short voyages on simple rafts or sailing vessels. The low sea levels during the Pleistocene enabled people to quickly colonise the islands of the western Pacific to the eastern end of the southern Solomon Islands chain. After this point was reached, all further movement into Remote Oceania ceased until around 3500 years before present (BP) (Kirch, 1997; Spriggs, 1997; Bedford et al., 2007; Clark et al., 2008).

The island groups of Vanuatu and New Caledonia were not colonised until around 3500 years ago (Summerhayes, 2001b). There are two theories of why colonisation in this area may have been delayed. The first is because ocean voyages were too long to suit the primitive class of sailing vessels and the second is because of subsistence restrictions (Green, 1991; Spriggs, 1997). In 1991, Green presented a review of the pattern of biogeographical impoverishment eastward from the Southeast Asian continent. He demonstrated that the declining number of taxa available to the early settlers of these islands would have significantly restricted the success of initial settlement (Green, 1991).

The archaeological evidence for subsistence and settlement in the Pacific prior to 20,000 BP is scarce but does suggest that little technological development occurred. Human populations remained small and subsistence was based on foraging from the wild plant and animal resources of land and sea. Some evidence suggests that cultivation of plants was developed in NG prior to the more widespread adoption of horticulture in the Pacific region (Golson, 1977; Denham, 2011).

1.1.2 Malaria and settlement of the Pacific Islands

The adoption of agriculture in the Old World heralded a period of rapid population growth which partly influenced technological advances, such as metallurgy and increasing social complexity, leading to the development of state systems (Cohen and Armelagos, 1984; Flannery and Marcus, 2012). In Oceania a paradox occurred where even with the adoption of agriculture, human populations in the islands of Near Oceania remained small. As a consequence of slow population growth, technological advancement and social stratification did not occur in Near Oceania (Kirch, 2000a). Conversely, when people reached the islands of Polynesia, around 3000BP, it was not long before these populations flourished and eventually developed into highly stratified chiefdoms with large and dense populations (Kirch, 1984; Kirch and Green, 2001; Kirch and Kahn, 2007).

It has been suggested that the presence of chronic infectious diseases in Near Oceania and their relative absence in Polynesia lent the early settlers in the Polynesian islands a strong advantage over people living in Near Oceania (Kirch and Hunt, 1997; Kirch, 2000a, 2010). The following chapters of this thesis will show that Kirch's suggestion that

Polynesia was free of the parasitic and infectious diseases which afflicted the people in the islands to the west is somewhat simplistic. However, there was one disease which was absent in Polynesia but present in most of Melanesia, a disease which has caused more death and illness than any other disease in the history of humankind, malaria (Bruce-Chwatt and de Zulueta, 1980).

Groube (1993) was the first to suggest that malaria may have negatively affected the growth and technological advancement of human populations in Near Oceania. He argued that the two less malignant *Plasmodium* species (*P. vivax* and *P. malariae*) were introduced into the Pacific region with the first human migrants where the mosquito vector was already established (Groube, 1993). As will be discussed further in chapter two, two of the possible sequelae of infection with all malaria species are an increased risk of foetal and maternal death. Therefore, as demonstrated by Groube (1993), population size could be significantly affected by the presence of such a disease. Conversely, a population released from such a disease and with the subsistence technology to feed a growing population, could conceivably begin to expand quite rapidly and archaeological evidence suggests that this was the case during the early settlement of Polynesia (Kirch, 1984). Groube's (1993) hypothesis was later expanded by Kirch (2000) who noted that the distribution of malaria in the Pacific islands corresponded to the distribution of different human population structures. For example, in areas of endemic malaria the population density is low and scattered throughout the landscape; while in areas where malaria is absent, the population reaches densities of up to 100 persons per square kilometre.

The idea that malaria influenced the course of human history is not a new one, and is not restricted to the Pacific islands. It has been suggested that with the manipulation of the environment associated with the first adoption of agriculture in Europe, the transmission of infectious diseases became increasingly favourable (Harper and Armelagos, 2013) and was particularly so for malaria (Brothwell and Sandison, 1967; Bruce-Chwatt and de Zulueta, 1980). In support of this, land clearance and irrigation systems required for subsistence agriculture are known to facilitate the transmission of malaria as these environments provide ideal breeding sites for the mosquito vector (De Zulueta, 1994).

Medical historians have also found references in written accounts from Greece and Rome for the impact of malaria on the success or failure of military campaigns and on the health of populations:

> The general opinion of those who have studied the effects of malaria on Greek and Roman history has been... that its damaging effects increased with passing time and that in the late Roman Empire and Byzantine period it became a much more serious health problem than it had been in early classical times (Bruce-Chwatt and de Zulueta, 1980, 22).

This increase in the effects of malaria over time has been correlated with the increase of deforestation and the build-up of coastal alluviation. These environmental changes were very desirable to local mosquito populations and over time saw the death of many a flourishing port city in Asia Minor and Greece (Bruce-Chwatt and de Zulueta, 1980). Similarly, from an extensive analysis of human skeletal material from the eastern Mediterranean, Angel (1984) found an increase in bone changes indicating anaemia in children after the adoption of agriculture. He attributes this characteristic bony change to a genetic disease which provides some protection against falciparum malaria Angel (1984). More recently, the relationship of malaria, palaeoenvironmental change and human health have been investigated (Gowland and Western, 2012).

1.1.3 Settlement of Remote Oceania

By 3500 BP the island groups of Vanuatu and New Caledonia were settled by what is believed by some to represent a new wave of people from the west (Bedford and Spriggs, 2008; Bedford et al., 2010; Sand and Bedford, 2011). New settlements in the Bismarck Archipelago and Solomon Islands also appeared shortly after this time (Summerhayes, 2001a; Sheppard and Walter, 2006). The Lapita cultural complex is recognised archaeologically partly by the elaborately decorated pottery that they left behind throughout the colonization sphere (Kirch, 2000a). Also, these new people are believed to have brought with them a different cultural tradition, language (Austronesian), and domesticated plants and animals (Bedford and Sand, 2007; Kinaston et al., 2014). The new animals and plants introduced with these people probably assisted with the settlement of previously uninhabitable islands (Green and Kirch, 1997).

During this period of island settlement, archaeological evidence suggests that the presence of malaria may have influenced the types and form of houses and geographic situation of the settlements of the initial Lapita inhabitants. The earliest Lapita settlements in Near Oceania were primarily small clusters of houses supported on stilts or poles over reef flats or small islands off the main islands (Kirch, 1997). This settlement pattern has also been observed in modern times throughout the malarial islands of Melanesia where 'artificial islands' were constructed offshore, possibly as a means of avoiding the malaria vector (Parsonson, 1966). As Kirch (1997) notes, it may be that the stilt-houses of the early Lapita colonists were also a means of avoiding the malaria mosquito. In support of this, the stilt-house form was abandoned when the Lapita people reached the malaria free islands of Polynesia (Kirch, 2000a). It is quite likely the Lapita people were aware of the dangers of sickness after visiting certain ecological zones, which may account for their avoidance of these areas when settling in the Pacific (Sheppard, 2011).

Over a period of only 200–300 years the Lapita people colonised the islands of Near and Remote Oceania and eventually reached Fiji and Western Polynesia (Anderson and Clark, 1999; Kirch, 2000a; Burley et al., 2001; Clark et al., 2008). The western group of Polynesian islands consisting of Tonga and Samoa was colonised by 3000 BP (Anderson and Clark, 1999; Burley et al., 2001; Clark et al., 2008). The distinctive Polynesian culture developed here and eventually spread through to the far reaches of the eastern Pacific islands. By 1000 BP the remote islands of New Zealand and Easter Island had been discovered and settled by Polynesian explorers (Spriggs, 1984; Kirch, 2000a; Walter et al., 2010).

A possible driving force behind the rapid dispersal of Lapita throughout the region may have been a desire to seek out lands where chronic infections were less pervasive (Sheppard, 2011). As described above, the original inhabitants of Near Oceania were probably quite severely affected by the morbidity and death associated with malaria and the Lapita people seemed to avoid areas where disease transmission was high. Groube (1993) and Kirch (1997, 2000a) have suggested that the Lapita people brought the more severe *P. falciparum* malaria with them when they first colonised Near Oceania. If this is correct, the Lapita people would have been well aware of the consequences of living in an area with endemic malaria, which would provide powerful incentive to seek new lands. A study of a Gm haplotype, a genetic component of the human immune response to disease that confers a genetic resistance to malaria, may provide some support for the concept that *falciparum* malaria was introduced by the Austronesian-speaking Lapita peoples (Clark and Kelly, 1993). These authors argue that these new arrivals carried an inherited resistance to malaria, as this Gm haplotype is absent among the inland and Highland inhabitants of NG. They do not go so far as to propose that the Austronesian speakers introduced *falciparum* malaria into the region, however if this were the case they do acknowledge that the Gm haplotype would have provided the new arrivals with an adaptive advantage over the original coastal inhabitants (Clark and Kelly, 1993).

As discussed in chapter two, the clinical effects of malaria in adults and older children are not so severe where malaria is stable. Stable malaria is characterised by a finely balanced continuum of constant transmission which imparts a relative immunity to indigenous peoples (Hendrickse, 1987). Therefore, if the Lapita people introduced a new and more deadly *Plasmodium* species, coupled with the fact they were relatively immune to that particular species (Clark and Kelly, 1993), clinical evidence suggests the original inhabitants would have been decimated as a result. Witnessing such dramatic ill-health in the local inhabitants would surely be an added incentive to build houses away from the mainland and to eventually seek new lands free of this disease. Therefore, it may be postulated that the Lapita even left their original homeland to be free of malaria and hence it may be argued that malaria played some role in the manner in which the Pacific Islands were settled and also affected the success of that initial habitation.

A further period of Pacific settlement relevant to this thesis is somewhat removed from the initial Lapita colonisation. This is the settlement of the so-called 'Polynesian Outliers'. These are islands dispersed throughout the eastern half of Near Oceania whose inhabitants speak a Polynesian language (Davidson, 1974; Spriggs, 1997; Leach and Davidson, 2008). While these people speak a Polynesian language and may have a more Polynesian social structure, whether they are more biologically Polynesian is uncertain (Kirch, 2000a; Feinberg and Scaglion, 2012). The majority of these islands lie to the east of the main islands of the Solomon and Vanuatu chains. The theories of the origins of these Polynesian speakers in Near Oceania are varied (Davidson, 1974; Bayard, 1976; Green and Kirch, 1997; Spriggs, 1997; Kirch, 2000a; Feinberg and Scaglion, 2012). They may be the direct descendants of the original Lapita colonists who remained on these islands, or some islands may have been colonised due to drift voyaging from western Polynesia, such as Tonga and Samoa. Other islands such as Bellona and Rennell west of the Solomons may have been colonised by subsequent waves from other outlier islands (Spriggs, 1997). The degree to which the Polynesian inhabitants assimilated with the indigenous people is also varied; ranging from annihilation in the case of Bellona (Spriggs, 1997) to an assimilation of language and culture, such as on Tikopia (Kirch, 2000). Green and Kirch (1997) suggested that both linguistic and archaeological evidence strongly indicates that the Polynesian influence on these 'outlier' islands is the result of a backwash of people moving west from the islands of western Polynesia. The island of Taumako is defined as a Polynesian outlier (Spriggs, 1997). The biological relationship of the Taumako people to Fiji and western Polynesia is discussed in chapter two.

1.2 Research Aims and Objectives

As mentioned above, the research aims of this thesis are three-fold:

1. Firstly, this thesis aims to assess the demographic structure of the two skeletal samples to test whether there are any differences in population health.
2. As will be discussed further in chapter four, one of the outcomes of infectious disease during childhood is an adverse effect on skeletal and dental growth. If an individual suffers infectious disease and/or inadequate nutrition during growth, the genetic potential for adult stature may not be attained. Therefore, by extending the first aim, the second aim of this study is to test whether there are any differences in individual growth between Taumako and Tonga. This second aim assumes a similar genetic potential for growth in both populations.
3. The third aim of the thesis is to test whether there were any differences in the skeletal manifestations of infectious disease and iron-deficiency anaemia between Taumako and Tonga.

Information of health and disease which aids in achieving the research aims is derived from multiple sources such as literature from wide ranging related topics including the analysis of human skeletal remains. The information derived from the analysis of skeletal material is then integrated with evidence from related literature. The aims of this research can be achieved by carrying out a number of objectives, based on the disciplines of skeletal biology and palaeopathology:

1. Firstly, the demographic profile of the skeletal samples is assessed by establishing the age and sex structure of each sample in order to provide comparative mortality rates.
2. Secondly, the analysis of non-specific dental markers of growth disruption is assessed. The stature of adults is also assessed for any evidence of environmental factors which may have restricted them from achieving their genetic potential for growth.
3. A third objective is to assess the relative levels of disease prevalence by observing and recording all evidence of pathological bony changes, either from an infectious origin or nutrition related. The skeletal pathology data are then integrated with clinical and historical literature of certain diseases in the Pacific Islands to construct a differential diagnosis of the disease(s) responsible for the skeletal lesions. The cultural and environmental constraints on nutrition, which may have influenced the course of infectious disease will also be considered.

Finally, the results from the preceding chapters are synthesised to consider the relative measures of health and disease between Taumako and Tonga.

1.3 Theoretical Concepts in Skeletal Biology

In this thesis bone tissue is the primary level of analysis. The skeletal samples used to address the aims outlined above were chosen because of their geographic provenance. The fact these samples were easily accessible for research in New Zealand was also an important consideration for their inclusion in this study. The following section will provide an outline of the theoretical basis of studying disease patterns in prehistoric skeletal material. It is by necessity brief. Where relevant, the concepts introduced in this section are expanded in subsequent chapters. In anthropology there is some variation in how basic theoretical concepts are applied when interpreting disease in prehistory. Therefore, it is considered important to outline the theoretical framework followed in this thesis.

1.3.1 Disease

Disease is defined in the Oxford Medical Dictionary as: 'A disorder with a specific cause and recognisable signs and symptoms; any bodily abnormality or failure to function properly, except that resulting from physical injury'. A disease may be due to infection with a microbial agent such as bacterium, virus or parasite that attacks certain tissues and leads to recognisable clinical symptoms. Disease may also be due to an immunological disorder that results in destruction of tissues. A lack of nutrients, vitamin and minerals can also lead to disease such as malnutrition, scurvy or iron-deficiency anaemia (May, 1958; Brickley and Ives, 2008).

The impact of disease on the health of an individual can be influenced by a number of factors. The type of disease, the immune status of the individual at the time of infection, the cultural response to disease, and the geographic location of the individual are all factors that influence the course of the disease (Ortner, 1992). In other words:

> Disease cannot arise without the convergence at a certain point in time and space of two orders of factors: factors that take the form of an environmental stimulus (a virus in the throat, poison in the food or in the air, or an emotional stress), and second, factors that condition the response of the tissues. These stimuli, these challenges to adjustment, are not the same in every environment. They vary with the geographical or cultural location. The response is conditioned by the genetic make-up of the recipient (May, 1958, 3).

When considering infectious disease, the circumstances under which both infection and disease occur are important to consider, as they may not be the same. Infection is the result of exposure to a potentially pathogenic organism and usually involves a host response. Disease on the other hand represents one of the possible outcomes of infection. As indicated above the development of disease depends on many factors, some of which are intrinsic to the host. The host response can include death, development of characteristic features of the disease, atypical expression, sub clinical infection, and a carrier state which may exist without a host response (Goodman and Armelagos, 1988). Also, infection does not always result in disease and a full understanding of the disease process requires consideration of the pathogenesis of infection (Evans, 1998). The pathogenesis of specific infections is outlined in chapters two and six.

Many studies of disease in prehistory ultimately include a consideration of how well a population has *adapted* to its environment (e.g. Mensforth et al, 1978; Walker, 1986; Cohen and Crane-Kramer 2007; Waldron, 2009). The concepts of adaptation and stress are used in varying ways. However, the way in which the researcher views the meaning of adaptation can influence the interpretation of the observed patterns of disease. It is frequently stated that patterns of disease may indicate either a successful or an unsuccessful adaptation to stress.

It will be evident from the following discussions of disease, stress, and adaptation that disease and stress have been conceptually separated. Disease is considered as an immediate causal agent of bone pathology that is recognisable. Stress on the other hand, is considered as an agent for non-specific markers that may be left in the dental or osseous tissue and are caused by a response to insults of infectious or nutrition-related origin.

1.3.2 Stress

The most often quoted definition of stress in relation to skeletal pathology is Selye's concept of stress as 'any extrinsic variable or combination of variables which cause the organism to react' (cited in Buikstra and Cook, 1980, 444) . The reaction of the individual to stress is mediated by the physiological status and may be completely buffered, with no skeletal record. If a disease state lasts longer than a few days, whether infectious or nutrition related, the growth of the individual may be interrupted producing an observable response in the osseous tissue or dentition. These responses may be analysed within the epidemiological context of a skeletal sample (Buikstra and Mielke, 1985).

The concept of Selyean stress emphasises a non-specific response to adverse environmental stimuli. This response involves several stages; initial alarm, resistance and adaptation or exhaustion (Goodman et al., 1988). Intrinsic in this view is that adaptation is dependant on maintaining a 'steady state' (Goodman et al., 1988, 173). This implies that natural selection would favour organisms that were able to maintain this steady physiological state. If stress is prolonged then the organism is unable to adapt and ill-effects can develop (Goodman et al., 1988). The outcome most relevant to this thesis is immune suppression. The occurrence of perceived threats are in most cases unmeasurable in skeletal studies although the palaeopathology of 'care provision' and agency are becoming very popular in the discipline (e.g. Tilley and Oxenham, 2011).

Goodman and colleagues (1984) integrated the concepts of Selyean stress within a framework applicable for bioarchaeology and extended by others (e.g. Domett, 2001; Halcrow et al., 2008) . The model illustrates that if stress is severe or prolonged enough it can cause a physiological response that may affect the evolutionary success of the individual and/or population. This model includes consideration of cultural buffers that may alleviate stress but also recognises that some cultural practices may increase stress. If stress is not buffered, then a physiological response is required to maintain equilibrium and this response may be left as a 'marker' in osseous or dental tissue. However, if the physiological response is not adequate then death of the individual may result, which in turn affects adaptation of the population. As with disease, the impact of stress varies with the age, sex, genetic susceptibility and immune status of the individual (Goodman et al., 1988). For a comprehensive and insightful review of the operationalization of stress and its sometimes misguided interpretation of health in bioarchaeology see Temple and Goodman (2014). The methodological issues of measuring stress in skeletal material are reviewed in chapter four.

1.3.3 Adaptation

In evolutionary terms, 'adaptations are characters shaped by natural selection to play a particular role or fulfil a particular function' (Harrison, 1993b, 11). These adaptations may be biochemical, morphological or cultural. The ultimate successful adaptation is survival of the vagaries of natural selection to reproduce and ensure that the offspring survives long enough to do the same. The ultimate failed adaptation is to die, but more importantly in evolutionary terms, to die before reproduction. Adaptation has been used more loosely by some anthropologists, particularly with reference to growth retardation and non-specific stress indicators. For instance, the use of the concept in relation to porotic hyperostosis, as a skeletal manifestation of iron-deficiency anaemia was critically reviewed by Goodman (1994), and the debate on the aetiology of these lesions has continued (Schultz, 2001; Walker et al., 2009; Oxenham and Cavill, 2010).

In the same paper Goodman (1994) reviews work examining growth retardation and concluded that evidence of growth disruption should be viewed as a successful adaptation to stress, simply because the individual has survived long enough to leave a marker on the bone. However, these types of explanations for evidence of ill-health 'could be viewed as benign reinterpretations of what are commonly referred to as 'stress indicators', reflecting a world view in which physiology is seen as no more complex than house plumbing' (Goodman, 1994, 171) and do not account for the physiological costs to the organism in staying alive. Any indicator that suggests disrupted growth of the individual is viewed in that study as a maladaptation. Adaptation in its purest form is to change the system that endangers survival, not to adjust to it. As a means of natural selection, disease varies in its importance. Some diseases may cause a quick death while others will elicit a physiological response and may aid in survival of the organism. The latter outcome, coined an adaptation by some (e.g. Stuart-Macadam, 1992), is not an adaptation but an adjustment (Goodman, 1994).

The term adaptability (Harrison, 1993a) is perhaps more appropriate to use when anthropologists are referring to the plasticity of human biology and culture in response to disease. In biology, adaptability refers to the responses of the individual or population to stress enabling change within a specific environment. This can refer to both phenotypic and genotypic changes. Adaptability, therefore, involves biological responses that include biochemical, physiological, and behavioural components (Harrison, 1993b). The field of molecular epigenetics and the intergenerational phenotypic plasticity are adding new layers to our application of the concepts of adaptation, health and stress (see Gowland, 2015 for a review).

For the purposes of this thesis, the term adaptation is used only in relation to evolutionary success, interpreted here as survival of the individual. I suggest that *response* is a more appropriate term for considering the impact of stress on an individual who survived long enough to exhibit a tissue response to that stress. Furthermore, a response to stress or disease may be positive or negative, depending on the disease and the immunity of the host.

1.3.4 *Environment*

The selective agent that determines the form of adaptation or response is the environment. The term 'environment' encompasses many different aspects. The *physical environment* is that in which people live and exploit for survival. The physical environment will determine what sort of subsistence is possible and which ecological niches can be exploited. Therefore, the physical environment can affect the health of the population. The *cultural environment* can be any culturally determined strategy that may affect the health of a sub-population or the population as a whole, irrespective of the physical environment. A sub-population may be defined by gender, social class, and/or age.

The *disease environment* has the most direct influence on the health of a population. Some pathogens are inescapable and all-pervading, while others may be avoided. The disease environment is also highly correlated to the physical environment. For example, malaria was not present above a certain altitude (Sayers, 1943) and yaws requires high temperatures and humidity to survive (Pirie, 1972). In turn, the cultural environment can either exacerbate disease transmission or cultural behaviours can alleviate the effects of certain pathogens.

In order to measure the health of a prehistoric population, it is best to consider the complex and dynamic unit that is the human body. Biology and culture are not mutually exclusive and a better understanding of prehistoric health in any region of the world is more easily achieved by a marriage of the two. This so-called Biocultural Perspective is well established in bioarchaeology (Buikstra and Beck, 2006; Larsen, 2015) and underpins the ethos of this thesis. A biocultural perspective evaluates the influence of ecology, such as diet, and culture on the expression of disease and was developed last century by researchers such as Angel (1966).

1.4 Palaeopathology and Differential Diagnosis

Palaeopathology is the study of prehistoric disease using skeletal and mummified material. It is essentially the study of ancient human suffering. A basic principle of palaeopathology is that while some kinds of illnesses and injury can leave their mark on bone, many do not (Ortner, 1992). The study of ancient disease is usually confined to chronic conditions that cause bone to respond. Therefore, interpretations of disease patterns may tell us more about chronic illness rather than the cause of death. Also, not all infections with the capability of affecting bone will do so (Lovell, 2000). This means that many studies of prehistoric disease patterns may underestimate the impact of a particular disease. Because of the limited way in which bone can respond to infection it may never be possible to know which particular disease was responsible, as in most cases only the effect of disease is preserved, not the pathogen itself (Lovell, 2000). Advances in the last 15 years on the genetic identification of pathogens affecting bone in archaeological samples have been increasing the accuracy of diagnosis for some diseases, such as tuberculosis (Mays and Taylor, 2002) but not so successful for others (Bouwman and Brown, 2005).

Despite the problems inherent in palaeopathology, by studying disease in ancient bones we seek to gain a better understanding of the stresses prehistoric people faced in their everyday lives. Certain questions are asked, such as: What diseases were present in the particular environment? Which age and sex groups were more commonly affected? How well did people cope with environmental pressures of infectious disease and food shortages? These questions follow the basic premise of modern palaeopathology that seeks to understand the impact of disease from a biocultural perspective.

The discipline of palaeopathology also provides information of disease in its natural state. With the modern hyper-prescription of antibiotics, we sometimes forget that there were thousands of years of human history when an individual's immune system may have been the only buffer between pathogens, illness and death by infectious disease. Without the advantage of modern medicine, millions of people are still at the mercy of infectious disease. Therefore, by studying disease patterns in prehistory it may be possible to provide an account of the natural course of diseases and the physiological responses to them.

There are a number of sources of data for studying the patterns of disease from prehistoric skeletal remains. The primary source of data is bone and, depending on the conditions, any soft tissues that may be preserved. Secondary sources of data are documents, such as early clinical reports from the research area. A knowledge of the sociocultural background and physical environment serves as a tertiary source. Tertiary sources may include data on climate, population density and other information that will aid the researcher in diagnosing the cause of skeletal changes in a population (Lovell, 2000).

The practical application of differential diagnosis differs between modern clinical cases and prehistoric skeletal material. In modern clinical cases the doctor has the enviable advantage of the patient history obtainable from a living person who can explain their symptoms. It is then the clinician's task to consider all diseases which may cause these symptoms and with the benefit of a raft of diagnostic techniques reach a diagnosis and embark on appropriate treatment (Waldron, 1994, 2009). When faced with skeletal lesions from individuals long dead, the task of diagnosis is not so simple. One of the primary limitations is that what is observed is a punctuated space in time of a chronic disease process, in an individual that is dead (Ortner, 1992).

The diagnosis of a disease in skeletal material is based firstly on assumption. By necessity this assumption is that: 'ancient disease...is sufficiently similar to a modern disease that a pattern fit means the diseases are related' (Buikstra and Cook, 1980, 440). Given our inability to identify the actual cause of disease in skeletal material (the pathogen), a certain level of uncertainty must be accepted. Similarly, it must be accepted that disease pathogenicity may change over time, and be expressed differently from descriptions in modern clinical literature. If it is assumed that the ancient disease is similar to a modern disease the researcher is dependant on clinical description that usually highlights the extreme expression (Ragsdale, 1996). Therefore, it must be considered that any abnormalities observed in the skeletal material might be an earlier non-symptomatic manifestation of that disease, or that the expression may be different from modern accounts.

The initial step in diagnosis of lesions in dry bone is careful observation and description of what has actually occurred, not what is assumed to have occurred. The bone changes should be described as whether the pathology has caused abnormal size, abnormal shape; or whether bone is being destroyed or produced in abnormal proportions (Ortner, 1992). Because diseases can affect the skeleton differently, the pattern of skeletal involvement is also important to record. For example, which bones are affected? Is it the appendicular or axial skeleton it is primarily affecting? Are the joint surfaces or the shaft of the bone affected? These patterns can then be interpreted to help in differential diagnosis of the disease (Lovell, 2000).

Similarly, if a diagnosis is to be correct it is dependant on the identification of variables that are characteristic of certain diseases. For instance, the skeletal lesions of leprosy are known to be primarily osteolytic in character, while secondary treponemal lesions are characteristically proliferative (Powell, 1991). The methodological issues surrounding differential diagnosis are reviewed in chapter six with reference to particular infectious diseases.

After the initial steps of observation and description of differential diagnosis are made, the next step is to: 'develop a model for differential diagnosis, interpreting evidence from samples of known clinical history, including autopsy series, specimens from museum collections, and literature surveys' (Buikstra and Cook, 1980, 441). Survey of the literature relevant to a particular subject should be limited to work that relates to cases reported prior to effective treatment of the disease. Therefore, the differential diagnosis will be based on the closest approximation of the disease in its natural state.

This thesis seeks to present a differential diagnosis of skeletal pathology in the two samples and to present an account of the relative health between the two Pacific samples. Non-specific indicators of ill-health are also assessed in this work. A full review of these indicators is presented in chapters four and five.

1.5 Is There an 'Osteological Paradox'?

The discussion of palaeopathology above was primarily concerned with the question; 'what is it?', when faced with skeletal pathology in dry bone. However, secondary to this and one of the main concerns of palaeopathology has been to correlate skeletal evidence of disease by asking the question; 'what does it mean?' (Ortner, 1991). For instance, does a high prevalence of periostitis in a skeletal sample indicate a more unhealthy population than a sample with little or no evidence of periostitis? It would seem intuitive to answer this question in the affirmative. However, as outlined above, any skeletal evidence of disease is an indication of a chronic condition and is therefore in itself a reflection of an immune system strong enough to combat an acute condition (Wood et al., 1992). Some researchers have long recognised that a paradox might exist between evidence of skeletal pathology and health (Ortner, 1979, 1991).

However, in 1992 Wood et al. published a paper whose primary conclusion has been included as the main caveat in many studies of prehistoric disease. The premise of the Wood et al. paper was that 'better health makes for worse skeletons' (Wood, 1992, 356). This premise was based on the concepts of selective mortality, hidden heterogeneity of risks and variation in frailty of individuals. Firstly, *selective mortality* illustrates an inherent problem in any skeletal study; that the cases which comprise a skeletal sample are dead and will never be representative of all the individuals in a population who were at risk of dying at a given age, only those that did in fact die. This is a problem that has been recognised by paleodemographers for some time (Jackes, 1992), but bears some relevance to palaeopathology because it must be recognised that individuals with skeletal evidence of disease will not be representative of the living population.

Secondly, *hidden heterogeneity in risks* means that the population from which the skeletal sample derives probably comprised a mixture of individuals with unknown variation in their underlying *frailty* to disease and death. Frailty is defined by Wood et al. (1992) as an individual's relative risk of death. They illustrate these concepts using an imaginary population with three subdivisions with respect to frailty. All of the subgroups are exposed to a condition which increases the risk of death and may eventually leave a marker on the skeletal tissue of survivors. Group A does not experience the stress so none of the individuals develop skeletal lesions. Group B experiences moderate stress enough to cause a skeletal response, but few deaths. Group C suffers heavy stress which result in many deaths quickly after the onset of disease and usually before a skeletal response is initiated. They argue that judging by the skeletal lesions left behind, only two groups would be recognised; one unhealthy, with skeletal lesions and one healthy, without skeletal lesions. This is the osteological paradox; that those individuals who survive long enough to develop skeletal lesions who may be interpreted as unhealthy are in fact the individuals whose 'health' saved them from early death.

This premise also suggests that previous conclusions in changing patterns of health over time are not simple (e.g. Cohen and Armelagos, 1984). Wood et al. (1992) suggest that the skeletal evidence in support of a decreasing health with the transition to agriculture indicated by more skeletal pathology, may in fact be interpreted another way; that the higher frequency of lesions in agricultural groups is an indication of better health and not worse health. This concept of the osteological paradox has been widely accepted, if somewhat grudgingly, and contains a well constructed cautionary note for researchers to consider when assessing the meaning of skeletal pathology.

While accepting the need for care when interpreting the meaning of skeletal disease in prehistory, Goodman (1993) valiantly defends the worthiness of measuring ill-health in skeletal samples:

> I suggest that what they consider paradoxical only appears so, because they focus on single rather than multiple indicators of health and misinterpret the goals of paleoepidemiology, and that the models they construct do not reflect biological realities and credible cultural contexts (Goodman, 1993, 281).

Goodman (1993) also states that Wood et al. (1992) somewhat erroneously assume that the primary goal of paleoepidemiologists is to understand the cause of death. To pinpoint the cause of death from skeletal disease is recognised as extremely difficult for a number of reasons and may not be essential to understand when considering the effect of disease processes in past populations: 'paleoepidemiologists find disease worth studying even when it has no measurable effect on mortality' (Goodman, 1993, 282). Furthermore, Goodman (1993) suggests that Wood et al. (1992) provide only enough data to make the situation appear paradoxical. For instance, if the biological and cultural processes which affect disease transmission are taken into account these can provide cross-confirmation which support a conclusion that skeletal lesions in prehistory do in fact represent ill-health.

Goodman (1993, 286) concludes:

> Their contribution to the literature is less than they assume. Their models ignore cultural processes, contradict known biological processes, leave out key information, and depend on false assumptions and ultimately prove mathematically possible but biologically improbable. Although it is useful to point out possible paradoxes and counterintuitive explanations, it is important to separate the probable from the suspect. Scientific snobbery is a poor substitute for a well-grounded critique.

Other authors have responded to the Wood et al. (1992) treatise defending specific aspects raised by them. For example, whether shorter stature is an indication of poorer health or better health (Saunders and Hoppa, 1993). These issues will be dealt with, where relevant, in subsequent chapters of this thesis. In relation to modern improvements in health due to the eradication of some infectious disease but with higher incidence of non-communicable disease such as cancers, Wood et al. (1992, 352) suggest:

> ...the shift in causes of death has resulted in greater longevity but also a higher prevalence of cancers and more years of life spent disabled, indicating that the very concept of "health" may be more complex than is often assumed.

This is a reasonable suggestion, however, it seems in conflict with their basic premise that better health makes for worse skeletons. For instance, if evidence of remodelled skeletal lesions caused by yaws (*Tr. pertenue*) leaves an individual with gross changes to the face and limbs, with disability, but indicates survival of the active stage of the disease, can we state that the person was healthy purely because they survived the active stage of a non-fatal disease? If the osteological paradox is to be applied, it must take into account the variation in disease expression and actual costs in pain and suffering to the individual and the population. To be left disfigured by a disease that one has survived may be viewed as an indication of relative health compared to those who died, but years of suffering may not be considered the healthy option for the individual. Furthermore, disfigurement by disease may eventually influence population survival if the reproductive success of individuals is inhibited.

Another factor to consider is the variation in disease biology and risk of death. If the development of a skeletal response to disease is an indication of an adequate immune response, what of diseases capable of leaving characteristic markers but evading recognition by the immune system during the acute phase of the disease? For example, the spirochete responsible for treponemal disease in humans is capable of evading the immune response by the nature of its cellular structure (Mims, 1993; Cotran et al., 1994). This survival strategy of the spirochete therefore contributes to the chronicity of the disease, not just the successful immune response of the host. Similarly, Ortner (1991) uses treponematosis as an example of how the development of skeletal lesions may be an indication of survival of the acute phase of disease and therefore better health. However, perhaps this concept should only apply to diseases which markedly increase the risk of death in affected individuals. In venereal syphilis this would have some validity, but not so for yaws. Yaws is known to cause to severe osseous changes in some people but does not in itself cause death (Powell, 1991). Therefore, does evidence of skeletal lesions characteristic of yaws indicate an effective immune response to acute disease or simply the pathogenesis of the disease?

The comments above do not presume to negate the insightful work of researchers such as Ortner, only to demonstrate that where infectious disease is concerned, it may be sensible to consider the variation of disease biology before interpreting all evidence of disease expression as an indication of better health.

1.6 Limitations of the Study of Skeletal Samples

The limitations of the samples and the study of pathology in archaeological remains need to be addressed in terms of the research aims. Firstly, as discussed in chapter two, the methods of archaeological investigation can significantly affect the size and structure of the skeletal sample. In terms of assessing the epidemiological patterns of stress and disease in these samples it is recognised from the outset that archaeological skeletal material introduces many biases which in most cases cannot be overcome (Waldron, 2007). These biases are discussed, where relevant, in subsequent chapters. Similarly, the methods applied for age and sex estimation can also skew the interpretation of patterns of pathology in skeletal remains.

A further limitation is the methods by which disease prevalence is calculated in archaeological skeletal material. The most significant factor which can affect the calculation of disease prevalence is missing data (Waldron, 1994, 2009) which is influenced by the preservation of the material and also by the method of archaeological investigation. With regards to comparability of results between palaeopathological studies, there is little standardisation in the methods used for calculating disease prevalence. For example, some studies may use the individual as the denominator, while others may use the skeletal element. Because of these limitations, it was decided to calculate prevalence of stress and disease indicators at the individual *and* skeletal element levels for this study.

Probably the most significant limitation of this study is the fact that malaria leaves no identifiable marker on dental or skeletal material. Therefore, the role of malaria as a primary causal factor of the evidence of stress and disease in these samples must be interpreted by inference alone. This approach has been applied with some success using detailed environmental inferential data (Buckley, 2006; Gowland and Western, 2012). Also, based on modern clinical literature, the presence of malaria on Taumako can only be assumed and it is not known to what extent malaria affected the health of the prehistoric Taumako people. Assuming that malaria was present on Taumako it is not known whether the artificial island of *Tahua* protected these people from the full onslaught of the disease, as Parsonson (1966) suggests. With these limitations in mind, it is assumed that malaria was a significant cause of ill-health on prehistoric Taumako, as it is in the greater Solomon Islands and most of Melanesia today.

1.7 Thesis Structure

Chapter two focuses on an overview of factors in the Pacific Islands which affected the health and survival of populations during prehistory. Where possible this is restricted to what is known of population health during the early periods of European contact and colonisation. The bone changes of disease included in the differential diagnosis of the skeletal lesions are discussed. A description of what is known of the archaeology of the sites is provided. This chapter has been updated to include studies on these skeletal samples and the archaeological sites that have been conducted and published since 2001.

Chapter three addresses the first aim of this study and provides a census of the two samples where age at death and sex are assessed for all individuals. This chapter aims to provide a firm base of the demographic structure of each sample which forms the basis of the analysis of age and sex related evidence of health and disease. The mortality rates of each sample are also considered in relation to health and disease.

Chapter four addresses the second aim of this thesis and examines the evidence of non-specific indicators of growth disturbance and adult stature.

Chapter five addresses the third aim of the study and reviews the literature concerning the impact of malaria on the health of Pacific Islanders and the genetic anomalies that have arisen as a selective response to malaria. This chapter also presents the results of the prevalence of the skeletal manifestations of iron-deficiency anaemia in each sample.

Chapter six also addresses the third aim of the thesis and investigates evidence of skeletal lesions in the two samples. This chapter aims to provide a picture of the epidemiology and impact of infectious disease on the populations.

Chapter seven synthesises and discusses the evidence from the analyses carried out in previous chapters in relation to the research aims of the thesis.

Chapter 2

The Pacific Context, Infectious Disease and the Skeletal Samples

This chapter includes a discussion of the physical environments of the Pacific Islands with respect to subsistence strategies. The disease environment of the Pacific is discussed with a review of early European observations concerning certain infectious diseases. The archaeological and environmental contexts and summary of work completed after this thesis was submitted of the two skeletal samples are also outlined.

2.1 The Island Environment

The Pacific Ocean covers one quarter of the globe's surface. The equator crosses this ocean and about half of the area is a tropical environment. Within this huge body of water, the land area is relatively small with islands which decrease in size from west to east. This decreasing island size is thought to have influenced the initial settlement of the Pacific world, with land-falls in small sailing craft becoming increasingly difficult as people ventured further east. The decrease in island size and biodiveristy may also have affected the success of initial colonisation depending on the food sources available (Green, 1991; Buckley et al., 2008).

The geology of the Pacific islands is divided by the 'Andesite Line'. The Andesite line runs to the east of New Zealand, Tonga, and Fiji, to the north of the Solomons, and Papua New Guinea, and to the north of most of the Micronesian islands. West of this line the islands are of 'continental' type, composed of volcanic sedimentary and metamorphic rocks with a mountainous terrain (Kirch, 2000a). The soils of these islands are generally very fertile with layers of volcanic ash and decaying limestone providing good gardening soil. Further east of the andesite line, the island geology is more simple consisting of basaltic rocks, corraline limestone or coral alone. These islands have less fertile soils for horticulture than the continental islands (Bellwood, 1978; Houghton, 1996).

The Pacific Islands are separated into groups based on their geology. These are the 'high' volcanic islands, 'low' coralline limestone islands and the ring like structures called 'atolls' (Cox and Banack, 1991). Atolls are the poorest of island environments, with little soil and are generally very small in size. As will be shown below, this difference in geology influences the type of subsistence strategy possible. The diversity of plant and animal species also diminishes from west to east. This greater diversity in the west is thought to be due to the early land connection of Australia and New Guinea (NG) with island Southeast Asia. This early land connection is reflected in the faunal and plant species of Australia and NG which are essentially of Indo-Malaysian origin and are almost as diverse and rich in species as the area of origin. This richness of diversity diminishes markedly further east and is associated with smaller land areas and greater distances between islands. As an example of this impoverishment in the east; Papua New Guinea has 520 species of birds, the Solomons 127, Fiji 54, Samoa 33, Tonga 20, the Society Islands 17, Marquesas, 11 and Henderson Island 4 (Green, 1991). This general impoverishment of species eastward would have had serious implications for the success of initial human settlement in some of the more remote island groups. Where it is relevant, the constraints of the island environment will be discussed in relation to population health and disease in subsequent chapters.

2.1.1 Pacific Island subsistence strategies

Knowledge of subsistence strategies in the regions under study will contribute to the interpretation of indicators of disease in the skeletal material. There is extreme variation in food economies throughout the region which makes any coherent synthesis of all areas in the Pacific prohibitive for this study. Therefore, the following outline will focus on the Solomon Islands and Tonga. The nutritional value of Pacific Island foods is not considered in any detail here but is covered in other relevant sections of the thesis.

The food economy of Pacific Islanders in general is primarily based on starchy root crops such as taro, yams, breadfruit and bananas. These are all perennial plants which are cultivated by various means. These cultivated foods are supplemented by the gathering of wild plants and animals, marine resources and domestic animals (Barrau, 1969). However, while there was no shortage of domestic animals, these are not consumed on a regular basis but are usually reserved for ritual feasts (Bellwood, 1978).

While the food economies of Oceania are broadly similar because they are based on root crop staples, there is a great deal of variation in the staples preferred among island groups. This variation is influenced by climate, soil type, and environmental differences such as altitude (Cox and Banack, 1991). In some cases the dominance of one staple over another may also be influenced by the personal preferences of the island inhabitants (Pollock, 1992).

Melanesia: Solomon Islands subsistence

The following discussion will concentrate mostly on information derived from nutrition surveys of non-westernised groups from the Solomon Islands. The Solomon Islands include a small group to the Southeast known as the eastern outer islands. This group includes Taumako Island in the Duff Group, and the Reef and Santa Cruz island groups (Willmott, 1969a).

Throughout the Solomon Islands there are several environmental zones which influence the type of subsistence economy employed. The first is the narrow strip of coastal land utilised for foraging the naturally occurring herbaceous plants. Horticulture was usually carried out in the foothills of the mountains, even if the settlement was situated on the coast. If the local rainfall is adequate, taro is grown near the coastal village. In the swampy lowlands, harvesting from wild sago stands is the predominant form of subsistence. Other food crops, such as sweet potato, taro and various greens, are grown on platforms with deep surrounding drainage systems. If the local environment does not favour gardening, the people of the swamps trade marine resources for root crops from gardeners further inland. Rain forest is the predominant vegetation on the foothills and mountain ranges in the Solomon Islands. In this environment taro is the predominant crop grown by bush clearing and long fallow periods (Barrau, 1958).

Breadfruit is the preferred staple of the eastern outer islands (Figure 2.1). In this region, this tree provides a resource with an almost continual fruiting season. The meat and juice of the coconut is also favoured by the coastal peoples of all the Solomon Islands and by the 'Polynesians on the Reef Islands' (Willmott, 1969a, 18). The Ngali nut (*Canarium mehenbethene*) is an important food source for the inhabitants of the eastern outer islands. Like the breadfruit the Ngali nut has no definite season and can be eaten all year round (Willmott, 1969).

In all of the Solomon Islands the green leaves from such plants as hibiscus, taro, and sweet potato are used as a relish for root staples and in soups (Willmott, 1969). Little fish is consumed by the inhabitants of the main islands: 'contrary to expectation, many islanders are neither fisherman by tradition or inclination' (Parsonson, 1966, 20), although women often gather shellfish and spear fish in the shallows of lagoons. On the smaller islands of the Solomons group including the eastern outer islands, the inhabitants are known as 'basically a maritime fisher folk, very different from the mainland peoples, the people 'bilong bush' ' (Parsonson, 1966, 9). However as mentioned above, the inhabitants of the swampy regions near the coast spend most of their time fishing, and the catch is used to trade with inland folk for tuberous crops (Barrau, 1958).

The consumption of animal meat is limited in most of the Solomon Islands. Domestic animals such as the pig and chicken are usually only consumed during feasts. Other sources of protein occasionally consumed are lizards, pigeons, opossum, flying foxes and wild pigs. Coconuts and other nuts are regular sources of fat for these island's inhabitants. The people of the eastern outer islands consume more breadfruit, coconut, nuts and fish than inhabitants of the main islands of the Western Solomons (Barrau, 1958; Willmott, 1969b).

The specific diet of the inhabitants of Taumako is not known. However, from the dietary surveys of the eastern outer islands reviewed above we can assume that the staple of Taumako was breadfruit with a heavy reliance on coconuts and other nuts. The people of Taumako probably also supplemented their diet considerably with marine resources (Parsonson, 1966). Minimal faunal material was recovered during the excavation of the burial mound on Taumako (Leach and Davidson, 2008).

2.1.2 Polynesia: Tongan subsistence

As with the Solomon Islands, there is also considerable variation in staples grown in Polynesia depending on the island environment and the preferences of the people. As mentioned above, there is a general impoverishment of indigenous flora and fauna in the eastern Pacific Islands. Therefore, it was necessary for the early migrants into Polynesia to introduce all the domestic plant and animal foods as part of the 'transported landscape' accompanying Lapita colonisers (Bellwood, 1978; Kirch, 2000a).

The yam is the primary staple in Tonga and is used in all important food related rituals. The predominance of yams is unusual for Polynesia where taro or breadfruit are usually preferred. In Tonga and Samoa the giant taro (*Alocasia macrorrhiza*) is also a staple food (Pollock, 1992) (Figure 2.2). The nuts of *Canarium commune L.*, of the same genus as those eaten in the eastern outer islands of the Solomon Islands, are also utilised in Tonga. In Polynesia breadfruit is more seasonal than in Melanesia and is more important as a food crop in the eastern islands. The coconut is an important condiment in Polynesian cooking and coconut cream produced from the ripe meat is an essential ingredient (Barrau, 1961).

Supplementary plant foods eaten by the Polynesians are more scarce compared to the great variety available in Melanesia. However, a source of fresh vegetable matter in Polynesia are certain marine algae and seaweeds (Abbott, 1991) and the young shoots from some fern species (Pollock, 1992). More importantly the young leaves of some taro species and certain parts of the coconut palm are also prepared as fresh vegetables (Barrau, 1961).

Seasonal fluctuations in food availability were often experienced in Polynesia. To guard against these shortages, various methods of preserving foods were developed (Leach,

Figure 2.1. *Breadfruit tree, Upolu, Samoa. Breadfruit is the main stable of eastern outer islands of the Solomons, including Taumako. Photo by author.*

Figure 2.2. *A taro garden on Uripiv island, Vanuatu. Taro is one of the main staple foods of the Pacific Islands. Photo by author.*

2003). For instance, in Tonga, yams, breadfruit and arrowroot were preserved. The vegetables were fermented in pits, resulting in a paste which could be preserved for years depending on demand and tastes (Pollock, 1992). These foods were also considered to be of 'higher status' (Leach, 2003).

The native terrestrial fauna of Tonga is scarce. There are however, numerous small species of lizards and insects that are possible food sources. The most noteworthy terrestrial species was the fruit bat and some birds, such as the fruit pigeon. The fruit pigeon was hunted with snares by the chiefly classes and many of the small mounds dotting the landscape were constructed for this purpose. The domestic animals brought into Polynesia were the pig and chicken. However, the pre-European presence of the dog is uncertain. The Polynesian rat was known to be present in pre-European times, however this was only consumed by the lower classes, although the aristocracy favoured hunting rats with a bow and arrow for sport (Poulson, 1987). As in Melanesia, pigs did not have a great importance in the traditional diet but were saved for feasting (Kirch, 2000b, 2001).

The sea provided a rich and abundant source of supplementary resources. The most plentiful fish, shellfish, and crustacean resources were found inshore around the reefs and shallow water. Fishing and gathering inshore can be done with ease and little risk. Deep water fish such as tuna, bonito, snapper and shark were also caught, but was by no means as regular as the exploitation of inshore species (Poulson, 1987).

Tongatapu is the largest and richest of the islands in the Tongan group. The soil is particularly fertile and the cultivated bounty of the island astounded early European explorers, most notably as described in Cook's observations on Tongatapu in 1773:

> I thought I was transported into the most fertile plains of Europe, here was not an inch of waste ground, the roads occupied no more space than was absolutely necessary and each fence did not take up above 4 Inches and even this was not wholly lost for in many of these fences were planted fruit trees and the Cloth plant, these served as a support to them....Nature assisted by little art, no were [sic] appears in a more flourishing state than at this isle. (Beaglehole, 1969, in Kirch 1984, 221).

Generally then, the Tongan subsistence pattern was heavily reliant on the cultivation of plant crops and supplemented by animal and marine resources.

Pacific peoples' attitude to food

The term 'food' is used in all societies, however the emphasis of what is considered food and what is just a substance that can be eaten may be different. Throughout the Pacific Islands the term 'food' is used only in relation to the starchy staples outlined above. For instance, taro is considered 'real food' which must be accompanied by an appropriate relish to enhance the taste (Pollock, 1992). The distinction between real food and condiments is found throughout Oceania. In Oceania a person is considered unfed unless the meal consisted of at least one of the starchy staples (Barrau, 1969). The accompanying food is the means by which meats and fresh vegetables are introduced into the diet. The accompanying dish is considered necessary for the person to feel totally satisfied with the meal (Pollock, 1992).

In Polynesia, the quantity of food eaten during meals depended on how much was available. In Tonga, early European observers noted that chiefs ate considerably more than commoners and more often (Mariner and Martin, 1827). A considerable amount of time and labour would be invested in feeding the chiefs at each meal time and a fat chief was a measure of a prosperous society (Leach, 2003). The availability of food was also affected by other cultural factors such as warfare (Poulsen, 1987). The habit of gorging when food was plentiful and fasting when none was available was frequently commented on. This attitude to food has been interpreted as an adjustment to seasonal variation in food availability. Therefore, the food habits of the Polynesians were, at times both irregular and variable in the amount consumed, depending on food availability and the social status of the person (Pollock, 1992).

It is uncertain whether this attitude to food was similar in the Melanesian islands to the west. The environmental constraints were probably similar to those of Polynesia and may have encouraged a similar feast or famine attitude to eating. However, the greater variety of plant and animal species available and less seasonality in the Melanesian islands may have alleviated the adverse effects of environmental disturbances in prehistory.

Nutrition and growth in the Pacific Islands

As further discussed in chapter four, patterns of infant feeding are of paramount importance to the growth and survival of children in developing countries (Jelliffe, 1968) (Figure 2.3). Some of the contents of this section have been published in Buckley and Oxenham (2016) as part of a review on disease and nutrition in the region. There are two main differences in infant feeding practices between Polynesia and Melanesia which may have affected the growth of children in prehistory. The first is the length of breastfeeding and second is the practice of food taboos. Polynesian weaning practices were reviewed in Jelliffe (1968) who described a regime in Hawaii which introduced a special supplementary food consisting of taro paste and shellfish viscera, rich in protein and vitamins, during the first four to six months of life. Similarly in Tonga, breast feeding was supplemented in the first few months of life by premasticated root vegetables and coconut milk (Jelliffe, 1968). In Samoa sole breastfeeding was continued for less than one year, when supplementary weaning foods were introduced (Wickes, 1953).

Figure 2.3. *Children from Tanna Island, Vanuatu. Photo by author.*

Figure 2.4. *Offering of food as a gift at a traditional wedding in NE Malakula, Vanuatu, including yams, pig, plantain and meat portions. Photo by author.*

In most tropical countries evidence of poor nutrition in the form of growth faltering and mild malnutrition is seen in children soon after weaning (Jelliffe, 1968). However, studies of Polynesian growth do not report this pattern. For instance, a study of growth in Rarotongan children under five years of age reported heights and weights which were comparable to Europeans (Faine and Hercus, 1951). The pattern of infant feeding described above is considered to account for the good nutrition and growth rates of Polynesian children. Early nutrition surveys carried out in a number of Polynesian island groups report consistent results of growth rates in Polynesian infants. Some studies have found that Polynesian growth rates are comparable and sometimes faster than Europeans (Bindon and Zansky, 1986). For instance, Tonkin (1966) reported birth weights in New Zealand Maori babies that were higher than their Pakeha counterparts during the first six months of life when thereafter the weight was the same. The height of Maori babies was also found to be consistently greater than that of European babies (Tonkin, 1966).

As mentioned above, one of the differences of infant feeding practices in Melanesian communities is prolonged breastfeeding without adequate supplementary foods. For example among the Bundi of the western NG Highlands a child is not fully weaned until two and a half years of age when a full deciduous dentition has erupted and the child is thought to be strong enough to survive the birth of the next child (Gray, 1982).

This correlation of weaning with the eruption of deciduous teeth was noted in other NG tribes (Thomason et al., 1986). Among the Bundi of the Madang Province in the Highlands weaning foods are introduced in small quantities around 6 months of age but a full adult diet is not achieved until four years of age (Malcolm, 1970a). In the Sepik region of NG, infants are fed breastmilk and sago paste until one year of age when solids are gradually increased, until complete weaning at two years of age. This pattern of prolonged breastfeeding with introduction of low protein supplementary foods is not restricted to NG and has been noted in the Solomon Islands (Willmott, 1969a) and Vanuatu (Jelliffe, 1968).

Probably the most significant difference between Melanesian and Polynesian child feeding patterns is the presence of food proscriptions connected to fish and meat as part of the weaning strategy in Melanesia. These food proscriptions are not necessarily determined by ecological constraints, such as limited food availability, but are found throughout the biodiverse tropical environments of Near Oceania. For example on Vanatinai island off the Southeast coast of NG, Lepowsky (1987) noted that protein rich foods were readily available to all members of the community except children from birth to around three years of age. The age-specific food proscriptions were explained by the mothers as necessary for ensuring the child's well-being or it would become ill and die (Lepowsky, 1987). Similar feeding patterns have been observed elsewhere in Melanesia. Among the Au people of the western Sepik the green leaves of the 'tulip' (*Gnetum gnemon*) are the main source of protein but it is denied to children until four years of age. This is explained as a means of avoiding sickness of the stomach and fevers in children which are linked to eating these foods (Thomason et al., 1986). A similar explanation is given by the Lumi people of the Sepik River as a reason for withholding protein rich foods from young children (Wark and Malcolm, 1969). The tribes of the northern coast of NG also follow this protein taboo where pregnant and lactating women and children are prohibited from eating fish and meat (van Der Hoeven, 1958). These northern NG tribes believe that if these taboos are broken the gods will be angered and intestinal disease and an enlarged spleen will result (van der Hoeven, 1958) (Figure 2.4). Feeding proscriptions have also been reported in the Solomon Islands (Willmott, 1969a).

Based on the indigenous peoples' explanation for why their children are not allowed foods which aid in adequate growth and well-being, Lepowsky (1987) proposed that this protein poor diet aided the survival of children living in endemic malarial areas. This model fits with the iron-withholding literature reviewed in chapter five, where mild malnutrition ameliorates the clinical effects of malaria. This hypothesis has some merit when the age-specific taboos from six months to around five years of age are correlated to the most vulnerable ages of malaria related morbidity in the Pacific Islands. Also in support of this is the absence of both malaria and food taboos in Polynesia.

The Lepowsky (1987) hypothesis has not been specifically tested but the long-term effects of chronic under-nutrition among children in Melanesian societies with food proscriptions is evident. The reported growth patterns of infants and children in NG and elsewhere in Melanesia are monotonous in their regularity. Nutritional surveys all report that prolonged breastfeeding, inadequate supplementary foods and maternal under nutrition have a cumulative effect of stunting growth in children (Jelliffe, 1968). These surveys also report growth rates which are comparable to Europeans up to the age of six months then a marked 'falling away' in growth velocity after this time (Malcolm, 1969; Wark and Malcolm, 1969; Willmott, 1969a; Malcolm, 1970a; Gray, 1982; Bogan and Crittenden, 1987). In some cases the degree of growth faltering among children in Melanesian societies is extreme. For example, in 1970 Malcolm reported that in the Bundi people of the Madang province of NG a child did not reach the height of a British seven year old until it was 12 years of age. A difference of five centimetres in height was found between Bundi and British children at three months old which had increased to 34 centimetres by 14 years of age (Malcolm, 1970b).

Most nutrition surveys in Melanesia comment that the starchy food staples were not sufficient to sustain growth in children and that frank malnutrition was prevalent (van der Hoeven, 1958; Wolff, 1965; Malcolm, 1969, 1970a). This is probably related to the prolonged period of breastfeeding without adequate supplementary foods and taboos which prohibit animal protein to be fed to young children. This regime of prolonged breastfeeding and food taboos is frequently practised in Melanesian and Southeast Asian societies. It is not known whether the inhabitants of Taumako followed weaning and infant feeding patterns similar to that of Polynesia or Melanesia.

Nutrition and infection

A significant cause of morbidity and mortality in developing countries is weanling diarrhoea which is related to the introduction of supplementary weaning foods (Gordon et al., 1963). The clinical impact of diarrhoeal episodes is somewhat dependant on the nutritional status of the infant at the time of infection. When the infant is well nourished, episodes of weanling diarrhoea are short, and the recovery is prompt with long intervals between infections. However, if a child is malnourished or already resisting other disease, episodes will be more protracted. In this setting of chronic weanling diarrhoea the infant may be in a constant state of illness and recovery, leaving the individual vulnerable to attacks from other pathogens and thereby compound any pre-existing nutritional deficiency or disease (Gordon et al., 1963; Scrimshaw, 1981). This concept of a synergistic relationship between infection and nutrition is developed further in light of the pathological findings in succeeding chapters.

2.2 Infectious Diseases of the Pacific Islands

> The tropics are dreamlands, released from the balance of northern things. Life down there moves between poetic loveliness and monstrous disgust. I could get used to the maimed adults, but the children always wrung my heart. (Lambert, 1941, 30)

While life in the Pacific Islands is usually considered idyllic, there are a number of infections that cause significant disease among the inhabitants. Many are non-specific or may have left no characteristic features on dry bone while others are diagnosed by pathognomonic skeletal involvement. The history and epidemiology of treponemal disease, tuberculosis, leprosy, malaria and hookworm are considered below. These particular diseases are reviewed because early European observations and clinical accounts suggest they were probably present in prehistory. This literature is reviewed in the appropriate sections. The pathogenesis of bone involvement in the first three diseases is also outlined, while the bone changes of anaemia are considered in chapter five as these are chronic conditions which may have contributed to the development of skeletal lesions observed in this study. The non-specific bone changes of osteomyelitis are also reviewed.

A review of the literature concerning erosive arthropathies is included in this chapter. These conditions are difficult to diagnose in skeletal material but are included because they may have adversely affected the expression of infectious disease in prehistory.

2.2.1 *Epidemiological and pathological concepts*

It is important to understand the biological basis of the clinical passage of some diseases before the skeletal response can be discussed with any degree of authority. The following terms are frequently used in palaeopathology and the definitions used in the current study are adapted from the work of Evans (1998) and Mims (1993).

Firstly, *aetiology* refers to the causal agent of disease, whether it be bacterial or nutrition related. Aetiology refers to the 'What' of disease in the prehistoric Pacific Islands. Secondly, *epidemiology* is the study of the distribution and causation of disease in specific populations. In a modern context the data from epidemiological studies are used to control health problems such as epidemics of infection (Evans, 1998). In the context of this thesis, epidemiology constitutes the 'Who' of disease. In other words, who is most affected with subperiosteal bone formation? Is it young children; or are more females affected than males? These questions are asked in order to understand the causation of disease by indicating the distribution of bone lesions in a skeletal sample.

In order to ask these questions the prevalence rate must be established. The *prevalence rate* is determined by the ratio of the affected individuals in a *defined population.* In the context of palaeopathology the prevalence rate is established by recording the evidence of skeletal changes in a defined population, which is the skeletal sample, in material that is available.

A disease is considered *endemic* when exposure and infection of a certain pathogen is constant, or the presence of the disease is usual in a given community (Evans, 1998). The levels of endemicity are measured differently for certain diseases but generally a high degree of endemicity is termed *hyperendemic* while an infection which causes disease early in life and affects most of the community is called *holoendemic*. An epidemic is differentiated from an endemic disease when an unusual number of cases of a disease arise compared to previous experience in a community (Evans, 1998). The causes of epidemics vary depending on the infection.

2.2.2 *Non-specific infections which cause bone changes*

The bacteria most commonly responsible for bone infection are staphylococci, streptococci and pneumococci. Infections with any of these bacteria can cause pathological changes in the bone tissue which are non-specific; meaning that infection with one species of bacteria is indistinguishable from another (Roberts and Manchester, 1995a). These non-specific pathological bone changes are reviewed below.

Osteomyelitis

The exact bacterial agent that leads to osteomyelitis is varied and in most cases is not able to be distinguished in dry bone. For this reason osteomyelitis is often classified as a 'non-specific' infection in palaeopathological texts (e.g. Roberts and Manchester, 2005). Osteomyelitis is an inflammation of the cortical bone (osteitis) and/or medullary cavity (myelitis) caused primarily by pus-producing bacteria (Aufderheide and Rodriguez-Martin, 1998). Fungus, parasites, and viruses can also infect bone (Resnick, 1995b) and some can cause distinctive bone changes. If strictly defined, osteomyelitis is an inflammation of the marrow cavity; periostitis, and osteitis can be associated with this primary infection but the localisation of infection in only one of these tissues rarely occurs (Steinbock, 1976).

Normally, bone is never exposed to the external environment, therefore pus-producing pathogens will ordinarily reach bone by indirect extension through the blood stream (Steinbock, 1976). Infection of bone tissue in this manner is known as haematogenous osteomyelitis, meaning the bacteria are blood-borne. The predominant aetiologic agent in haematogenous osteomyelitis is bacteria originating from a primary infection of the throat, the upper respiratory tract, the gastrointestinal system, or another remote bacillary source (Silverman, 1985; Resnick, 1995b). In some cases bone may become infected directly following a severe trauma such as a compound fracture or burn. These primary sites of infection may form the focus for indirect haematogenous spread to other bones (Steinbock, 1976). A further mode of infection is from a contiguous source, such as extension into the bone from an adjacent site. Infection of bone by this mechanism can develop from cutaneous, paranasal sinus and dental infections in which the skeletal tissue can become involved (Resnick, 1995b).

Clinically, osteomyelitis presents in three different stages; acute, subacute and chronic. The last two stages may indicate an inadequate immune response to the infection or, in modern cases, resistance by the pathogen to treatment (Aufderheide and Rodriguez-Martin, 1998). Use of these terms should not imply definitive divisions between one stage or another, nor do they signify that all cases will progressively advance through each of the stages (Resnick, 1995b).

When infection is focused in the medullary cavity, areas of radiolucent sites denote rarefaction in the spongiosa and indicate foci of pyogenic fluid (Silverman, 1985). Intramedullary pressure caused by the associated oedema of infection forces the pyogenic fluid to infiltrate Haversian systems. Cortical bone underlying a subperiosteal abscess becomes sclerotic, appearing denser radiographically indicating bone necrosis and the formation of a sequestrum (Silverman, 1985). Sequestra are sections of necrotic bone which may reside in the marrow for long periods of time and harbour infective organisms, sometimes evoking an acute resurgence of disease (Resnick, 1995b). Ordinarily, sequestra are resorbed by osteoclasts, or expelled through the soft tissue to the surface of the body. The apposition of new bone around a sequestrum results in the formation of an involucrum which is a shell-like casing of new bone. This new bone formation indicates the beginning of the reparative stage of bone involvement. Small defects can develop in the involucrum which function as drains for the purulent fluid or cloacae (Ortner, 2003).

Infants and Children

Although osteomyelitis can occur at all ages, infants and children comprise most cases (Krugman and Katz, 1981). In children, the metaphysis is most commonly involved and bacteria are introduced into the medullary cavity via the metaphyseal vascular system. Capillaries on the metaphyseal side of the growth plate are the terminal end of a nutrient artery which turn sharply to join with large sinusoidal veins occupying the intramedullary portion of the metaphysis where blood flow is slow relative to the medullary space. This slow blood flow and factors such as the inability of the vessels to penetrate the open physeal plate, contribute to the focus of haematogenous osteomyelitis in the metaphyses of the tubular bones in children (Resnick, 1995b).

The pathogenesis of haematogenous osteomyelitis is essentially the same in infants. However, because the metaphysis and epiphysis still communicate through a remnant foetal vascular system penetrating the growth plate, the epiphyses can become involved. The involvement of the epiphyses may lead to septic arthrosis. Because of the particularly loose attachment of the periosteum of this age, the infantile pattern of haematogenous osteomyelitis can result in the profuse formation of involucra and extensive sequestration (Resnick, 1995b). Due to the extreme vascularity of the infantile periosteum, an involucrum may rapidly remodel into the cortex and eventually leave little trace of infection (Resnick, 1995b).

In infants and children, osteomyelitis mainly affects the tubular bones. Involvement of the tibia and femur predominates and these bones, together with the fibula, humerus, and radius, comprise about 80 per cent of cases in young children. In children, infection of more than one bone occurs in only approximately 15 per cent of cases. However, in infants, multiple bones are more commonly involved (Krugman and Katz, 1981).

Adults

In adults the ossification of the growth plate leads to a vascular connection between the metaphysis and epiphysis. Therefore, a subchondral focus of bone infection, which may lead to joint destruction, is not unusual in this age group. The firm attachment of the periosteum to the cortex in adults, it resists displacement by purulent fluids, therefore, the formation of a subperiosteal abscess and extensive periostitis is unusual in this age group. However, if a subperiosteal abscess does develop, the consequent elevation of the periosteum initiates the laying down of new bone.

Haematogenous osteomyelitis in adults does not commonly affect the tubular bones but is more common in the spine, pelvis, and small bones of the hands and feet (Resnick, 1995b). The intimate association with the cortex of the adult periosteum ensures an adequate blood supply in most sufferers of osteomyelitis and large sequestra do not commonly develop. However, the infection usually destroys the cortical bone disseminating throughout the Haversian systems, which results in atrophy and weakening of the osseous structure. Therefore, pathological fracture is a common sequel to adult osteomyelitis (Resnick, 1995b).

2.2.3 *Distinctive characteristics of osteomyelitis*

The most distinctive characteristic of osteomyelitis which may aid in differential diagnosis is the array of bone changes associated with the process of bone destruction and repair. These bone changes have been outlined above and have been assigned labels denoting pathological processes particular to osteomyelitis; the sequestrae, involucrum, and cloacae (Ortner, 2003). However, because of the particular anatomy and function of the infantile periosteum outlined above, it should be noted that cases of osteomyelitis in young subadult skeletal remains may not be clearly distinguishable from other conditions that lead to diffuse subperiosteal bone production.

2.2.4 *Treponemal disease*

The pathogenic treponemes are highly virulent microorganisms responsible for chronic human disease. The four treponemes that cause disease in humans are: pinta (*Treponema carateum*), yaws (*Tr. pertenue*), endemic syphilis (*Tr. pallidum endemicum*), and venereal syphilis (*Tr. pallidum pallidum*) (Arya, 1996). These are usually differentiated entirely on the basis of epidemiology and clinical manifestations (Arya, 1996). Pinta is a disease that affects only the skin, so is excluded from further discussion. Endemic syphilis or bejel is a form of treponematosis that is restricted to more temperate and arid climates than the research area covered in this study. Yaws is specific to tropical climates, while the transmission of venereal syphilis is not affected by climatic factors (Hackett, 1976). Because of these climatic considerations, endemic syphilis is mostly excluded from further discussion, while yaws and venereal and congenital syphilis are considered in more detail.

The origins of treponemal disease is a subject that has been hotly debated for decades. This debate has concentrated more on the origin of venereal syphilis than the other treponemes. There are three main hypotheses concerning the evolution of treponemal disease. Firstly, the Colombian hypothesis argues that syphilis originated in the Americas and was brought to Europe by Columbus' crew on their return from the Americas in AD1493. Coinciding with their return was an epidemic of the disease around AD1500. The champions of the Columbus hypothesis state that the rapid spread of the disease throughout Europe attests to its New World origin (Baker and Armelagos, 1988). However the proponents of the pre-Columbian hypothesis argue that syphilis was always present in Europe but was not distinguished from ''leprosy' which encompassed a whole raft of infections that affected the skin (reviewed in Baker and Armelagos, 1988). The epidemic of a 'new' disease was purely the result of syphilis finally being recognised as a separate entity. The pre-Columbian hypothesis asserts that geographic isolation led to speciation of the treponemes. As living conditions and hygiene improved in Europe, those strains which favoured transmission through sexual contact were selected for and strains which relied on skin to skin contact for transmission were selected against. The pre-Colombians argue that better living conditions and improved hygiene in Europe had developed to such a standard that by AD1493 the venereal form predominated (Cockburn, 1963). Finally, the Unitarian hypothesis argues that the four treponemal treponemes are essentially gradients of the same disease which have evolved according to such environmental aspects such as climate, hygiene, and socio-cultural factors (Baker and Armelagos, 1988).

It is beyond the scope of this thesis to consider the merits of each hypothesis further. Recent work on evidence of treponemal disease in prehistory is more concerned with attempting to differentiate between the treponemal conditions and assessing the impact of disease on past communities than engaging in the debate over its origins. However, studies of skeletal evidence for treponemal disease have revealed pre-European occurrence of an endemic form of the disease in the Americas (e.g. Powell, 1988) and there is mounting evidence of pre-Columbian treponemal disease in Europe (Stirland, 1994). This evidence would support the pre-Colombian and Unitarian hypotheses more than the New World Hypothesis.

Further discussion of studies concerned with treponemal disease in Europe and the Americas will be limited to their value in aiding in a diagnosis of skeletal pathology in the samples used in this study. As Trembly (1996, 398) points out: 'To listen to, and read all these papers and discussions, one would think the world ended there, with North and South America, Europe, the Mediterranean, and Africa. Literally half the world is left out...'. Trembly (1996) is referring to the fact that patterns of treponemal disease in the Pacific Islands may contribute to the understanding of this disease in the rest of the world, but is largely ignored by many theorists. What is relevant to this thesis is the origin and antiquity of yaws in the prehistoric Pacific Islands. For a review on the antiquity of treponemal disease in this region and recent genetic data, see Oxenham and Buckley (2016) and Buckley et al. (2008).

Based on early European accounts there is a universal agreement that a treponemal infection was present in the tropical Pacific Islands prior to European contact and that this infection was yaws (van Der Sluis, 1969). Yaws is dependant on a warm moist climate for effective transmission and is therefore thought not to have been present in the more temperate islands of New Zealand. The New Zealand Maori's first experience with a treponemal disease was venereal syphilis introduced by European men (Houghton, 1996). There is also some evidence that not all island groups had yaws in prehistory. There is no clear evidence of skeletal changes of a treponemal disease from Hawaii (Miles, 1997). However, historical accounts negate this, where florid descriptions abound from early European observers from the first half of the nineteenth century which van Der Sluis (1969) cites as evidence of the presence of endemic yaws in Hawaii.

The early descriptions of treponemal disease in the tropical Pacific Islands attest to the widespread presence of a chronic endemic disease that afflicted most people with very early transmission of the disease. Some of these descriptions are particularly graphic. For example (van Der Sluis, 1969, 111) cites the work of a Dr. Fox working in Fiji in the early nineteenth century:

> It usually attacks children from two to nine years of age and according to the natives and white men's experience, none escape....The mouth, arms and umbilic, ulcerate around the whole circumference....Very large and extensive ulcers, at the same time, exist in various parts of the body, some having the appearance of a fungous mass. In adults the pericranium is oftener affected than in children....Cases are by no means rare of the loss of the bones of the palate and nose. In several instances we observed the upper lip entirely gone....The females, in particular, are very often seen with deep cicatrices about the lips ...The natives say this disease has always prevailed among them, and always speak of it as a feejee disease. We have observed something of a similar nature on the other islands which I have heretofore mentioned.

The other islands that Dr Fox is alluding to are the Samoa and Tonga groups (van Der Sluis, 1969).

Traditionally yaws is described as the benign cousin of venereal syphilis that does not have a grave impact on the health of individuals and is never directly fatal (Powell, 1991).

> ...it is improbable that septic infections of the bones is responsible for the changes seen. Untreated septic infection of the extent necessary to produce the widespread changes seen in some cases would be accompanied by grave general symptoms and high mortality; whereas the patients showing these bone lesions were not severely ill, although they suffered considerable discomfort. (Hackett, 1951 cited in: Powell, 1991, 174)

Despite this reasonable description of yaws, it should be remembered that while yaws is not as dangerous as its fatal cousin this was a disease that contributed to a considerable amount of suffering for people afflicted with the disease. The following quotes illustrates the experience of yaws prior to effective medical treatment:

> The most loathsome of diseases prevailed through the tribes, nor were the youngest infants exempt from them. Indeed so young were some, whose condition was truly disgusting, that I cannot suppose they must have been born in a state of disease; but I am uncertain whether it is fatal or not in its results, though most probably it hurries many to a premature grave. (Sturt, 1833, 96–97).

> During its presence [the disease yaws] the patient is generally so enfeebled as to be unable to procure food, and is in fact totally helpless (Gason, 1879, 284).

By all accounts yaws was an accepted part of life among the inhabitants of the Pacific Islands (Marples and Bacon, 1953). Infection with the disease was considered inevitable and in some cases was precipitated by intentional infection. 'Native mothers expose their babies in the hopes of 'getting it out of their systems' ' (Lambert, 1941, 31). A similar practice was noted in West Africa where mothers would expose their children to the disease early because the primary stage of the disease was believed to be less dangerous in childhood than as an adult (Hackett, 1946).

Yaws has been largely eradicated in the Pacific Islands. Although resurgence of the disease has been known to occur (Gershman et al., 1992). With few exceptions yaws was widespread and endemic throughout the Pacific Islands (Geizer, 1986). Prior to eradication programmes during World War II '[i]t has been said that no native grew to adult life without showing signs of past or present yaws infection' (May, 1958, 226).

An eradication survey carried out on Simbo Island in the Solomon Islands in 1953 found that 30 per cent of the children under 12 years of age had active yaws. Of the children from Simbo, 78 per cent had been infected with yaws, the transmission of which was prior to two and a half years of age. In light of the clinicians attitude to yaws as a benign disease, it is interesting that yaws was considered by May (1958) to be primarily responsible for the high mortality rate among infants in some of the island groups of Micronesia and on Simbo Island (May, 1958).

Yaws was thought to have been eradicated in the Solomon Islands, however an outbreak of the disease in the early 1980s revealed a high incidence rate among inhabitants of three islands in the western province of the group. The peak incidence of yaws was seven years of age, and again at 13 years of age (Alemaena, 1986). This age of infection is much older than earlier accounts of the disease where children under three years of age were said to have been more affected. A similar situation occurred on Karkar island off northern Papua New Guinea where prevalence was very high prior to eradication and a resurgence of the disease had occurred (Backhouse et al., 1998). Over 50 per cent of the tertiary cases in the Solomon Islands were children under ten years of age (Fegan et al., 1990). In an early survey of yaws in Vanuatu the population prevalence of yaws was found to be 50 per cent, and children under five years of age were predominantly affected by primary or secondary infection. The highest prevalence of early yaws was found in children between one and five years of age (Mills, 1955).

Prior to eradication programmes in Fiji and Tonga, yaws was one of the leading causes of morbidity and hospitalisation (Geizer, 1986). In 1933, 50.9 per cent of the total population of western Samoa had yaws. In a later survey on Manona Island in Samoa 13 per cent of the population was found to have yaws (Marples and Bacon, 1953). Seventeen per cent of one year olds and 23 per cent of children less than ten years old had primary and secondary yaws. The primary yaw was recorded in subadults aged from ten months to eight years old, while secondary yaws was found to be most prevalent between the ages of two and 15 years. No cases of tertiary yaws were recorded for subadults under two years of age (Marples and Bacon, 1953). While no early clinical reports of yaws in Tonga have been located for this study, accounts by early European explorers note its existence. For example, Mariner and Martin (1827) noted its extreme prevalence amongst all young Tongan children he encountered.

Generally then, it can be said that where yaws was endemic in the Pacific every child was exposed at some point, in some populations beginning in the first year of life. Transmission then increased in toddlers, with new cases declining after adolescence. From these early clinical reports it may be accepted that yaws was highly prevalent amongst some Pacific Islanders prior to European contact. The early infection of children under one year of age in Samoa and other Pacific Islands differs from results of research carried out in Uganda, west Africa, where infection under two years of age was very unusual (Buxton, 1928; Hackett, 1946). This may be a reflection of more intense endemicity of yaws in some Pacific Islands.

Palaeopathological studies in the Pacific have lent credence to the conclusion that this was a disease endemic in the prehistoric Pacific Islands (Brothwell, 1976; Hanson, 1990; Houghton, 1996; Heathcote et al., 1998; MacKay, 1938; Pietrusewsky, 1976; Rothschild and Heathcote, 1993; Rothschild and Rothschild, 1995, 1996; Roy, 1989; Stewart and Spoehr, 1952; Stodder, 1997; Stodder and Martin, 1992; Trembly, 1996; van Der Sluis, 1969). Papers published reporting treponemal disease since this thesis was completed include Buckley and Tayles (2003b) and Buckley and Tayles (2003a).

Most written accounts of skeletal evidence of treponemal disease prior to European contact are contained within unpublished 'grey literature' (Trembly, 1996). However, recent work carried out in Micronesia has increased our knowledge of the paleoepidemiology of yaws in this region (Stodder and Martin, 1992; Rothschild and Heathcote, 1993). Earlier work conducted on skeletal material in other regions of the Pacific have also found evidence of treponemal disease before European contact (Stewart and Spoehr, 1952; Pietrusewsky, 1969, 1976; Houghton, 1996).

Factors which influence disease prevalence

The transmission of yaws has been found to vary considerably between island groups and even within islands. Accordingly, the distribution of yaws is patchy. For example, areas of high yaws endemicity may be found close to areas where the disease is absent (May, 1958; Pirie, 1972). A number of factors are thought to influence this variation in prevalence and also the clinical expression of the disease. For instance, it has been demonstrated that yaws is most prevalent wherever rainfall is heavy and the soil is fertile and moist, supporting a lush vegetation. Poor sanitary conditions and a high population density are also factors which favour yaws transmission (May, 1958).

Furthermore, a high temperature and humidity have been shown to influence the course of the disease, where more florid lesions develop as the temperature rises and lesions decrease during the cooler months. Where the soil remains warm and moist for longer periods, the treponeme is thought to survive outside the body long enough to cause new infections (May, 1958).

Where vegetation is more abundant, a correlation with higher prevalence of yaws has been found. This is thought to be caused by the greater risk of skin abrasions which aid transmission of the treponeme. 'In the Solomons twice as much yaws is found in the bush as on the coast' (May, 1958, 229). In summary;

> Cultural factors seem to be the most significant in the epidemiology of yaws. The disease is the expression of miserable living conditions, of densely populated areas, of crowded village markets, of a primitive way of life in dark, windowless huts. These cultural factors that expediate the transmission of yaws are probably the only ones among all factors listed to which no exceptions have been found (May, 1958, 230).

May (1958) goes on to list examples in where the transmission of yaws is highest due to cramped living conditions and where people, especially children, are forced to huddle together for warmth at night. The high prevalence of yaws on Simbo Island was also explained by cramped living conditions.

The following account quoted in (Hackett, 1936b, 736) describes the association of close proximity with ready transmission of the disease in Malaysia:

> During winter they, 'tis true, are very much subject to a kind of scurvy, which, from its prevalence, might be deemed contagious, ...each breakout is due to a lack of nutritious food, combined with cold, wet lodgings. As the mild spring advances, and food becomes plentiful this distemper gradually leaves them, and by summer their skins have returned to their normal sleekness....

This reference to poor nutrition exacerbating the course of the disease has been made in clinical literature (Hackett, 1936b, 1946; May, 1958). The synergism between infection and nutrition has long been recognised in relation to other diseases, such as tuberculosis (Scrimshaw et al., 1968) but the anthropological literature rarely recognises this relationship in yaws.

2.2.5 *The pathogenesis of yaws and syphilis*

Yaws is a chronic granulomatous infection that usually results from direct contact of broken skin with an infective yaws lesion (Dooley and Binford, 1976). The transmission of yaws by way of fomites or flying insects has also been suggested (Aufderheide and Rodriguez-Martin, 1998). The 'mother yaw' or initial skin lesion usually develops after two to four weeks' incubation at the site of inoculation. A papule forms and then gradually enlarges forming a granuloma from which the epidermis erodes. It is usually located on the lower extremities and buttocks (Woodruff and Wright, 1987) because this is the area of skin that is most exposed to dirt, the risk of abrasion, and subsequent inoculation. In venereal syphilis the usual mode of transmission is by sexual contact. The initial lesion in syphilis is known as a 'chancre' (Kampmeier, 1954). In the following section, congenital syphilis will be discussed separately from yaws.

The initial yaws lesion is usually associated with bone pain and mild fever. The bone pain is an indication of inflamed lymph nodes as a response to infection (Hackett, 1946; Jelliffe, 1970). In syphilis, before the initial chancre has developed, the spirochetes migrate to the regional lymph nodes from the inoculation site and from there haematogenous spread begins, resulting in bacteraemia (Krugman and Katz, 1981; Resnick, 1995b). Within a few months the infection will escalate to the secondary stage, characterised by a diffuse papillomatous skin rash, in both syphilis and yaws. In yaws the number of lesions is variable (from one to hundreds) and they usually cluster on the limbs leaving the trunk relatively free (Hackett, 1946). The distribution of the papillomatous rash is probably influenced by microclimatic conditions, as it has been observed they cluster in the warm moist areas of the body, such as the armpits and groin (Woodruff and Wright, 1987). In syphilis, the skin rash is variable in its distribution on the body and in the numbers present. It does not seem to be as strongly influenced by the microclimate of the body as the skin rash of yaws. The skin lesions develop in many different forms and, like the papillomatous rash of yaws, they are never pustular (Kampmeier, 1954).

It is during the secondary stage of infection that bone involvement can occur. The periosteum can become inflamed at this stage with subsequent new bone formation (Woodruff and Wright, 1987). The periostitis of syphilis is associated with intimal thickening and an accumulation of lymphoid cells in the walls of medium sized arteries (Resnick, 1995b). The subsequent ischaemia, together with the damage caused to blood vessels, triggers inflammation in these sites resulting in the hypervascularity and exuberant formation of new bone in syphilis. It is implied that this or a less destructive process occurs in the pathology of other treponemes. However, the intimal thickening and obliterative endarteritis of syphilis do not occur in yaws (Arya, 1996). Anterior bowing of the tibiae and other bone changes that are distinctive to all treponemal diseases are discussed below.

Tertiary Disease

In yaws, usually after two to three years, the secondary skin and bone lesions will resolve spontaneously. After this period many sufferers will overcome the infection completely or enter the latent secondary stage in which no cutaneous lesions persist. However, some active residual bone lesions may persist and many patients complain of bone pain that may indicate a chronic state of bone involvement (Hackett, 1946). In both yaws and syphilis, this tertiary stage of infection marks the development of permanent and grossly destructive cutaneous and bony changes (Jelliffe, 1970; Sengupta, 1983; Woodruff and Wright, 1987). The most featured lesion of the tertiary stage is the gumma, a type of granuloma. Granulomatous inflammation is a specific type of chronic inflammation characterised by accumulations of granulation tissue. This process is initiated by a variety of infectious and noninfectious agents. The presence of poorly digestible irritants (such as a treponeme) or a cell-mediated immune response to the irritant appears to be necessary for granuloma formation (Cotran et al., 1994).

The formation of gumma is actually a hyper-immune response to re-exposure which may cause a reactivation of the infection, upsetting the host-parasite relationship (Steinbock, 1976; Mims, 1993). The presence of dead spirochetes in a degrading gumma would conceivably initiate the 'hyper' destruction of tissues characteristic of tertiary treponemal disease.

In syphilis, no organ or tissue is immune from gumma formation, including the mucous membranes. Involvement of these can lead to destruction of associated cartilage and the nasal and palatal bones can also be destroyed (Woodruff and Wright, 1987). Gummatous lesions are frequently found around the joints such as the elbow, and become fibrous when healed (Jelliffe, 1970). This distribution of gummatous lesions around the joints may relate to the site of affected lymph nodes. The contribution of the lymph nodes to the pathogenesis of bone lesions is discussed below. However, if the bone becomes involved and the gumma has originated from within the bone tissue itself, the subsequent ulcer will eventually discharge through the skin. Sometimes bony involvement is secondary to subcutaneous gummatous destruction (Jelliffe, 1970; Woodruff and Wright, 1987).

Congenital syphilis

If a pregnant woman is infected with venereal syphilis the spirochete may pass through the placenta to the foetus resulting in congenital syphilis. In congenital syphilis, bone death is usually due to trophic or nutrient disturbances caused by the spirochetes blocking the blood supply to bone tissue, rather than inflammation (Jaffe, 1972). Two kinds of trophic changes occur, mostly at the metaphyses. The first trophic change is due to the non-specific effect of a chronic disease on bone formation and is observed radiographically as transverse bands of decreased density parallel to the growth plate (Silverman, 1985).

The second trophic-type change is osteochondritis. The osteochondritis specific to congenital syphilis is characterised by a disturbance of endochondral ossification at the metaphyses by the accumulation of granulation tissue as a response to spirochetaemia. Radiographically, osteochondritis appears as irregular osteolytic foci along the contour of the metaphyseal-growth plate junction (Resnick, 1995b). Osteochondritis may be present in the first six months of life but resolves after this age even without treatment (Krugman and Katz, 1981; Silverman, 1985). Osteochondritic changes can occur in any bone but are found most commonly at the fastest growing metaphyses of the proximal tibia and distal femur, known as Wimberger's sign (Jaffe, 1972). This trophic change is considered by Jaffe (1972) to be pathognomonic of early congenital syphilis and are predominantly symmetrical (Ortner, 2003).

In congenital syphilis, diffuse periosteal new bone deposition is less common than osteochondritic changes. However, after six months of age reparative subperiosteal new bone can develop during the remodelling of affected metaphyses. This is representative of a 'callus' formation rather than a response to spirochete infiltration of the periosteum. Syphilitic periostitis often develops during infancy, after resolution of the osteochondritic changes, is generalised, diffuse, and affects multiple bones (Resnick, 1995b; Ortner, 2003). The subperiosteal new bone of early congenital syphilis is likely to be irregular in form and can envelop the entire diaphysis (Krugman and Katz, 1981; Steinbock, 1976). The cortical bone under the subperiosteal new bone either remains the same or becomes demineralised (Jaffe, 1972). Distinctive disturbances in the formation of enamel in the permanent teeth are also considered pathognomonic changes of congenital syphilis (Hillson et al., 1998). The bone changes of late congenital syphilis are essentially the same as those in acquired syphilis (Ortner, 2003). The anterior diaphyseal 'bowing' of the tibiae that is characteristic of yaws and syphilis can also develop in congenital syphilis.

The relationship between yaws and syphilis

The following statement typifies the prevailing view concerning the relationship between the skeletal lesions of the endemic treponematosis, yaws and venereal syphilis:

> Hackett and many others have repeatedly emphasised the close similarity of the bone lesions in both yaws and syphilis. Indeed, except for the osteochondritis and dental stigmata of congenital syphilis, no lesion is found in one disease that may not be found in the other. The differences are merely quantitative.... (Steinbock, 1976, 143).

The general pathology of yaws and syphilis is the same. However, there is less endothelial proliferation in yaws, the obliterative changes to the blood vessels (endarteritis) do not occur in yaws, and there is a more marked acanthosis (cellular change in the epidermis related to profuse papillomatous rash). Furthermore, in secondary yaws, spirochete activity is confined to the epidermis, whereas in syphilis all the skin layers can be affected (Arya, 1996). Probably the most fundamental difference between venereal syphilis and yaws is the capacity of syphilis to attack the central nervous and cardiovascular systems (Resnick, 1995b). This may explain why syphilis can be fatal while yaws is not. It is implied that the histological mechanism for bony change is similar, but different by degrees, in all treponemal conditions.

Expected distribution of bone changes in treponemal disease

Based on the type of bone change and the distribution of skeletal involvement, the diagnosis of treponemal bone lesions is well established (Hackett, 1976; Steinbock, 1976; Ortner, 2003). Yet there is some confusion as to the actual mechanism producing this distribution. This confusion might lead to inaccurate diagnosis of treponemal disease in prehistory. The following discussion critically reviews current ideas of skeletal involvement in treponemal disease and proposes an alternative explanation for its pathogenesis.

The published frequency of skeletal distribution of bones involved in treponemal disease largely depends on the conclusions of the particular author and is therefore dependant on the sample from which the study is based. However, there seems to be general agreement that certain bones are essential to consider in any differential diagnosis that includes a treponematosis. These bones, in order of importance, are: tibia, clavicle, femur, ulna, hands, and feet (Steinbock, 1976; Powell, 1988; Stodder, 1997; Aufderheide and Rodriguez-Martin, 1998; Ortner, 2003) (Figure 2.7). Any bone may be affected, but generally those bones distal to the knee and elbow are most commonly involved (Aufderheide and Rodriguez-Martin, 1998). Lesions are predominantly bilateral and symmetrical (Hackett, 1976). Although it is mostly the diaphyses and metaphyses of the bones that are involved (Steinbock, 1976; Powell, 1988; Stodder, 1997; Aufderheide and Rodriguez-Martin, 1998; Ortner, 2003), severe joint deformities have been recorded in the tertiary stages of yaws (Sengupta, 1983). It has been suggested that the joint involvement in yaws may be useful for differentiating this condition from other treponemal diseases in skeletal samples Ortner et al. (1992). As discussed below cranial lesions also develop in treponemal disease.

The involvement of the tibiae, with concurrent involvement of multiple bones, is often used as a minimum criteria for the diagnosis of treponemal disease in skeletal samples (Stodder and Martin, 1992; Rothschild and Heathcote, 1993; Rothschild and Rothschild, 1995). Although the periostitis of yaws is characteristically exuberant and diffuse, subperiosteal new bone formation of the tibia alone should not be considered pathognomonic of yaws (Heathcote et al., 1998). In 1998, Heathcote et al. presented a critique of some previous work concerning treponemal disease in the western Pacific Islands. The following discussion is an attempt to extend the arguments of Heathcote et al. (1998) concerning the biological basis of bone lesions of treponemal diseases.

While there is general agreement on which bones are affected in treponemal disease the actual mechanism of the distribution of bony involvement is not as clear.

> Tertiary skin involvement tends to occur in places that frequently are subject to trauma or irritation such as the elbows, but the process by which trauma contributes to the development of the lesions is obscure (Bogdan and Weaver, 1992, 158).

Many anthropological accounts of the bone lesions of yaws suggest that they are the result of a direct extension from the skin rash and lesions at all stages of the disease (Powell, 1988; Bogdan and Weaver, 1992). Hackett (1946) reported frequent bone involvement in the secondary stage of yaws in Uganda, however, these were not associated with skin ulcers (Hackett, 1946). Similarly, 'Dactylitis affecting several fingers is commonly seen in children...[but] They are never related to supra-adjacent skin lesions' (Jelliffe, 1970, 758). The distribution of bone lesions is largely uniform and the papillomatous rash is random, therefore the direct extension of irritation of the periosteum from skin lesions seems unlikely. Furthermore, the papilloma of yaws do not penetrate the epidermis (Arya, 1996), therefore the periosteum is unlikely to become directly involved from this pathway. Where bone involvement is secondary to skin involvement it is usually in the inner nasal bones and palate (Keyes, 1908); that is the mucous membranes and not the 'skin'.

A further explanation for the skeletal distribution of treponemal disease is trauma. The sentiment of local trauma causing the production of new bone on commonly affected bones stems from a statement made by Steinbock (1976):

> Apparently the increased incidence of syphilitic involvement in these bones is due to a greater incidence of constant irritation and minor trauma to these more exposed bones. Minor but constant trauma to the bones mentioned may produce similar conditions for syphilitic periosteal inflammation (Steinbock, 1976, 115)

There are a number of questions that the above statement raises, but first, Keyes (1908), whom Steinbock is citing on the subject of why gummatous lesions are commonly found on the tibiae in syphilis, makes the following statement:

> ...[the tibia is the] bone most commonly subject to trauma, easily first, and such subcutaneous bones as the parietals, frontal, sternum, clavicles etc., following, there is doubtless some connection between the position of a bone, subjecting it to the possibility of trauma, and the occurrence of syphilis in it. Add to this the fact that the lesions are chiefly periosteal, that they usually occur upon the surface of the bone most exposed to trauma (e.g. anterior surface of the tibia, exterior surface of the cranial bones) and the picture is complete (Keyes, 1908, 426)

However, Keyes (1908) goes on to suggest: 'But in most instances the direct connection between trauma and syphilis cannot be established, and surely of the many bruised shins few become syphilitic' (Keyes, 1908, 426). This second excerpt from Keyes (1908) demonstrates not only a lack of understanding of the pathogenesis of syphilis, but also the fact that Steinbock (1976) has ignored the more cautious statements of Keyes (1908).

It is doubtful that trauma alone will precipitate the formation of gumma at a bone site as Keyes (1908) implies, however, any major trauma to a local site would likely produce some periosteal reaction. From the clinical descriptions above, it would seem clear that gumma form as part of the immune response to treponemal infection, and not to external forces such as trauma. However, local trauma at a site where a gumma is forming, may *accelerate* its development.

Based on the comments of Keyes and Steinbock, certain questions are raised. For example, if the causative mechanism of the characteristic skeletal pattern is *local* trauma, then why are the lesions predominantly bilateral and symmetrical in their distribution? This implies that all treponemal sufferers whose bones were affected were unusually clumsy, managing to induce major trauma or constant irritation to both limbs, and at the same time. Accidental trauma is by its nature random, unilateral, and asymmetrical and the symmetrical and bilateral pattern of treponemal skeletal lesions reflects a systemic infection, not local trauma. Constant irritation of any bone lying close under the skin, such as the clavicle when carrying heavy loads, might eventually produce a bony reaction, regardless of whether there is any underlying treponemal infection (Resnick, 1995b).

Finally, it would be reasonable to assume that trauma patterns will differ according to cultural activities such as modes of subsistence and patterns of interpersonal violence (Roberts and Manchester, 1995a). Yet, the reported pattern of skeletal involvement in treponemal disease is largely universal, regardless of the temporal or spatial origin of the skeletal or clinical sample from which the study is derived.

2.2.6 *An alternative explanation for the distribution of bone lesions in treponemal disease*

An alternative explanation for the distribution of bone lesions in treponemal disease has been proposed (Buckley and Dias, 2002). It suggests that the distribution of the skeletal lesions of treponemal disease is related to the anatomical distribution of the lymph nodes and vessels that are intimately associated with these bones (Figure 2.5). The lymphatic system consists of a network of vessels and nodes that act as a second line of defence against microbial infection when the local inflammatory reaction fails to combat microbial invasion. Microbial organisms are taken up by the lymph vessels during this period and transported to the regional lymph nodes, where the lymph is filtered. Once the microorganism reaches the regional node, the node will become inflamed and enlarged (Cotran et al., 1994). This is known to occur in treponemal disease (Mims, 1993). Once within the node the treponeme is able to multiply. This proliferation is characteristic of primary treponemal disease, while in the secondary stage further multiplication of the organism within the node causes tissue damage (Kampmeier, 1954; Mims, 1993).

As discussed above the distribution of bone lesions in treponemal disease is characterised by bilateral and symmetrical lesions of multiple bones with the almost universal involvement of the tibia. The inflammation of the regional lymph nodes would spread to the adjacent bones and could incite increased vascularity of the periosteum and eventually the production of new bone. The general lymphadenitis of treponemal disease at all stages would further support this hypothesis.

If the anatomical distribution of the deep lymph nodes and superficial lymphatic vessels that lie near or directly over bones is considered (Figure 2.6) a pattern emerges that is strikingly similar to the skeletal involvement of treponemal disease (Figure 2.7). In such places where bone lesions are infrequent, such as the proximal femur, a mass of thick muscle and fascia protect the periosteum from direct association with the lymph nodes and vessels. This pattern is seen throughout the body where bones are infrequently involved, unless by direct extension from gumma where the soft tissues are destroyed.

The characteristic *caries sicca* lesions of treponemal disease develop in the tertiary stages of the disease (Hackett, 1946). These lesions are mixed destructive and regenerative and result from clusters of small gumma formation in the soft tissues of the pericranium (Steinbock, 1976). The same principle might apply to the cranial bones as it does with the postcranial skeleton, where the bones most intimately associated with the lymph nodes and vessels are affected. The frontal and parietal bones are commonly affected, whereas the temporals and occipital are less frequently involved (Steinbock, 1976). This might be explained by the location of the temporalis muscle protecting most of the temporal bones from contact with inflamed lymphatic structures. Similarly, the occipital bone is protected by the trapezius and sternomastoid muscles and the sheath of the posterior epicranial aponeurosis. However, it should be noted that there are no lymph nodes superior to the mid-facial region, and lymph drainage of the face and scalp occurs by gravity. Therefore, it seems less likely that the association of inflamed or gummatous lymph nodes could be the mechanism for the characteristic osseous lesions of treponemal disease.

Alternatively, if the anatomy of the bone tissue in the cranial bones is considered, some clues to the mechanism for the development of tertiary lesions may be found. The diploë of the flat bones of the cranial vault is highly vascular and involved in haemopoiesis into adulthood (Jacobs, 1982). It is possible that during haematogenous spread of the spirochetes, some may become lodged in the diploic bone, eventually forming gumma. If it is accepted that the high vascularity of the flat cranial bones provides an explanation for their involvement in treponemal disease, the more frequent involvement of the frontal and parietals might be explained by their higher proportion of diploic tissue, compared to the temporals and occipital bones.

The mechanism for the development of lesions of the facial bones fit more comfortably with the hypothesis of the lymphatic system as directly relating to bone involvement. An exuberant formation of new bone on the nasal bones that produces a 'horned' appearance is characteristic of secondary yaws (Hackett, 1946). The involvement of the nasal bones might be explained by the presence of the nasolabial nodes that lie directly over this area.

2.2.7 Distinctive lesions of treponemal disease which aid in differential diagnosis

The previous discussion has outlined the pathogenesis of the bone changes associated with yaws, syphilis, and congenital syphilis. These bone changes are characterised by diffuse apposition of subperiosteal new bone in the secondary stages of the disease, due to an inflammatory response to chronic infection. The postcranial skeleton is most commonly affected at this stage of the disease and the apposition of new bone on the anterior aspect of the tibiae, known as 'sabre shin' or 'boomerang leg' is the most pathognomonic (Hackett, 1936b; Ortner, 2003). New bone may also be produced on the superior nasal bones and inferior palate from inflammation of the mucosa (Hackett, 1976). The diffuse apposition of new bone on the outer aspect of the nasal bones is believed to be particular to yaws, and African individuals with this particular lesion were described as 'Horned Men' (Hackett, 1946).

In the tertiary stages of treponemal disease the pathognomonic gummatous lesions may develop. These can affect any bone, however, the characteristic *caries sicca* lesions of the cranial vault are most distinctive. The lesions of the skull will develop from within the diploic bone and progress outwards and inwards, occasionally resulting in perforation of the inner table. The cranial vault is reportedly more commonly involved in syphilis than in yaws (Roberts and Manchester, 2005), while perforations of the inferior palate and resorption of the nasal area and alveolar bone of the maxillae, known as 'gangosa', are more commonly associated with yaws (Hackett, 1946; Roberts and Manchester, 1995a).

In syphilis, the presence of resorptive bone changes associated with a syphilitic aneurysm may aid in differentiating the two treponemes (Roberts and Manchester, 2005). Similarly, syphilis may also result in destructive changes of the joints, particularly the knee. These changes may result from infection within the joint itself, or they may become damaged and ultimately ankylosed due to the loss of sensation from nervous system involvement: the so-called 'Charcot's joint' (Ortner, 2003). Gummatous destruction of the joints have also been proposed as a means of differentiating venereal syphilis from yaws in skeletal material (Ortner, 1992).and have been reported in clinical (Sengupta, 1983) and skeletal samples (Buckley and Tayles, 2003a).

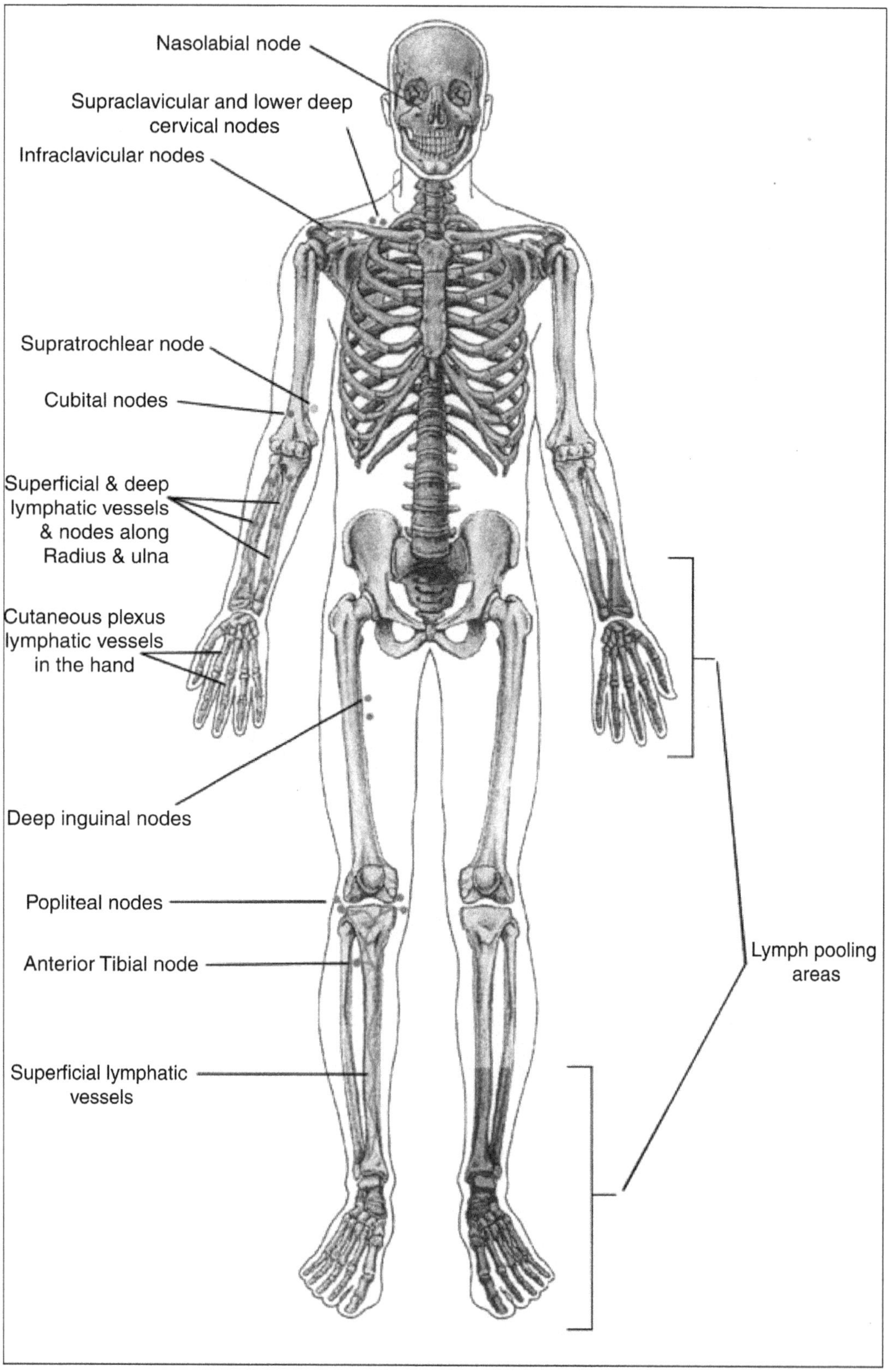

Figure 2.5. *Illustration of the association of the deep and superficial lymph nodes and vessels with bone. The nodes which are indicated by dots are those which lie directly against the bone with no soft tissue intervention.*

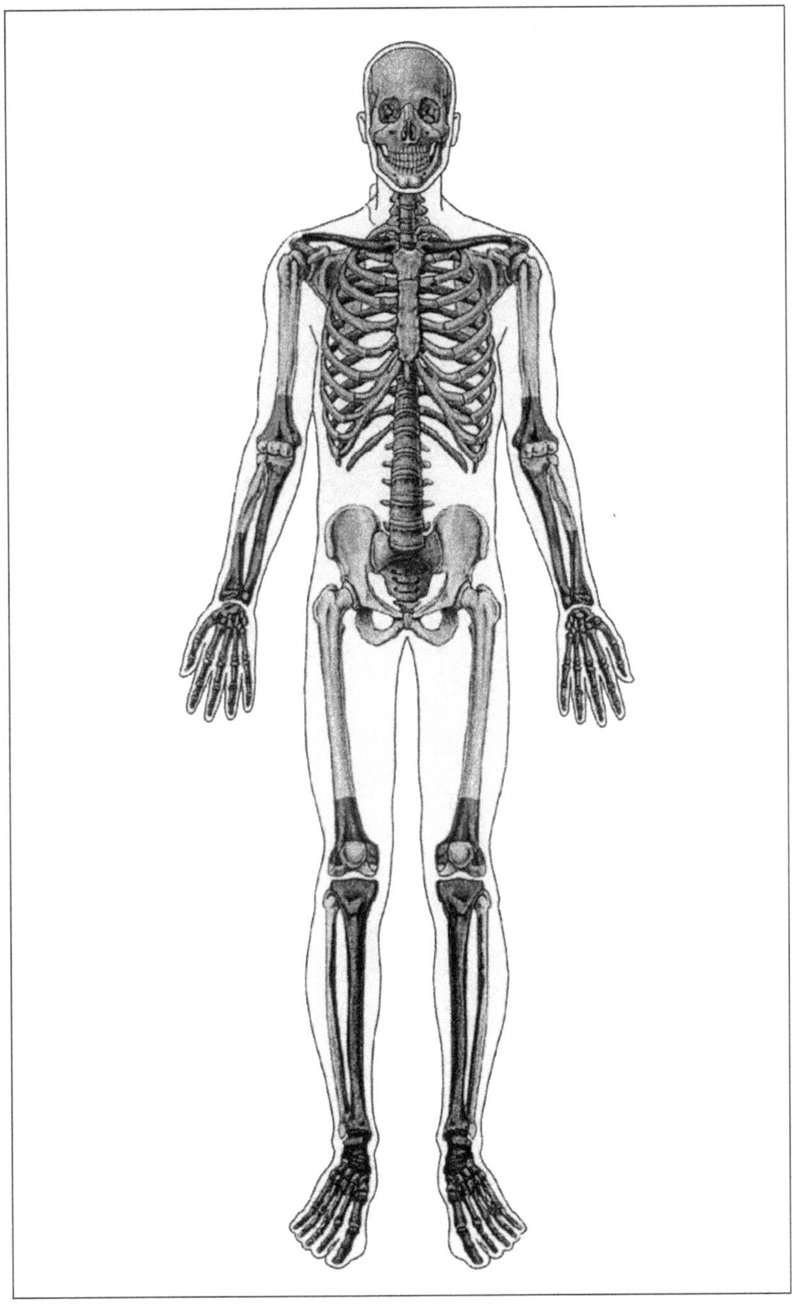

Figure 2.6. *The dark shaded regions illustrate the areas where lymph nodes or vessels are directly associated with bone tissue.*

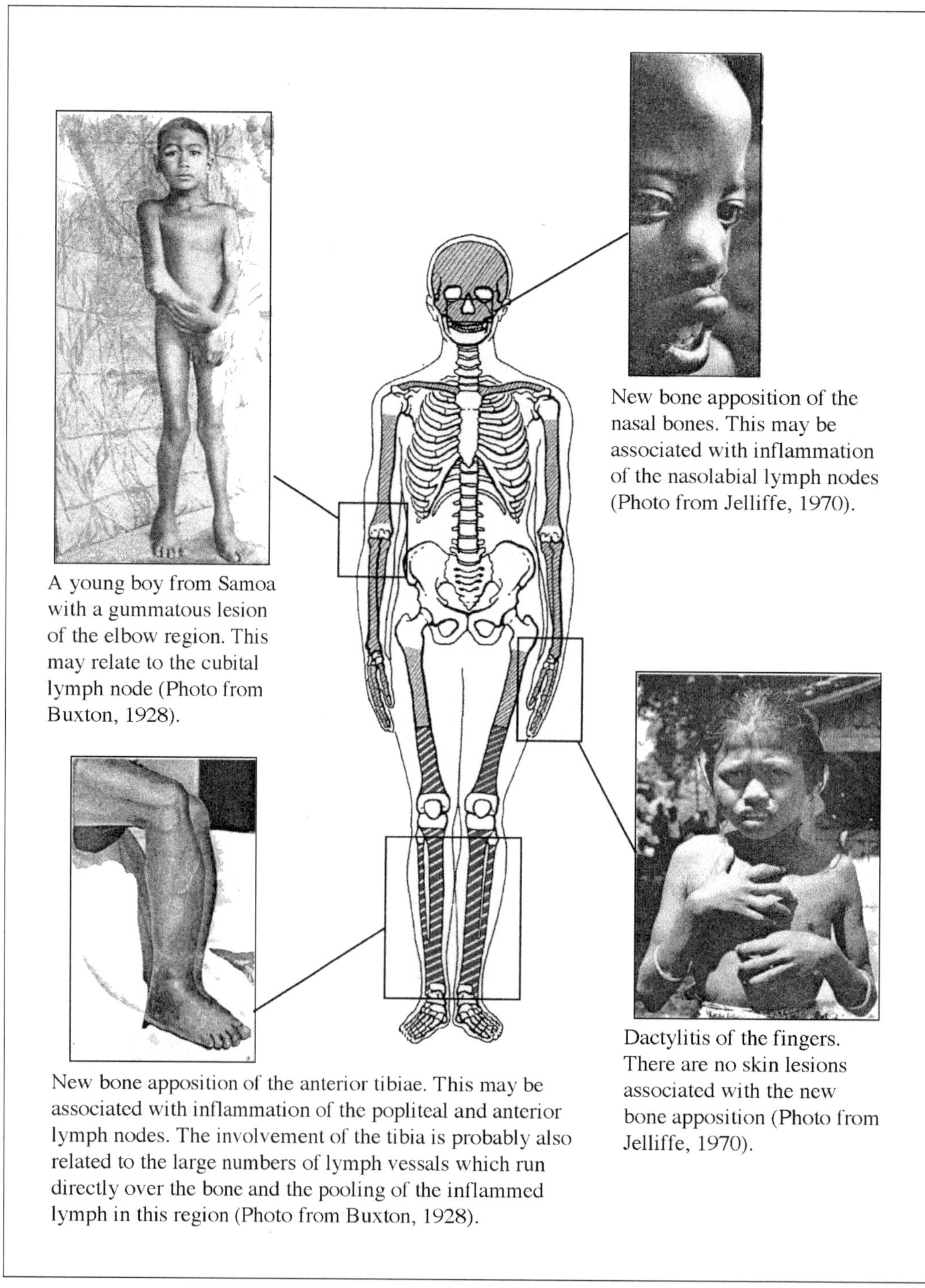

Figure 2.7. *The central skeleton which is adapted from Steinbock (1976) shows the skeletal involvement of yaws. This illustrates the association of bone lesions with lymph nodes and lymph pooling areas as shown in Figure 2.6.*

In both yaws and syphilis, the skeletal distribution of bone lesions is distinctive, with an overwhelming predilection for the tibiae and to a lesser extent the other lower limb bones (Powell, 1988). The involvement of multiple bones, including the tibiae, with bilateral and symmetrical distribution of lesions is often used by researchers as a minimum criteria for the diagnosis of treponemal disease in a skeletal sample (Stodder et al., 1992). Powell (1988) has suggested that the absence of gummatous lesions should preclude a diagnosis of treponematosis, even if multiple bones are affected with diffuse subperiosteal bone apposition. Powell (1988) argues that those individuals with gummatous lesions may be assigned a diagnosis of 'probable' treponemal infection; while those without gummatous lesions may represent a portion of the populations with secondary treponematosis or a non-specific infection and should therefore be assigned a diagnosis of 'possible' treponemal infection, a method which has been applied in the Pacific (Buckley and Tayles, 2003b).

Other researchers were not as cautious as Powell (1988) and developed methods that trumpeted a solution to not only the diagnosis of treponemal disease in prehistoric skeletal material but also as a definitive means of differentiating between the treponemes themselves. The most notable of these is a series of papers by Rothschild and colleagues in which they have developed the SPIRAL method (Rothschild and Heathcote, 1993; Rothschild and Rothschild, 1995, 1996); where S= Saber shin without periostitis in venereal syphilis but not in yaws or endemic syphilis; where P= Prepubescent bone involvement is present in yaws and endemic syphilis but not in venereal syphilis; I= Involvement of the tibia unilaterally occurs in venereal syphilis but not in the other two conditions; R= Routinely affected hand or foot in yaws only; A= Average number of bone groups affected are greater than three in yaws only and L= Lacking periostitis but flattened in venereal and endemic syphilis only. While this method has some merits, the simplistic style in which it was developed has been strongly criticised. In most of these criteria the SPIRAL method does not take account of the biological basis of bone involvement or the variation in treponemal disease expression (Heathcote et al., 1998; Powell and Cook, 2005).

2.2.8 Mycobacteria

The question of the pre-European presence of tuberculosis and leprosy in the Pacific Islands is somewhat more confused than yaws. These diseases are caused by bacteria of the genus *Mycobacterium* and have been responsible for more suffering and misery than any other bacterial disease. While they have been successfully eliminated from Western Europe, they are still causes of significant amounts of disease in the tropics and subtropics (Manchester, 1991).

Leprosy

Leprosy (*Mycobacterium leprae*) is a chronic condition caused by invasion with *M. leprae* bacterium. Leprosy is able to infect people of all ages from early infancy through to old age. In most parts of the world males are affected more frequently than females (Noordeen and Pannikar, 1996).

The successful transmission of leprosy is dependant on consistent and close contact. Therefore the risk of infection within families is high. Transmission of the disease is also influenced by hygiene, nutrition, and what has been described as a genetic susceptibility among families. A high population density and crowding of people in small, badly ventilated dwellings is also thought to influence the prevalence of the disease. These, and other epidemiological factors, suggest that leprosy is a disease of both village and city life. It is therefore not necessarily dependant on large populations for survival (Manchester, 1991).

The palaeopathology of leprosy is best known from Europe (Möller-Christensen, 1967). However, putative skeletal evidence of leprosy has been reported from the islands of Guam and Saipan in Micronesia (Stodder and Martin, 1992; Trembly, 1995). The origin of leprosy in Micronesia is not certain, however Trembly (1995) argues it was carried into this area after contact with China or Japan. Archaeological evidence of Chamorro pottery sherds in Japanese sites supports this hypothesis. No other evidence of leprosy has been reported from skeletal collections from Melanesia or Polynesia. However, it was a significant health problem in many tropical environments in the early period of European colonisation (Russell, 1960; Glasse, 1965; Noordeen and Pannikar, 1996). Van Der Sluis (1969) recounts descriptions of 'leprosy' on first European contact in Tahiti, but argues these were misguided diagnoses and most probably yaws.

In a survey of 19th and 20th century clinical work in other Melanesian and Polynesian Islands groups, Miles (1997) concludes that leprosy was probably in Fiji and the Solomon Islands pre-European contact. Oral traditions in other island groups can name the date in which it was introduced, usually in the late 1800s (Miles, 1997). Te Rangi Hiroa argued for its presence in New Zealand prior to European arrival (Buck, 1910) but there are no published accounts of this in skeletal material to date.

From the brief outline of evidence given above it seems there is some confusion concerning the presence of leprosy in Melanesia and Polynesia prior to European contact. Despite this, it is not ruled out as a contributor to the skeletal pathology in the samples used in this study. There is archaeological evidence and oral traditions that contact between Micronesia and some parts of Polynesia was frequent in prehistory (Spriggs, 1997) and contact between Tonga and Fiji was continual up until recent times (Kirch, 1984). Similarly, the Taumako people were known for their active role in the trade of feather money and as experienced navigators (Collins, 1944; Parsonson, 1966; Spriggs, 1997). Therefore, it is reasonable to assume that if leprosy was present in some Pacific Islands prior to European contact, its existence was possible on both Taumako and Tongatapu.

Pathogenesis and bone changes of leprosy

Leprosy (*Mycobacterium leprae*) is a chronic condition caused by invasion with M. leprae bacteria. It mainly affects the peripheral nerves and has the capacity to attack the vital organs of the body (Andersen et al., 1994; Roberts and Manchester, 1995a). The type of clinical expression and extent of skeletal involvement depends on the immunological status of the affected individual (Andersen et al., 1994) and possibly the route of infection (Noordeen and Pannikar, 1996). Based on this extent of clinical expression the condition can be divided into two groups which are actually extremes of a continuum. These are termed tuberculoid and lepromatous leprosy. Tuberculoid leprosy is a self-limiting condition characterised by few skin lesions holding minimal bacillus in which host resistance is high while lepromatous leprosy is generalised in its distribution with more diffuse skin lesions (Andersen et al., 1994) and low host resistance (Noordeen and Pannikar, 1996). Clinical and paleopathological accounts of the disease attest to the fact that only lepromatous leprosy will affect the skeleton (Faget and Mayoral, 1944; Andersen et al., 1994; Resnick, 1995b; Cook, 1996; Noordeen and Pannikar, 1996; Ortner, 2003).

In lepromatous leprosy the bacilli will mass in intracellular and extracellular material in encapsulated clumps known as globi (Noordeen and Pannikar, 1996). The exact mechanism for disease transmission to a healthy person is not known. Experimental evidence suggests that droplet infection via the respiratory tract may be the most probable route. Following infection, incubation has been reported as ranging from three weeks to 30 years before clinical expression (Noordeen and Pannikar, 1996).

Lepromatous leprosy is characterised by widely disseminated bacilli throughout the skin, nerves, and reticuloendothelial system. There may be bacillary invasion of the eyes, testes, bones, and mucous membranes of the mouth, nose, pharynx, larynx, and trachea (Noordeen and Pannikar, 1996). Motor function and sensation are affected by nerve damage, which causes anaesthesia of the plantar and palmar regions of the hand and foot and paralysis of muscle action leading to 'drop foot', longitudinal plantar arch collapse and 'claw hand' deformities. The continual pressure and shearing of the affected areas, without the limiting factor of pain, can result in trauma-induced periostitis and osteomyelitis of the hands and feet (Cooney and Crosby, 1944; Noordeen and Pannikar, 1996).

Significant to the development of bone lesions in leprous infection is the permanent dilation of arteriole systems (peripheral hyperaemia), a sympathetic component of nerve damage (Andersen et al., 1994). Continued minor injury to anaesthetised areas can lead to local capillary damage and haemorrhage followed by an aseptic necrosis of affected tissues, thereby providing a site for the establishment of bacterial pathogens able to induce bone changes. An individual may suffer from superficial cutaneous ulcers for many years before pain sensation is lost and ulceration spreads to the deep tissue, bones and joints testifying to the insidious nature of the disease (Noordeen and Pannikar, 1996).

There are three broad types of bone changes in leprosy. The first is a non-pyogenic form of lesion caused by the formation of a leprotic granuloma. The second is necrotic bone and joint lesions that develop as the result of nerve damage by the bacillus. The third type of bone lesions are non-specific periostitis, osteomyelitis, and septic arthritis due to secondary infection. An infected individual may develop lesions of all these types either in combination or isolation. Yet, the neurotrophic changes to the hands, feet, and face are considered the most diagnostic of leprosy Möller-Christensen (1967).

Bone lesions resulting from direct invasion by *M. leprae* bacilli are characteristically cystic lesions small in size and restricted to a small area. The epiphysis is most frequently affected and the lesions may take decades to develop (Ortner, 2003). Development of bone cysts may be due to formation of globi clusters in the medulla, periosteum, and nutrient vessels, becoming large enough to produce bone necrosis (Faget and Mayoral, 1944; Noordeen and Pannikar, 1996). A leproma (discrete leprous granuloma) may cause erosion of the outer cortical bone from direct infection of the periosteum, or multiple lepromae may cause extensive erosive lesions. The granulomatous lesions can also affect the subchondral region of the bone leading to instability of the bone architecture and joint collapse. If lepromae develop in the vertebrae, the process may lead to the formation of 'cup joint deformity', which is characterised by vertebral body collapse. There is usually no sclerosis or periostitis associated with these lesions (Ortner, 2003). The cysts usually affect the hands, feet, legs, and forearms but will never invade the proximal joints (i.e. the knee and elbow) and changes to the lower limbs or forearms will only ever be present in conjunction with changes to the hands and feet (Möller-Christensen, 1967; Andersen et al., 1994).

Neurotrophic changes to the bones of the hands and feet are the result of remote nerve damage that causes resorption of the cortex known as 'concentric bone atrophy' (Andersen et al., 1994; Faget and Mayoral, 1944; Noordeen and Pannikar, 1996). In the hands, concentric cortical atrophy begins at the distal phalanges with eventual 'collar-button' shortening of the diaphysis. The phalanges will eventually resorb completely. Bone atrophy may progress to the proximal phalanges, while the metacarpals are usually spared. In the feet, concentric bone atrophy commences in the diaphyses of the proximal phalanges and in the heads of the metatarsals (Faget and Mayoral, 1944). The distal ends of the affected bones become thinned and pointed (Faget and Mayoral, 1944; Noordeen and Pannikar, 1996). This process leaves the foot dramatically shortened with the remaining distal phalanges attached by soft tissue (Faget and Mayoral, 1944; Noordeen and Pannikar, 1996). Concentric bone atrophy of the distal finger phalanges, metatarsals and toes, can also develop as a result of circulatory alteration (hyperaemia) when osteoclastic activity becomes greatly increased. Consequently, resorption occurs at the more vascularised regions of bone (Andersen et al., 1994; Resnick and Niwayama, 1995).

Damage to the motor nerves can cause wasting of the muscle tissue in leprosy. For example, the distinctive 'claw hands' of leprosy are the result of denervation of the ulnar nerve (Noordeen and Pannikar, 1996), which leaves the medial metacarpophalangeal joints vulnerable to trauma (Faget and Mayoral, 1944). Disuse of a limb as the result of muscle paralysis can lead to bony ankylosis of the interphalangeal, metacarpophalangeal, and metatarsophalangeal joints. Osteoporosis is also a common consequence of muscular atrophy (Noordeen and Pannikar, 1996) .

Septic arthritis and osteomyelitis may develop as a result of secondary infection from cutaneous lesions. Deeper bone invasion may lead to a cystic cavity in the cortical bone that may be indistinguishable from the 'Brodies abscess' of osteomyelitis (Andersen et al., 1994). A 'Brodies abscess' is an encapsulated, radiolucent abscess that can develop during sub-acute or chronic stages of osteomyelitis, possibly as the result of increased host resistance or reduced organism virulence (Resnick, 1995b).

Specific leprous arthritis, although uncommon, can occur. Joint involvement results from intra-articular extension of an infective focus, or haematogenous contamination of the synovial membrane by bacilli. Neuropathic osteoarthropathy can also be considered as a secondary mechanism of joint involvement in leprosy (Resnick, 1995b).

The pathognomonic *facies leprosa* lesions of the rhinomaxillary region were described by Möller-Christensen (1967) following detailed investigations of burial material from Medieval leper hospitals in Denmark. The atrophic *facies leprosa* changes include a blunting and widening of the normally sharp nasal aperture due to resorption of the bone from the inferior aspect. Atrophy of the anterior nasal spine may also occur. Destruction of the nasal cartilage is a common sequel of lepromatous leprosy, and the dissolution of the cartilage produces the characteristic 'saddle nose' of late leprosy (Faget and Mayoral, 1944). The maxillary alveolar margin may also resorb horizontally, beginning centrally, and leading to premature loss of anterior dentition (Möller-Christensen, 1967; Roberts and Manchester, 1995a; Ortner, 2003).

Möller-Christensen (1967) stresses that inflammatory changes to the superior palate are pathognomonic of leprosy, because the former rhino-maxillary changes may not always be present. The nasal septum and hard palate are usually left intact (Faget and Mayoral, 1944; Ortner, 2003), however perforation of the palate beginning from the superior aspect can occur (Möller-Christensen, 1967; Roberts and Manchester, 1995a). The atrophic and inflammatory changes of the rhino-maxillary region of the skull are considered the result of direct invasion by *M. leprae* pathogens and secondary infection from cutaneous tissue (Andersen et al., 1994; Resnick and Niwayama, 1995). Changes to the cranial vault are not believed to occur in leprosy (Möller-Christensen, 1967).

2.2.9 Distinctive lesions of leprosy which may aid in differential diagnosis

Probably the most diagnostic characteristics of leprosy are the almost exclusive involvement of nasal regions and the bones of the hands and feet (Steinbock, 1976). The involvement of these bones is further characterised by lesions that are purely resorptive in nature. Atrophy of the anterior nasal spine is not known to occur in any other disease. The perforation of the hard palate can occur from leprosy and treponemal disease. However, in leprosy the palate is resorbed from the superior aspect while in yaws resorption begins from the inferior palate, reflecting the mucosal involvement of leprosy (Ortner, 1992).

In yaws and syphilis the nasal region can also be involved. However, in leprosy this is restricted to resorption of the bones around the inferior nasal aperture, while in treponemal disease, the more superior aspect of the nasal bones are usually affected by a proliferative rather than resorptive reaction (Ortner, 1992). The resorption of the alveolar bone of the maxillae, beginning centrally and resulting in loss of the anterior dentition, might be confused with adult scurvy. However, in cases where the other distinctive resorptive changes of the nasal bones and concomitant hand and food involvement in leprosy are present, scurvy can be ruled out (Aufderheide and Rodriguez-Martin, 1998).

Tuberculosis

Tuberculosis (TB) is either an acute or chronic infection of the skeletal and/or soft tissues caused by *Mycobacterium tuberculosis*. There are two species of *Mycobacterium* that cause disease in humans. One is *M. bovis* where cattle are the reservoir for disease. Humans are infected by ingesting the milk from infected animals. This species is obviously not relevant in the pre-European Pacific Islands where cattle was not present, so only *Mycobacterium tuberculosis* is considered in the following review.

There are a number of risk factors in transmission of TB. For example, infants and young children are more susceptible to infection because of an underdeveloped immunity compared to adults. Malnutrition is also correlated with high infection rates. In children, the risk of infection is higher in closed rooms or huts with little ventilation. Prior to effective treatment of TB, the highest mortality rate was in children under five years of age, then again in the third decade (Manchester, 1991). Tuberculosis has been described as a density dependant disease of urban life (Cockburn, 1963; Buikstra and Cook, 1980).

The existence of TB in the pre-European Pacific is again confused. Equivocal cases of skeletal TB have been reported from Hawaii (Snow, 1974; Trembly, 1997). Tuberculosis lesions have been reported from the Marquesas and NG (Pietrusewsky, 1976), but these have not held up to subsequent scrutiny (Trembly, 1997). Skeletal lesions consistent with TB have been reported in early New Zealand Maori (Buckley et al., 2010) but as these authors acknowledge, aDNA is required to confirm diagnosis. A further up to date review of this issue can be found in (Buckley and Oxenham, 2016).

As to the origins of TB in Hawaii, Trembly (1997) argues for an American origin by way of multiple Polynesian voyages to American soil. Although no previous evidence of the presence of TB has been found in skeletal material from Tonga or Taumako, it is not ruled out as a possible contributor to lesions observed in this study.

Pathogenesis and bone changes of tuberculosis

Tuberculosis is a disease of two phases: 1) the primary phase and 2) the re-infection or reactivation phase. Human infection with the bacillus is usually caused by inhaling infective moisture droplets in the air from a person with pulmonary tuberculosis. As a result the primary infection of tuberculosis is commonly a respiratory disease. A period of bacterial multiplication follows and some bacillus escape into the extravascular fluid, to be picked up by the lymphatic system which eventually carries them to the lymph nodes of the lung hilum (Aufderheide and Rodriguez-Martin, 1998). Both of these lesions together form the 'primary complex' of tuberculosis. Less common is the intestinal pathway of a primary complex in the intestinal wall and mesenteric lymph nodes (Ortner, 2003).

The bacilli are able to proliferate freely within the immune cells' cytoplasm and resist destruction by macrophages because of a waxy coating. However, an immune response by macrophages and lymphocytes at the infected site secretes chemicals which cause injury to tissues. In the lung, this process results in necrosis of the tissues affected. If this immune response is effective, the necrotic caseous tissue will become walled off in an envelope of scar tissue. This same process occurs in the infected lymph nodes. These caseous scars may become partially or completely calcified. Trapped within this shell of scar tissue and calcium the bacilli survive and represent a potential for reactivation of the infection if conditions in the environment of the lung or immunological status of the person changes (Resnick and Niwayama, 1995; Aufderheide and Rodriguez-Martin, 1998).

In most cases primary infection with the bacillus will run its course without any clinical effects. In some cases a compromised host immune status which may be caused by factors including malnutrition and co-infection, may impede the normal immune response and allow progression of the primary infection (Aufderheide and Rodriguez-Martin, 1998; Ortner, 2003). If this occurs, galloping consumption, or rapidly progressive pneumonia will follow with erosion into a bronchus and seeding into the rest of the lung. If this occurs, a pulmonary vein may be eroded and dissemination of the bacteria into the blood stream results in infection of any or all organs. Skeletal tuberculosis is mainly the result of haematogenous dissemination of the bacillus (Ortner, 2003). Although Steinbock (1976) stresses a pathway of secondary infection from either the lungs or lymph system.

Secondary tuberculosis

Secondary tuberculosis may develop after a period of many years, as the result of re-infection with the tuberculous bacteria. This may occur by the breakdown of the encapsulated primary lesion releasing bacteria or because the individual is reinfected by the bacillus (Aufderheide and Rodriguez-Martin, 1998). A vigorous cell-mediated immune response is activated, because the hosts immune system is already 'primed'. This results in a larger area of tissue death and an accompanying granulomatous inflammatory reaction. A larger area of tissue necrosis ensues and an accompanying granulomatous inflammatory reaction. With extensive tissue necrosis, structures bordering these foci may become involved. For example, infection of a lymph node adjacent to a bronchus may erode the bronchial wall and the liquefied caseous material containing bacteria will then discharge into the bronchial lumen and be coughed up into the oral cavity and trachea. With the bacteria transported into the trachea by this pathway the infected sputum may be swallowed and infect the intestinal tract (Aufderheide and Rodriguez-Martin, 1998). All organs and lymph nodes may become involved by many different pathways.

The tubercle bacilli spread by haematogenous dissemination will affect the areas of skeleton that have a high circulatory and metabolic rate; namely the haemopoietic (red) marrow. Therefore, it is the areas of cancellous bone rather than the cortex or medullary cavity that are usually affected (Ortner, 2003). In the long bones of adults, the epiphyses and metaphyses are affected to a greater extent because of the concentration of cancellous bone in these areas (Resnick, 1995b). However, because the distribution of haemopoietic marrow is greater in infants and children, the tuberculous bacteria can be much wider spread than in adults (Ortner, 2003).

Initially tuberculous infection will permeate the marrow space leading to devitalized areas of cancellous bone and thus the formation of central sequestra of the cancellous tissue. In the secondary granulomatous stage of infection the process leads to local destruction and cavitation of the cancellous bone. In both stages there is very little, if any, production of new subperiosteal bone. An exception may be in the small tubular bones of infants and children where the destruction of the cortical bone initiates the production of reparative bone over the necrotic area (*spina ventosa*) (Ortner, 2003). The term spina ventosa is derived from *spina* meaning 'a spine-like projection', and *ventosa* meaning 'full of air'. This descriptive term evokes the appearance of these lesions very well, where the affected bone looks to be hollowed out and covered with a thin shell of reparative bone.

Spina ventosa changes to the diaphyses are most commonly observed in the phalanges, metacarpals, and metatarsals. Similar changes can develop in other bones, such as the ulna, radius and humerus (Resnick, 1995b). The formation of new subperiosteal bone may also be observed in association with an extraosseous abscess (Ortner, 2003). A rare form of tuberculous bone lesion is of a cystic type affecting one or more bones which is more frequent in children than adults and usually affects the metaphyseal regions of the tubular bones. The cystic lesions may be symmetrical and are not accompanied by any sclerosis (Resnick, 1995b).

Joint involvement in tuberculosis is intimately linked with infection of the adjacent bone, where the process may begin in the synovial membrane, in the bone itself, or in both simultaneously (Ortner, 2003). The vertebral column, hip, and knee are most frequently affected (Steinbock, 1976). Joints of the lower extremities are more commonly involved than those of the upper limb (Resnick, 1995b). When lesions of the arm or shoulder develop, they are usually found in adults (Steinbock, 1976). Joint involvement resolves after destruction of the joint space often ending in bony ankylosis of the affected bones (Ortner, 2003).

Tuberculosis of the spine (Pott's disease)

The spine is the most frequent site of skeletal involvement in tuberculosis, constituting 25–60 per cent of lesions (Resnick, 1995b). This is because the vertebrae house the largest amount of cancellous bone in the skeleton and are very well supplied with oxygenated arterial blood on which the bacillus thrive (Aufderheide and Rodriguez-Martin, 1998). The bacillus are deposited within the vertebral bodies by way of the arterial blood supply and proliferate adjacent to the cartilaginous end plate, forming an abscess. Eventually the abscess will perforate the end plate, through the disc, and encroach on the neighbouring vertebra. Lesions of the vertebrae predominantly affect the anterior portion of the bodies, although why this is the case is unclear. Lesions of the spine are sometimes considered the result of haematogenous dissemination of the bacteria, but the vertebrae may also become involved by direct extension from visceral and lung lesions (Aufderheide and Rodriguez-Martin, 1998).

The reporting of the regions within the spine most affected varies considerably. However, there is general agreement that the lower thoracic and upper lumbar are more commonly involved than other regions (Aufderheide and Rodriguez-Martin, 1998; Resnick, 1995a; Ortner, 2003). The bacteria spread from an infected site in the trabeculae of the vertebrae and extend beneath the anterior longitudinal ligament. The adjacent vertebra may also become involved by direct spread of necrosis of the intervening tissues. Distribution of lesions in the vertebrae may be anterior (cortical destruction under the anterior longitudinal ligament), paradiscal (erosion of the cartilaginous end-plate with narrowing of joint space and kyphosis caused by destruction of the metaphyseal region) and central, where destruction begins in middle of the vertebral body and spreads, leads to kyphosis and collapse (Bullough, 1992; Aufderheide and Rodriguez-Martin, 1998). Kyphosis is a sharp anterior angulation of the spine due to vertebral collapse and is most common in the thoracic vertebrae (Ortner, 2003). The paradiscal involvement is the most frequent change, constituting more than 50 per cent of lesions (Aufderheide and Rodriguez-Martin, 1998). The pattern of resorptive lesions of the spine is not particular to tuberculosis only. Many other diseases can cause these changes, which makes a diagnosis of tuberculosis in skeletal material particularly difficult (Steinbock, 1976; Ortner, 2003).

Involvement of the ribs has also been suggested as a possible means of diagnosing tuberculosis in skeletal material. Kelley and Micozzi (1984) found that 9 per cent of individuals who died of pulmonary tuberculosis had either diffuse periosteal lesions and/or localised areas of resorption on the pleural aspect of the rib shafts. They suggest these lesions are a likely sequel of adjacent infection of the pleura and lungs (Kelley and Micozzi, 1984). Further issues concerning the diagnosis of tuberculosis in skeletal samples will be covered in the discussion at the end of this chapter.

Distinctive changes of tuberculosis which may aid in differential diagnosis

Generally the skeletal changes of tuberculosis are characterised by solely resorptive changes. If new bone is produced it is minimal in comparison to diseases such as treponematosis (Ortner, 2003). The spina ventosa lesions are similar in appearance to a large area of reparative bone as seen in osteomyelitis and congenital syphilis. However, lesions of the latter two diseases usually involve a single bone and are less severe than the spina ventosa lesions (Ortner, 2003). The destruction of the vertebral bodies, leading to kyphosis and subsequent fusion (Pott's disease) are considered pathognomonic of this disease (Ortner, 2003).

2.2.10 Diseases which may cause erosive arthropathy of appendicular joints

Firstly, all of the bacterial diseases discussed above have been known to cause joint destruction. A further infectious disease that is worthy of consideration is Ross River Fever (RRF) or epidemic polyarthritis. This viral disease has been known to cause epidemic outbreaks of disease in the Solomons Islands, NG, and Australia for decades. More recently epidemics have been reported in American Samoa (Tesh et al., 1981), Fiji, and the Cook islands (Rosen et al., 1981). The virus is transmitted by a mosquito vector, usually the same vector for filariasis. It is a self-limiting disease which is characterised by acute joint pain and effusions of the knees, ankles and wrists. The initial attack may last only a few days but pain persists in affected joints for up to a year. The pathology of joint involvement is thought to be from the virus multiplying within synovial cells causing local inflammation and pain (Adebajo, 1996). The arthritis is known to affect adults more than children and older adults more frequently than the young. Males and females are affected in similar proportions (Adebajo, 1996). Arthritis seems to develop in joints which have been previously diseased or injured (Tesh et al., 1981). No long term sequelae are reported in association with this virus (Resnick and Niwayama, 1995), so it cannot be argued for as a strong contender of the cause of the joint lesions observed in this study. However, the fact that chronic joint pain continues for months in some affected people suggests that inflammation could lead to some bony change.

Erosive arthropathies

A group of conditions which can affect joints are non-infectious in origin; these are rheumatoid arthritis and the seronegative spondylarthropathies (ankylosing spondylitis, psoriasis and Rieter's syndrome) and gout. Rheumatoid arthritis (RA) is the articular expression of a chronic systemic disease (Jaffe, 1972). The aetiology of RA is unknown, although a hereditary link has been proposed (Jaffe, 1972). RA also has many characteristics suggestive of an autoimmune disease including the presence of an abnormal immunoglobulin in the blood known as rheumatoid factor (Aufderheide and Rodriguez-Martin, 1998). An association with infectious disease has also been proposed (Silman, 1991). In European populations, about 2 per cent of adults are affected and three quarters of them are women (Jaffe, 1972; Aufderheide and Rodriguez-Martin, 1998).

RA affects the synovial membrane of joints which becomes inflamed, thickened and hypervascular. Long-term synovial inflammation produces a chronic state in which the affected synovium is referred to as a pannus (Rogers and Waldron, 1995). Pannus will eventually destroy articular cartilage causing subchondral destruction and erosion of bone marginal to the joint which is unprotected by articular cartilage. The affected joint can eventually be completely destroyed resulting in fibrous and bony ankylosis (Resnick and Niwayama, 1995). The small joints of the hands and feet are invariably involved in the early stages of the disease and later, the large joints such as the knee, shoulder, elbow and wrist (Aufderheide and Rodriguez-Martin, 1998). The synovial joints of the vertebrae may also be affected, particularly those of the cervical spine (Rogers and Waldron, 1995).

The bone changes of RA are characterised by purely erosive lesions beginning at the margins of the joints (Rogers and Waldron, 1995). The pattern of involvement is usually polyarticular and symmetrical (Aufderheide and Rodriguez-Martin, 1998; Rogers and Waldron, 1995). Large cysts can also develop in the subchondral bone where increased intrarticular pressure forces the synovial fluid and pannus into the bone. The development of these articular cysts are found in people who remain physically active while affected by the disease (Resnick and Niwayama, 1995). Contrary to other erosive arthropathies, the sacro-iliac joint and cartilaginous joints of the spine are spared in RA (Rogers and Waldron, 1995).

RA is generally thought to be an urban disease of Europeans (Rogers and Waldron, 1995; Blondiaux et al., 1997). However, it has been found in African and Asian populations (Kilgore, 1989; Silman, 1991) and also reported in modern Pacific Islanders (Wigley, 1987). The clinical expression of RA in individuals from tropical climates is less severe than in temperate climates. This is thought to be due to a protective advantage of infectious disease, such as malaria (Adebajo, 1991). The actual mechanism for why this protective relationship might exist is not known.

RA has been diagnosed in skeletal material from Europe (Rogers and Waldron, 1995), the Americas (Woods and Rothschild, 1988) and Africa (Kilgore, 1989). RA has not been reported in prehistoric skeletal material from the Pacific Islands. It was assumed to be absent from this region, however, its presence in Asia (Adebajo, 1991) and a report of definite cases from Tokelau (Wigley, 1987) would suggest that rheumatoid arthritis should be considered in the differential diagnosis when evidence of an erosive arthropathy is recorded in this region of the world.

Seronegative spondylarthropathies

This group of diseases includes ankylosing spondylitis (AS), psoriasis, and Reiter's syndrome. AS affects the cartilaginous joints more than the synovial joints (Rogers and Waldron, 1995) so will not be considered further. The two latter conditions are characterised by primarily erosive lesions of the joints so will be considered in more detail here. Psoriasis has not been reported in the Pacific Islands, so will not be considered any further in this discussion.

Reiter's syndrome is an uncommon disorder is associated with concurrent infection, usually of a venereal origin. It is much more common in men than women. This syndrome has been reported in association with venereal disease in modern NG (Adebajo, 1996). The joint lesions of Reiter's syndrome are usually asymmetrically distributed and are marginal to the joint with subchondral destruction sometimes evident. The lesions are more common in the lower extremities and in the small joints of the foot (Resnick and Niwayama, 1995). This condition is distinguished from RA by the association of proliferative new bone with the erosive lesions and sacroiliac and spinal joint ankylosis (Rogers and Waldron, 1995). RA is also characterised by osteoporosis and uncommon involvement of the axial skeleton (Resnick and Niwayama, 1995).

Gout

Gout (hyperuricemia) is a disorder of the metabolism in which urate crystals accumulate as granulomatous masses in and around the joints (Resnick, 1995a). These masses are called tophi which can affect all joint structures. Like RA, gout is a disease of the synovium. A single joint is usually involved, but multiple joints can be affected (Rogers and Waldron, 1995). The bony lesions of gout can be situated on any aspect of the joint or even remote from a joint space. The lesions are described as 'punched out' in appearance and typically have a sclerotic margin with overhanging edges indicative of bony proliferation (Rogers and Waldron, 1995). The distribution of lesions is typically asymmetrical and in 75 per cent of cases is restricted to the metatarsophalangeal joint of the big toe (Rogers and Waldron, 1995). More men than women are affected with gout in Europeans (Resnick, 1995a).

An unusually high incidence of gout has been noted in Pacific Island populations which is considered to be the result of a genetic predisposition for hyperuricemia (Klemp et al., 1997; Gosling et al., 2014). Skeletal evidence for gouty lesions has been reported from Micronesia (Rothschild and Heathcote, 1995). More recently, gout has also been reported in Lapita skeletons from Vanuatu (Buckley, 2007) and in early New Zealand Maori (Buckley et al., 2010). The study of this disease in prehistoric Pacific peoples is a potential future avenue for understanding the origins and antiquity of metabolic syndrome related diseases in modern populations (Buckley, 2011; Gosling et al., 2015).

2.3 The Skeletal Samples

An outline of the archaeology and environmental context of the skeletal samples used in this study will be provided below. This is important because of the different cultural and physical environments of each site and the effect this may have had on the health of the populations. Where relevant I will also give an update on publications and other research on the samples since the completion of the thesis.

2.3.1 Taumako

Taumako is the largest island in the Duff group, southeast Solomon Islands, Melanesia. There are also other smaller islands in the Duff group. Taumako is a high volcanic island with 16.2 square kilometres of land area on the main island (Terrell, 1986). A fringing reef around the island forms a lagoon rich in marine resources. The main island of Taumako is heavily forested and the report from the Spanish explorer Quiros suggests at least one good source of water in the form of a small river (Kelly, 1966). Early in the 20th century all inhabitants of Taumako lived on an artificial island named *Tahua* in the lagoon, as they do today. The population of Taumako was said to number 143 persons in 1931 (Collins, 1944).

The climate of Taumako can be extrapolated from information concerning the Solomon Islands as a general group. In the Solomons the climate is consistently hot and humid with a heavy rainfall throughout the year. The prevailing wind is the southeast trade which is strongest from the months of April to November. Hurricanes are rare in this region, but when they do occur it is predominantly in the Southeast Solomon Islands around the Santa Cruz group (Collins, 1944).

The Excavation

The sample is derived from an excavation by Janet Davidson and Foss Leach in 1977–78 on Taumako. Two burial mounds about 200 m inland from the beach were identified at Tetoli Bay, Taumako (Fyfe et al., 1993; Redvers-Newton, 1995). Namu, the largest of the two mounds, was chosen for excavation. The name Namu means either mosquito or smelly in Polynesian and it was noted that the large numbers of mosquitoes made living conditions difficult on the main island (Leach and Davidson, 1978). It is interesting that the method of burying the dead in large raised mounds was commented on by Quiros' men in 1606 (Kelly, 1966). However, the local modern inhabitants have no tradition concerned with this mound or any of the other mounds in the area (Spriggs, 1997).

At the time of the submission of this thesis the archaeology of the sample had not been published, therefore much of the information pertaining to this site was gleaned from the excavation diary (Leach and Davidson, 1978), personal communication with the archaeologists and published accounts of some aspects of the site. In 2008 the archaeologists published the monograph on their expedition to Taumako, including a detailed account of the excavation of the Namu burial mound on the island (Leach and Davidson, 2008). The monograph included numerous appendices compiled by various experts on the human skeletal remains and faunal and cultural materials. These appendices were published as they had been prepared in the 1980s.

The mound was about 8 meters in diameter and 70 cm high and was formed by the activity of burial itself. There were no graves dug for a burial but the individual was lain on the ground with coral rubble heaped over the body. Thus the mound was formed through the act of placing multiple burials in one area (Figure 2.10).

An 8x8 metre square excavation area was initially opened, but as time grew short, the emphasis was shifted to the north east quadrant of the square. This part of the site was excavated to natural sediments. The first archaeological layer of the site consisted of mixed cultural deposits, with some 'shadows' of graves where the bone material had been destroyed. However, the second layer contained many burials within the coral matrix. The bone material of the burials was well preserved, however in many cases skeletal elements were fragmented. Unfortunately, there was little information pertaining to the stratigraphic levels of individual burials. Therefore, no temporal assessments of changes in health could be carried out in the present study.

However, some interesting information about patterns of burial could be gleaned from the excavation diary and later described in the site report (Leach and Davidson, 2008). Most burials in the more recent levels were supine and extended. Many were buried with grave goods. Some of the grave offerings were very rich and consisted of jewellery such as elaborate shell bead necklaces, *tavi* (a large disc-shaped shell pendant) (Figure 2.8), and knee rattles with bone toggles (Figure 2.12). Young children and babies were buried with equally rich grave offerings, such as 'aprons' of beads sewn together like a skirt (Figure 2.9).

From the way in which the burials were grouped in lines, the archaeologists suggested these were family plots that had been used for generations (Figures 2.10 and 2.11). These groups consisted of successive burials which were in some cases up to four layers. One group held three males, all with knee rattles. These family plots included some double buri-

Figure 2.8. *Burials 36 and 37 showing a Tridacna shell breast plate (tavi). The smaller disk is not part of the burial. Photo courtesy of Foss Leach.*

Figure 2.9. *Beads associated with Burial 129, a 1.25 year old child, Taumako. The beads were overlying the lower limbs in the form of an apron. Photo courtesy of Foss Leach.*

als where males and females were buried together. The men were always placed on the right of the women. Similarly, the archaeologist identified several burials where a female and young child or baby were buried together (Leach and Davidson, 2008).

The earlier levels of the excavation revealed some patterns different from the more recent layers of burials. For instance, individuals were buried in a crouched or flexed position, people were buried with shell necklaces and knee rattles which were smaller in size. Also, many of the knee rattles in the lower levels were made of white cowrie shell, rather than the more common black and brown type. The archaeologists also felt the individuals in the lower levels were younger at death and more young women and children were represented. They also felt there were more individuals with evidence of disease than in the later levels.

Without any information on the stratigraphy of the site, or an analysis of grave goods and site layout, none of these impressions could be tested at the time of the thesis completion. The conclusions of the archaeologist relating to sex and age of the individuals may differ from the estimates in this thesis. However, these impressions provide tantalising clues of the culture and ideals of the people buried in the mound. It would seem the people of Taumako had a rich material culture and recognised infants as members of the community. Also, the difference in burial position and grave goods between the early and later levels may indicate a changing cultural tradition.

A radiocarbon date was obtained from a piece of charcoal near the base of the site which provided a single date of about 340 years BP (NZ4639, CRA = 338±55) (Davidson and Leach, 1991). Electron Spin Resonance (ESR) analysis of a series of bone samples suggested that the burial mound was in use for about 170 years, from about AD1530 to 1700 (Whitehead et al., 1986).

European contact at Taumako

The Santa Cruz Islands were first discovered in 1595 by the Spaniard Mendana, while the first European contact on Taumako was in 1606 by his countryman Quiros (Kelly, 1966). Quiros and his sailors stayed on the island for eight days to collect water and firewood. The inhabitants of Taumako were reported to be most accommodating. The chief Tumai ordered that the houses on the artificial islet of *Tahua* be vacated by the locals for the duration of their stay. It was noted that *Tahua* was densely populated, with one hundred dwellings and small openings for entrances:

> The dwellings are laid out in good order with streets and squares, adorned with very tall palm trees and other fruit trees, which made the place very pleasant (Kelly, 1966).

A more recent impression of the islet off Taumako is from Parsonson (1966) who describes the village area as covered with fine yellow sand and kept immaculately clean. The people of Taumako were described in Quiros' chronicle as:

> mullattos in colour, being not very dark; their hair is frizzy like that of the negroes; in fact there are some negroes among them. Some of the natives seen were white and ruddy, with fair hair like Flemings (Kelly, 1966, 184).

Tumai was said to be a strong well-muscled man with fine features and one of his wives was described as: 'of good height, half-caste in colour and clothed in mats' (Kelly, 1966, 188). There was little contact with other locals during Quiros' stay and the protests of Tumai that the Spaniards were overstaying their welcome suggests a certain amount of fear on the part of the Taumako people. The only mention of the health of the local inhabitants was in relation to wounds acquired from raids on their neighbours in the Santa Cruz Islands:

> Those natives seemed *corsairs*, and they must raid all those islands, where they are accustomed to make captives of their neighbours....They had feuds, because many of them had wounds, some of which had not healed, on their left side, from combats fought with bows and arrows (Kelly, 1966, 167).

Possible evidence of interpersonal violence in this population has been reported. The evidence was in the form of a bone spear or arrow point lodged within a lumbar vertebra of a child from this site 2.13. A healed depression fracture of the frontal bone of an elderly male may also indicate interpersonal violence (Buckley, 2000a; Scott and Buckley, 2010). Considering the lack of contact between the Spaniards and the locals during their stay it is not surprising that little else of the health of the people was remarked upon.

The Taumako inhabitants were considered accomplished sailors (Kelly, 1966; Parsonson, 1966) and therefore may have had indirect contact with European pathogens from AD1595 onwards after Mendana visited Santa Cruz. Certainly, when Quiros reached the Duff Group, the local inhabitants remembered Mendana's stay at Santa Cruz, because one of the local men had been killed by the Spaniards. This and the description of taking captives attests to some contact between these islands. The dates of this site indicate the burial mound was in use during and after the time both Spanish expeditions landed at the island. Therefore, European pathogens cannot be ruled out as contributing to the pattern of pathology found at the site.

Biological and cultural affinities of Taumako

Taumako Island is considered to be a 'Polynesian Outlier' (Bellwood, 1989). As mentioned in chapter one, this term applied to the 18 islands in the western Pacific whose inhabitants speak a Polynesian language. One of the research objectives of the excavation at Taumako was to test whether these people were more biologically similar to Polynesians than Melanesians (Leach and Davidson, 2008). Houghton (2008) suggested they exhibited morphological characteristics that were indicative of Polynesian ancestry, with some

Figure 2.10. *An overhead view of the Namu excavations in progress. The combination of square and burial excavation methods is evident. Photo courtesy Foss Leach.*

Figure 2.11. *The Namu excavations in progress at an early stage. Note the lines of burials and also lines of vertically placed stones. Although some of these may have outlined the placement of burials, many of these stones appear to have been re-used for subsequent burials, resulting in considerable jumbling of stone features. Photo courtesy of Foss Leach.*

Figure 2.12. *Shell knee rattles with Burial 92, a young child (two years old) from Taumako. Photo courtesy of Foss Leach.*

'Melanesian' admixture. This suggestion was based on comparative metric analysis with other Pacific Island skeletal material. This analysis of the possible biological relationships of the Namu skeletal material was conducted in the 1980s and published without modification in 2008 in the Leach and Davidson (2008) monograph.

The following quote from Parsonson (1966, 20) lends further support to a more Polynesian origin of the Taumako people:

> It would...seem from the early Spanish accounts that the original culture of Taumako, I think the first artificial islet in these seas and the archetype of the rest, was Tongan and that the artificial islet itself was derived from the Tongan *esi* mound, a dryland structure, adapted to a specific function, that of warding off malaria.

The artefacts associated with burials in the Namu mound were generally similar to ethnographically observed materials from the wider area. However, some ornaments, such as dugong and whale tooth beads were cut in a reel-shape, reminiscent of examples from western and eastern Polynesia (Spriggs, 1997). Excavations of other sites on Taumako attest to a period of habitation as long as 3000 years (Kirch, 1984). The artefacts recovered from these sites might suggest that a small population has lived on this island with sporadic contact with Polynesian peoples for at least this amount of time. Various sources have suggested that the traditions of the people of Taumako are more closely affiliated with neighbours on Santa Cruz than with any source in Polynesia (Davenport, 1968; Terrell, 1986; Leach and Davidson, 2008).

Therefore, the biological ancestry of the Taumako people is unclear. While some argue that the Taumako people were more closely affiliated with Santa Cruz (Kirch, 1984), contact with the islands of western Polynesia may have been frequent with an exchange of genetic and cultural material. The chronicle of Quiros' stay on this island describes a people of 'mixed race' and the study of Houghton (2008) would seem to support this conclusion. For the purposes of the present study it is assumed the Taumako people who formed the mortuary sample were Polynesian speakers with cultural and genetic affinities with their close neighbours as well as their Polynesian relations.

Previous work on the Taumako sample

An account of the human skeletal material was compiled by Philip Houghton for inclusion into the anticipated monograph on the archaeology of the Namu site (Houghton, 2008). Houghton (1996) included much of the metric data for comparative analysis in his book *People of the Great Ocean*. This book also included an account of the health and disease of the people, but failed to mention any pathology

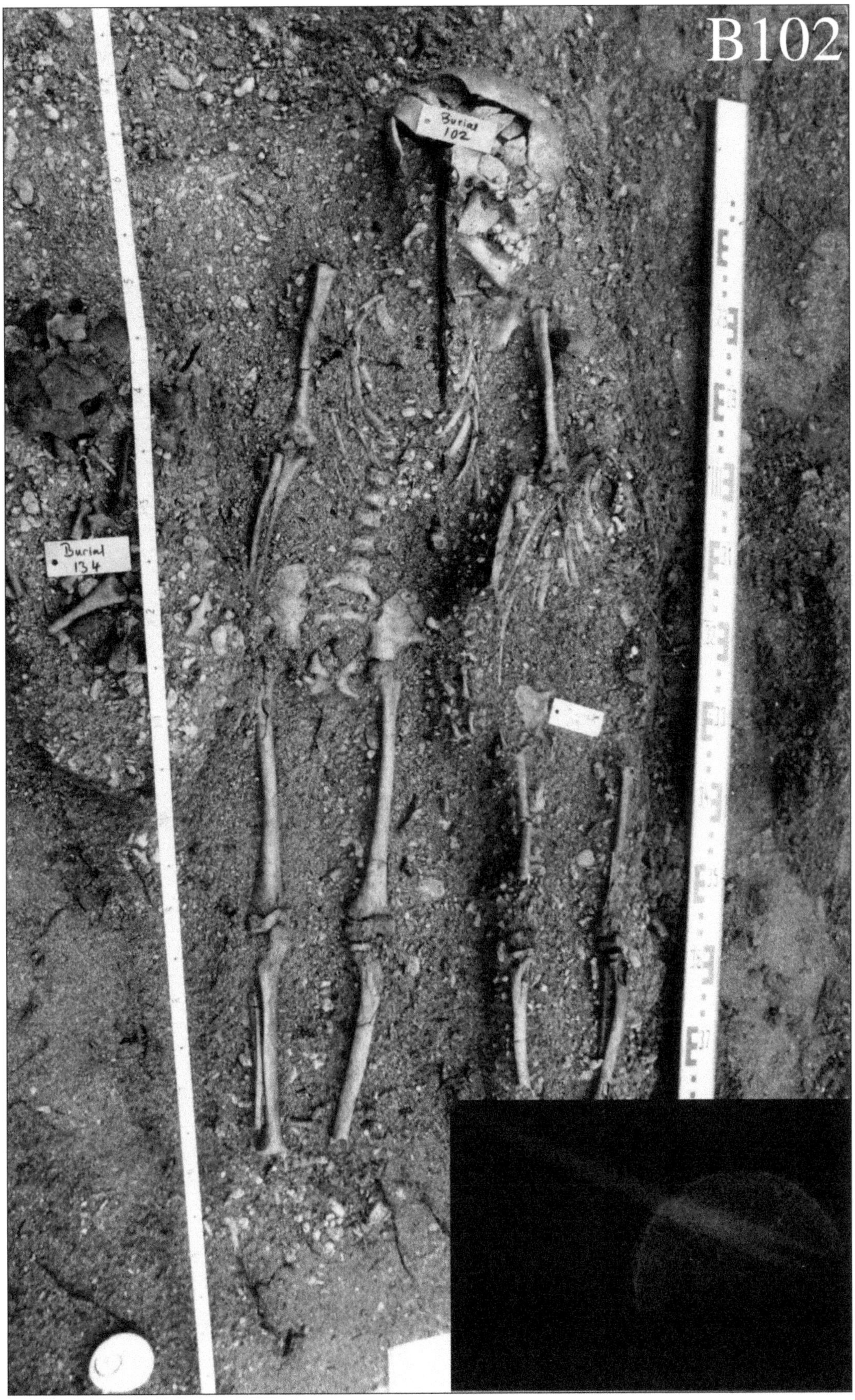

Figure 2.13. *Burial 102, a ten year old child, shown in situ (photo courtesy of Foss Leach). The inset is a radiograph of the bone point found lodged in the vertebra of this individual.*

of the infants or young children.

A comparative study on the dental health of some Pacific Island skeletal samples was undertaken by Evans (1987) which included data from the Taumako collection. Fyfe et al. (1993) analysed the status of alveolar bone in some of the adult individuals from Taumako. Simpson (1979) included data from the Taumako sample for comparison with certain health parameters of prehistoric Maori samples. One study of trace elements in human bone and prehistoric diet included some material from Taumako (Horwood, 1988), and another examined the relationship between stable isotopes and diet in the prehistoric Pacific Islands (Quinn, 1990). Finally, a study on environmental effects on the human body form included some material from this collection for comparative data (Prestney, 1997). Since the completion of this thesis there have been numerous publications and student projects on this sample. Buckley and Tayles (2003b; 2003a) published on aspects of infectious disease in the sample. Buckley (2006) included data from this sample in a chapter exploring the different prevalence rates of non-specific indicators of stress in several Pacific samples.

Student projects on the sample include an assessment of degenerative joint disease in the axial (Stiles, 2006) and appendicular skeletons (Robb, 2005). Earlier Rodney Jansen 2002 had investigated the relationship between circular caries and dental developmental defects in the deciduous dentitions. Foster (2007) conducted a detailed assessment of the hand entheses in this sample and the Teouma Lapita sample which she later extended as a doctoral study of the upper limbs and hands in several Pacific and Southeast Asian samples, including 'Atele (Foster, 2011). Based on information from the field diary, Sarah Mann (2007) made an assessment on the relative health and wealth of individuals investigating variation in the sample based on age, sex and certain health parameters. Robb (2009) included the Taumako sample in a study on cortical thickness in several Pacific island samples and aspects of this study were later published (Robb et al., 2012). Rachel Scott (2009) included this sample in a comparative study of skeletal trauma in several Pacific island samples and later published on this (Scott and Buckley, 2010). In 2010, Dudley made an assessment of the burial positions based on field photos given to Buckley by Foss Leach. In 2010 Rebecca Kinaston completed her PhD on the stable isotope and oral health evidence for diet throughout the life history in several Pacific samples and later published on a the diet and relative wealth of people from the Namu burial ground (Kinaston et al., 2013). Some individuals from this sample had also been included in an earlier study on stable isotopic evidence of diet but variation in sex, age and wealth was not assessed in this study (Leach et al., 2003). Other reporting of research on this sample includes many conference presentations that are not listed here.

As will be shown in chapter three, the present study identified 226 individuals which is somewhat larger than Houghton's estimate of 200. The main reason for this discrepancy is that several boxes containing skeletal remains included individuals that had not been previously recorded (e.g. Burial 104 which contained four individuals). Twenty-two other subadults were assigned burial numbers after their identification from large paper bags of bone fragments within a numbered burial that showed little evidence of having been opened and sorted previously. These subadults were given a burial number if they had at least three identifiable skeletal elements and/or enough dental material to provide an age estimate. Those which were given a burial number were allocated a suffix of the burial from whose box they were identified. For example, Burial 1 (2.14)included the cranial and dental remains of a two year old child. The child was then allocated the burial number of 1.2 and Burial 1 became 1.1. Many other fragmentary remains of infants and children were found in the bags of bone fragments in this collection. However, because minimal material was present, they have been treated as non-burial material. At the request of the Solomon Islands National museum the skeletal sample from Namu was returned to Honiara in 2009 on the understanding they would be reburied on Taumako. The collection remains in Honiara to this day. With the permission of the then Director of the museum, samples of bones and teeth were retained for future isotope and aDNA analyses.

2.3.2 Tongatapu

The Tongan archipelago consists of 150 islands comprising a total land area of about 700 square kilometres (Poulson, 1987). The Tonga islands consist of three groups, the southern Tongatapu group, the central Ha'apai group and the northern Vava'u group. Like the 'Melanesian' islands Tonga is of continental type and is the only island group in Polynesia situated west of the Andesite line (Poulson, 1987). Generally the soils of Tonga are rich and fertile and cultivated land comprises a large part of the islands today. Tongatapu is among one of the most fertile islands with volcanic ash layers mixed with decaying limestone which has produced this fertility (Collins, 1944).

The climate of Tonga is typically tropical with a dry and relatively cool period from May to November and a hot and humid period from December to April. The prevailing wind is the south east trade but westerly and northerly winds can sometimes reach extreme speeds particularly between February and March which is the hurricane season. The average rainfall in the southern group is 150–180 cm while in the northern groups it is considerably wetter. Periods of drought in the southern islands are not uncommon in the dry season. The average temperatures are between 21 and 27 degrees Celsius (Collins, 1944).

The supply of natural fresh water is scarce in Tonga. On Tongatapu there are several open-air pools and some cave pools. However the usual method of collecting fresh water is to dig pits into the clay soil to collect rain water. A major feature of Tongatapu is the extensive lagoon, which occupies most of the central part of the island. The lagoons are generally very shallow and an abundance of shellfish can be collected from all over the lagoon (Collins, 1944).

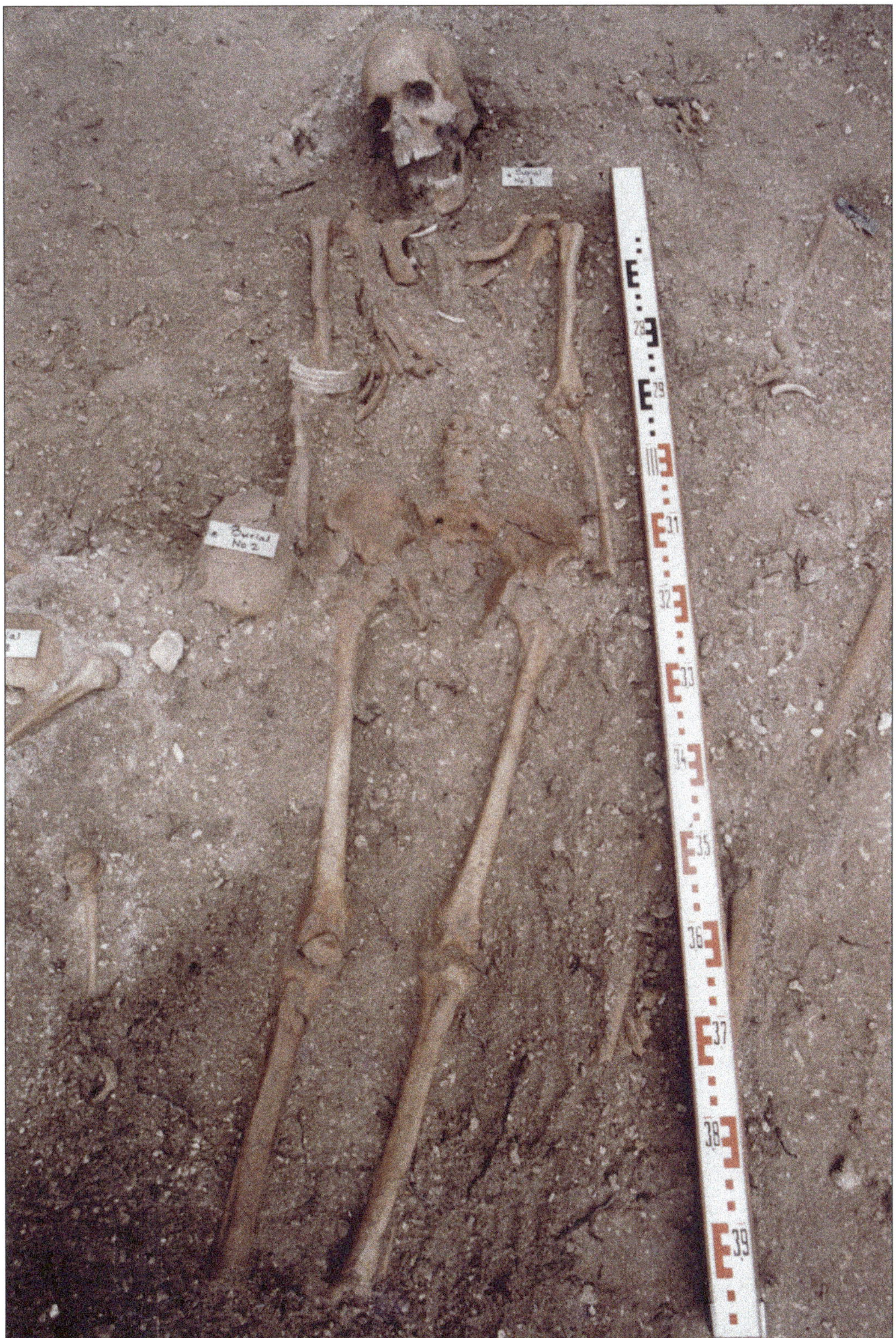

Figure 2.14. *Burial 1 with* Trochus *arm rings. Photo courtesy of Foss Leach.*

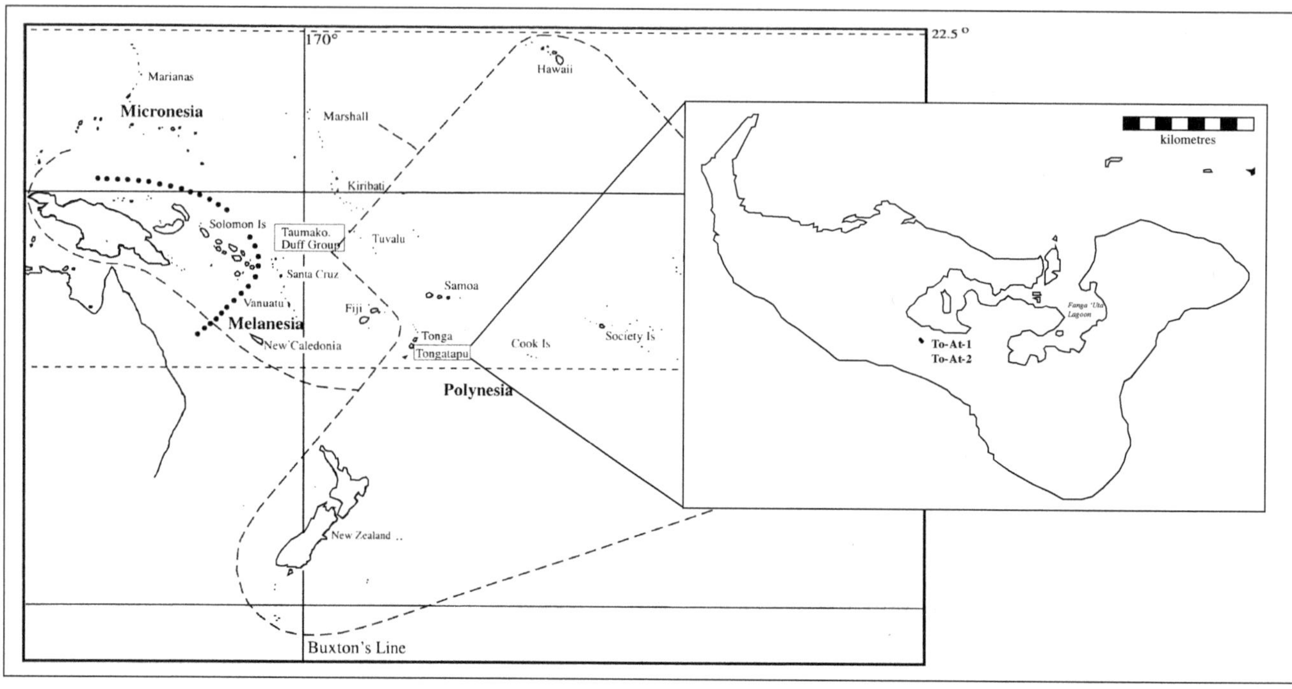

Figure 2.15. *Tongatapu island, Tonga. The 'Atele region is situated at the head of the lagoon.*

Archaeological investigations of Tongatapu

In 1964 Davidson conducted the excavation of two burial mounds from the 'Atele region around the lagoon of Tongatapu Island (Figure 2.15). These mounds were both situated in the grounds of the Tonga College and were 200 meters apart. The skeletal remains from these excavations form the second sample used in this thesis. The research objectives of the archaeological excavation of the mounds were to test the assumption that white sand was indicative of burial and to investigate the physical structure of the mounds. Because the excavation strategy was designed to investigate the structure of the mounds and not for recovery of burials, the samples are less than ideal for skeletal analysis. This is particularly so for Mound 2 (ToAt-2) where the 1 meter trench excavated through the center of the mound (Figure 2.16) meant that in most cases only half of the skeleton was recovered. Despite this, the bone material from both mounds was very well preserved and less fragmented than Taumako. The modern inhabitants of this region have no oral tradition concerning the origins of these particular mounds (Davidson, 1969).

Excavation of Mound 1, 'Atele, Tongatapu

Mound 1 (ToAt-1) is a small low mound approximately 40 m in diameter, with a height of only 80 cm above the ground level. Altogether 27 square meters of the mound was excavated (Figure 2.16). The central portion of the site was used intensively for burial activity requiring painstaking interpretation of a complicated series of intercutting burial pits. In the western and northern squares, five main layers were identified based on differences in soil matrix. These were interpreted as relating to four periods of use of the site.

The sequence of use of this mound was divided into four periods from five stratigraphic layers:

1. A number of holes were initially dug into the subsoil and were filled with Layer 5 material. These holes may have been postholes for construction of a dwelling, however the excavated area was too small to distinguish any pattern. These holes were likely to represent yam planting holes (Davidson, 1969).

2. After this initial evidence of human alteration at the site, a more definite period of domestic occupation ensued. Further postholes were dug and filled with the black midden material of Layer 4. Layer 4 was rather thin and therefore the period of occupation it represents is probably of short duration.

3. After the cessation of the build-up of Layer 4, the first burials were interred in the site. This group of burials precedes the construction of the mound itself. Graves were dug from the top of Layer 4. The bodies were partially covered with white coralline sand, then the pits were filled in. As the result of successive grave digging and filling in with soil and white sand, the mixed soil of Layer 3 built up on top of the midden Layer 4.

   During this period at least one structure was built on the site and possibly several successive structures. These structures were represented by postholes which did not penetrate the subsoil but were visible on the top of Layer 4. A domestic dwelling may have been present before the first burials were made, in which case the burials were laid under the house floor but it is probably more likely that the postholes repre-

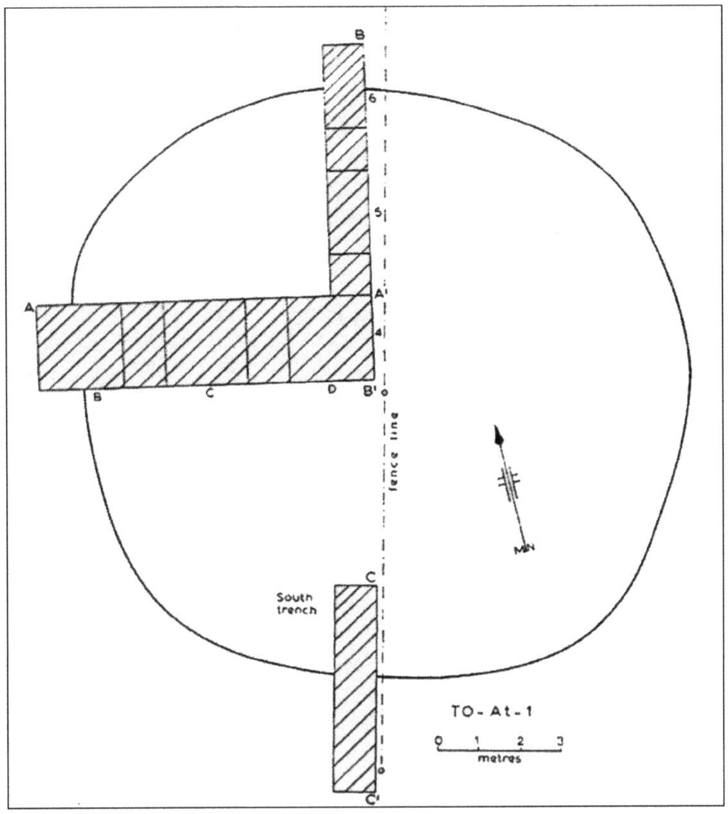

Figure 2.16. *Excavation plan of Mound 1, Tongatapu. (Adapted from Davidson, 1969).*

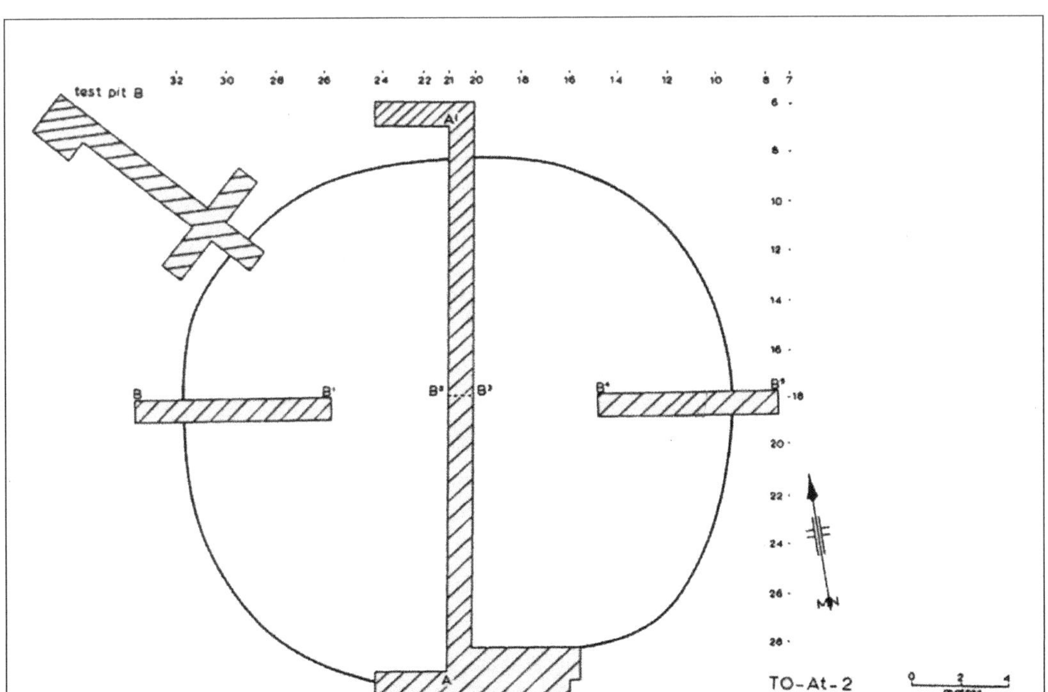

Figure 2.17. *Excavation plan of Mound 2, Tongatapu. (Adapted from Davidson, 1969).*

sent a special structure built over the burial ground (Davidson, 1969).

4. The final phase of use at the site related to further construction of the mound and subsequent use of the mound for burial. The low mound was constructed by spoil derived from the circular ditch surrounding the site. This mound appeared to have been built over the centre of the existing burial ground. The new surface was then used for burial. During the process of burial, a layer of soil and sand very similar to Layer 3 built up. A number of postholes dug into this surface probably indicate a structure erected over the burial area.

The midden content of Layer 4 was crudely analysed in the field. A number of shellfish species were identified with an overwhelming representation of one species (*Gafrarium gibbosum*). No faunal remains were mentioned in the excavation report. The midden remains probably represent a few meals consumed at a gardening site rather than a habitation area. Very few artefacts were recovered from this site and no inorganic grave goods were present.

The burial activity of the site was divided into two distinct groups by the archaeologist. Group A was interred prior to the construction of the mound. These groups were further subdivided into four subgroups. Subgroup 1 is the oldest and consisted of Burials 34, then 30 and 29. Burial 29 contained the disarticulated remains of multiple adults and subadults placed in an oval pit. Most other pits were rectangular, of varying depths, and some graves were little more than scooped depressions. Group B was interred from the surface of the newly constructed mound. Three subgroups were identified by the archaeologist within Group B. In total, 38 interments were excavated from this mound, consisting of 42 individuals. Despite the subgroups identified by the archaeologist, no temporal divisions are employed in this thesis. This was primarily decided because the sample sizes are so small that further subdivision would render subsequent analyses difficult.

A C^{14} date based on bone collagen from one burial yielded an age of less than 1200 years old (Davidson, 1969). While this is not an ideal description of the date for this site, it is all that was provided in the publication concerning the excavation (see below for an update on dating of this sample).

Excavation of Mound 2, 'Atele, Tongatapu

The second mound excavated in the 'Atele region was considerably larger than the first. It was also encircled by a ditch that was clearly visible from the ground surface. The excavation strategy of this mound consisted of opening up a one meter wide trench over the entire mound (Figure 2.17). Other trenches were opened up at the edges of the mound in order to determine the type of ditch construction. During the excavation of the mound six layers were identified. Layer 6 was the lowest and therefore earliest of the stratigraphic layers. It consisted of a thick layer of dark brown soil which filled a large shallow pit and several postholes dug into the subsoil. Five other layers were identified. The stratigraphy was very clear in the centre of the mound, but the merging of these layers towards the mound fringes made interpretation difficult. The construction of the mound was similar to that employed at Mound 1, except that one more period of ditch digging was carried out with the fill thrown on the top of mound. This additional period of construction probably contributed to the larger size of this mound compared to Mound 1.

Four phases of burial activity were identified at Mound 2 (Phases 2–5). Phase 2 is the oldest, and Phase 4 is the most recent. Most of the burials were found in Phase 4. Phase 5 consists of four adults buried in the fill of the encircling ditch, and is considered to be contemporary with, or a little younger than, Phase 4 (Davidson, 1969). In total, 41 interments constituting 52 individuals were excavated at this site. Again the small size of this sample negates any assessment of temporal changes in health.

The two samples of human skeletal material are kept separate in this thesis despite the small sample. This was decided primarily because different patterns of health and disease may be observed between the sites. Also, each site has its own sampling problems which may be compounded by merging the two samples.

Some differences between the burials in the mounds were noted by the archaeologist which support this decision to keep the two samples separate. Firstly, the archaeologist noted a more intensive use of Mound 1 during both periods of construction which tended to correlate with a greater number of subadult burials (Figure 2.18). At Mound 2 however, the more intensive period of use was during Phase 4 (Figure 2.19). This impression may in part be influenced by the different methods of excavation at both mounds. Another difference was an apparent greater use of black tapa cloth in wrapping the burials at Mound 1 than at Mound 2. This may be due to different burial practices between the two groups or an indication of different status groups buried at each mound (Davidson, 1969). These differences in burial practice, although admittedly not dramatic, may indicate certain cultural practices that have influenced the health at each mound.

The overall aim of this thesis is to assess any differences in health and disease between Taumako and Tonga. Therefore, the data from the Tongan samples will also be combined and then compared with Taumako as a complete sample. This is considered to be valid because while there might be differences in health between the two mounds the Tongan populations lived in the same physical environment, creating similar influences on health when compared with Taumako. Therefore, with the data from the two mounds merged, more solid comparisons can be made between the impact of the environments of Tonga and Taumako on the health of the people.

Figure 2.18. *Lower limbs and thorax of a burial from Mound 1, Tongatapu. Photo courtesy of J. Davidson.*

Figure 2.19. *Skull and upper limbs of a burial from Mound 2, Tongatapu. The association of white sand with the burial is clearly illustrated here. Photo courtesy of J. Davidson.*

2.3.3 *The context of the 'Atele mounds in Tongan prehistory*

Tongatapu was first settled by Lapita colonists 2840 cal BP (Burley et al., 2015). Archaeological investigations suggest that the first thousand years of habitation was centred close to the shores of the inland lagoon of the island with intensive exploitation of shellfish resources (Spennemann, 1987). The Talasiu site recently discovered on Tongatapu shows direct evidence of domesticated plant use while retaining intensive shellfish resource use persisting at least 200 years later (Clark et al., 2015).

A highly visible indicator of Tongan cultural change is found across the landscape in the form of mounds of various sizes and forms. These mounds include communal burial mounds where local commoners were buried, mounds for chiefs to sit upon (*'esi*), mounds for snaring pigeons which was a chiefly amusement (*sia heu lupe*), and mounds for the burial of the highest ranking aristocracy and their relatives (*langi*). The *langi* were very large structures with facings of coral and limestone blocks built by specialist stone cutters (Kirch, 1984).

It is believed that the explosion of construction of these mounds in the second millennium AD was an indication of increased population and social complexity, with intensified agriculture. This period also saw a shift of settlement patterns away from the lagoon shores and into the interior of the island. The mounds at 'Atele were probably in use during this period of intense social and economic change (Kirch, 1984).

Burley (1998) synthesised Tongan culture history using a four phase chronology as follows:

- Eastern Early Lapita Ceramic Period (2850–2650 BP)
- Polynesian Plain Ware Ceramic Period (2650–1550 BP)
- Aceramic Formative Development Period (1550–750 BP)
- Complex Centralised Chiefdom Period (750–150 BP)

Based on the form and size of the 'Atele burial mounds, they are thought to have been in use during the latter part of the Formative Period and the first part of the Chiefdom Period (around AD1100–AD1250) (Davidson, 1969). However recent AMS dating of human from the beginning and end of the burial sequences in both mounds have revealed a much more recent age at 500–150 cal BP (AD 1500–1850) placing this well within the Cheifdom period of Tongan history (Stantis et al., 2015).

European contact

The first documented contact with Europeans in southern Tonga was during an exploratory voyage, lasting only a few days, by Abel Tasman in 1643. Tasman anchored in the extreme north-west of Tongatapu, a considerable distance from 'Atele (Burley, 1998). Therefore, the radiocarbon dates from Mound 2 indicate that contact with Europeans was possible in the latter part of the sequence.

There was no evidence of European contact in either of the 'Atele mounds, although this would not necessarily have been expected, given the general paucity of artefacts at the sites. The possibility that at least some of the burials date to the period of European contact cannot be excluded on present evidence, but it is considered unlikely (J. Davidson, pers. comm.).

Ecology of prehistoric Tongatapu

In the early 1960s Poulsen carried out several archaeological excavations on midden sites from the Lapita period, at the base of the inner lagoon. Midden sites were selected for investigation in the hope that these would yield enough pottery sherds to construct a pottery sequence for Tongatapu. The shell and faunal remains from the middens also provided some ecological information. The district selected for the most intensive work was in and near the villages of Pea, Tokomolo, and Ha'ateiho. This district is bounded on three sides by uplifted terrain and on the fourth side by the lagoon.

It was concluded by Davidson (1969) that the 'Atele mounds and the surrounding area were used primarily for burial and cultivation while the Pea and Ha'ateiho areas previously excavated by Poulsen were the main habitation areas. The archaeological evidence would seem to support this. The 'Atele mounds were probably in use a little later than the occupation sites at Pea and Ha'ateiho.

Some ecological information for the 'Atele mounds could be extracted from the Poulsen monographs (Poulsen, 1987). For example, shellfish from the lagoon constituted a substantial amount of the remains from the middens. Other marine resources such as fish, turtles, and sharks were also exploited. Of the fish, reef and shallow water species predominated. Indirect evidence of agriculture in the form of fermentation pits and domestic animal remains were also excavated (Poulsen, 1987). It is not known whether foraging or agriculture predominated in the economy of the sites Poulsen excavated.

There are several environmental constraints that would have influenced the socio-political change in the later periods of Tongan prehistory (Kirch, 1984). These are worth noting, because they would also have influenced the health of the people: Firstly, the islands are small, with fixed boundaries and therefore restrict the size of the population. Also, there are no permanent sources of water, which restricted agriculture to dryland cultivation. Finally, environmental disturbances were common, particularly cyclones and drought.

These environmental constraints did not restrict the development of an agricultural system that utilised every available tract of land which has prompted such florid narratives as relayed above in the section on subsistence. Agriculture was so successful that it has been estimated that the island was very densely populated not long after initial colonisation (Kirch, 1984). The intense land usage and consequent competition over land is believed to have precipitated the rise of ever increasing social complexity, so much so that by the beginning of the second millennium AD one particular chiefly lineage had achieved supremacy over several smaller chiefdoms within Tongatapu. To this day the head of this family line is known as Tui'i Tonga: meaning Lord of Tonga (Kirch, 1984). The possible influence of the social stratification of Tongan society on the health of the people buried at the 'Atele mounds will be considered in chapter seven.

2.3.4 Previous work on the 'Atele samples

Prior to the completion of this thesis Michael Pietrusewsky (1969) published an extensive paper on the biological affinities, health and disease of these samples. Like Houghton, this study did not identify any pathology in the infants and children (Buckley, 2000b). The general prevalence of non-specific stress indicators from this sample were published alongside the Taumako data in (Buckley, 2006) and aspects of skeletal trauma were addressed by Scott and Buckley (2014). Differences in the diet of the people from the two mounds were found in a stable isotope study of diet and migration by Stantis (2015) as part of her doctoral study and also published (Stantis et al., 2015). The oral health of these people were also compared with Fijians and Cook Islanders with respect to differences in island ecologies (Stantis et al., 2016).

2.4 Conclusions

This chapter has outlined the environmental factors of the Pacific Islands which may have affected health in prehistory. The infectious diseases which may have contributed to the development of skeletal lesions have been outlined and the archaeological context of the samples used in this study has also been reviewed. Based on the ethnohistorical literature reviewed in this chapter it is assumed that all of the diseases included in the differential diagnosis present on Taumako in prehistory were probably also present on Tonga, except malaria.

Chapter 3

Census of Samples and Mortality Patterns

The first aim of this is to test whether the mortality patterns of the samples reflect any differences in population health. In order to address this aim, the objective of this chapter is to provide a census of age and sex of each of the samples and to compare mortality between the samples.

An accurate estimation of age and sex from skeletal remains is an essential initial step in any study of the epidemiology of disease patterns in prehistoric populations. Bone is a living tissue that grows and remodels at a known rate. It is this known rate of skeletal and dental development that is used by anthropologists to estimate the age at death in subadults. After skeletal growth and development is complete, the rate of the degeneration and remodelling of bone tissue can be used as a relative measure within a population to aid in age estimation of adults. The relative rate of attrition in dental tissue can also be used to estimate adult age (Saunders, 1992; Buikstra and Ubelaker, 1994). Below is a brief outline of normal bone anatomy and remodelling processes of bone to aid in understanding age estimation methods.

3.1 Normal Bone Anatomy, Growth and Remodelling

Generally, the structure of a mature bone consists of a cortex of compact bone, an outer shell which encloses areas of mesh-like trabecular bone. This cortical bone is encased by the periosteal membrane which communicates the nerve and blood supply to bone tissue (Resnick and Niwayama, 1981). The periosteum is intimately associated with the cortex on all surfaces except for the intra-articular portion where cartilage covers the bone. In young children, a typical long bone is comprised of four elements during growth. The diaphysis or shaft, the epiphyseal plate, made of hyaline cartilage, the epiphysis, usually at the extremities, and the metaphysis, the bone adjacent to the epiphyseal plate (Boyce et al., 2013). The growth of the bone begins as a cartilaginous model called a primary centre of ossification. Within these primary centres, growth and remodelling of the bone occurs which allows growth of the diaphysis. The linear growth of the diaphysis occurs through the epiphyseal plate where the cartilage model grows longer and is progressively replaced by bone. The shape of the bone is moulded by appositional ossification and resorption of the periosteal and endosteal surfaces, respectively (Kieser, 2012). When linear growth of the bone is complete, the epiphyseal plate is replaced by osseous tissue and union of the epiphysis to the shaft gradually commences (Jaffe, 1972; Krogman and İşcan, 1986). This type of growth is called endochondral, meaning within cartilage, and occurs in all skeletal elements except the cranial vault where bone growth is intramembranous.

The normal growth of bones is dependant on the activity of certain cells. The osteoblast is the bone cell that is responsible for the processes of bone growth. These cells decrease in number and size when the skeleton reaches maturity (Ortner, 2003). However, when a pathological process is present, the osteoblast is capable of responding with the same intensity as during foetal bone growth. The osteoblast spends some time on the surface of the bone and is eventually incorporated into compact bone when it matures into an osteocyte. The osteocyte may also be involved in bone resorption through a process called osteocytic osteolysis. A third cell, the osteoclast, is responsible solely for bone resorption (Jaffe, 1972; Boyce et al., 2013).

3.2 Age Estimation

As stated above, the estimation of biological age of individuals in skeletal populations is a crucial initial step in any analysis of prehistoric health. One of the main difficulties of estimating age in a burial sample is correlating biological age with chronological age (Eveleth and Tanner, 1990). The estimation of a chronological age is problematic in skeletal samples because while an individual is growing, environmental stressors and/or population variation in growth may affect the growth rate (Angel et al., 1986; Hillson, 2006; Eveleth and Tanner, 1990).

The methods used for age estimation are different for subadults and adults. For subadults, the methods most commonly used are dental calcification, for foetal and infant ages (Hillson, 1986); and dental eruption, epiphyseal appearance/union, and diaphyseal length, for children, adolescents and young adults (Krogman and İşcan, 1986). In adults the degree of reliability in age estimation decreases with increasing age of an individual. After skeletal maturation is complete, degenerative changes in skeletal and dental tissues can be utilised for estimating the age of adults from mid to old age (McKern, 1970; Hillson, 2006). The following is an outline of some of the methods used in subadult and adult age estimation.

3.2.1 Subadult age estimation

Dental development

The normal processes of skeletal and dental development have been extensively studied to compile various standards of estimating age in subadult skeletal material (Johnston and Zimmer, 1989). The calcification and eruption of teeth are considered to provide the closest approximation of chronological age in young subadults because the development of the teeth are less affected by environmental disturbances than linear growth of the diaphyses (Saunders, 1992). Each individual has three dentitions during life which correlate with developmental stages; a deciduous dentition in early childhood (infancy and up to six years), mixed deciduous and permanent teeth through late childhood and a complete permanent dentition during adult life (Hillson, 1986). The calcification of deciduous dentition begins early during foetal life and continues until birth, while the permanent teeth begin to calcify from birth to around 15 years of age (Costa, 1986).

Eruption of the deciduous dentition begins from six to nine months of age until around 18 months, while the first permanent molar erupts at around six years old (Hillson, 1996). Female dentition develops and erupts earlier than males which provides some error in estimating age in skeletal samples where the sex of subadults is not known (Saunders, 1992).

Various standards for estimating age from dental material have been developed. Probably the most widely used of these methods is Schour and Massler (1941). The Schour and Massler standard is based on histological sections of teeth and documents the calcification and eruption ages of individuals from five foetal months to 35 years old (cited in Costa, 1986; Hillson, 1986) . The simple cut-away view in which the chart is presented probably accounts for its popularity amongst field bioarchaeologists. However, the chart of Gustafson and Koch (1974) has an advantage over the Schour and Massler chart because it accounts for some variation in eruption times that is not acknowledged in the latter.

However, the diagrammatic format of Gustafson and Koch chart is not as easy to interpret and is therefore less widely used than the chart of Schour and Massler (1941). The Moorees, Fanning and Hunt method (1963a; 1963b) is based on two longitudinal studies of the development of deciduous and permanent teeth in Bostonian children (Costa, 1986). The ages of crown formation, closure of root apices and later resorption of roots during exfoliation of deciduous teeth are presented. These developmental stages were then correlated to the known chronological age of the subjects. The study of Moorees et al. (1963b) is useful because it provides the researcher with a relatively simple means of estimating the age of young children complete with standard deviations for each tooth. The second study of Moorees et al. (1963a) includes ten permanent teeth, and served to fill in the gaps of the original Schour and Massler chart (Miles, 1978). Despite the usefulness of these standards, radiographs of subadult jaws are ideally required to employ them which is not always possible in field situations (Moorees et al., 1963a,b).

Subsequently, Merchant and Ubelaker (1977) compared the ages yielded by the Schour and Massler chart to those of Moorees et al. (1963a,b), and found the former to yield higher ages which may introduce error in further analysis of growth rates in the sample population. These fundamental problems have since been addressed, most notably by Ubelaker in 1978. This modified version of the Schour and Massler chart allows for sexual and racial variation between populations. However, the standard is specifically tailored for American Indian samples and should therefore be used with caution on other populations (Ubelaker and Grant, 1989). A number of histological methods for estimating dental are available (Costa, 1986). These were not used in the present study so are not reviewed further.

Skeletal growth

Diaphyseal lengths can also be used for age estimation in subadults aged from birth to around 12 years old, before the epiphyses begin to fuse (Saunders, 1992). Linear growth of long bones is considered more susceptible to environmental stress than dental development (Eveleth and Tanner, 1990). However, coupled with a 'reliable' dental standard, age estimates based on diaphyseal lengths can provide a base with which to construct age estimations that are specific to that sample (Saunders, 1992).

A number of standards based on diaphyseal lengths are available. One of these is a comparative study of populations in which the prehistoric Indian Knoll population of North America and the population from the Alterneding site of West Germany (sixth to seventh century AD) were analysed by dental and long bone criteria (Sundick, 1978). Another diaphyseal length standard was based on a longitudinal study of 334 European Americans from middle to upper socio-economic backgrounds (Hoffman, 1979).

It is important to note that the methods introduced above supply age estimation standards that are only a reflection of the skeletal growth of infants and young children of the populations from which the standard is derived. However, some other methods have been developed that provide ageing standards which amalgamate dental and diaphyseal data from varying sources. For example, Ubelaker and Grant (1989) provide extensive charts on epiphyseal union and also standards of ageing by diaphyseal length. The data of diaphyseal lengths presented in the Ubelaker (1989) volume were derived from a paper co-authored with Merchant in 1977. The 1977 paper established an age estimation standard based on diaphyseal lengths, but also addressed the problems of using standards based on different dental methods (Ubelaker, 1979).

Skeletal maturation

The appearance of primary centres of ossification of the hand and wrist can also be used for estimating the age of infants and young children. Eveleth and Tanner (1990) state this is the most favourable technique for analysing the growth rates of living children. The two most widely cited techniques are the Greulich-Pyle standard (GP) and the Tanner and Whitehouse standards (TWI and TWII) (Eveleth and Tanner, 1990). These methods have provided researchers with comparative data of skeletal maturation. However, they are not particularly useful for archaeological samples. The small epiphyses of the hand and wrist are more likely to be lost post-deposition from taphonomic processes or overlooked during excavation. These small amorphous bones are also difficult to orient correctly out of their anatomical context (Saunders, 1992).

However, epiphyseal union is the most appropriate method for estimating the age of adolescents and young adults (Saunders, 1992). Krogman and İşcan (1986) supply charts on the stages of epiphyseal union, based on the original Todd Atlas (McKern, 1970). Subsequent to publication of the Todd Atlas, McKern and Stewart published a report in 1957 of a comprehensive study on American soldiers killed in the Korean war (McKern, 1970). As would be expected, the study is based on data solely from men aged 17–50 years at death. McKern (1970), proposed the lower limit of reliability using this technique as 'puberty' (13–15 years old), and the upper limit as 28–30 years. The lower range of epiphyseal union reliability can be supplemented by standards for dental development and the upper range may be aided by degenerative changes of the skeleton and teeth.

While this is a useful method there are a number of limitations associated with age estimation in subadults. With all stages of bone and dental development in humans there is significant variation between the sexes and union of the epiphyses are no exception (Saunders, 1992). The sex discrepancy in skeletal maturation can be as much as two to three years and this margin of error can be further increased by the lack of accurate methods for estimating sex in subadult skeletal material. Variation in skeletal maturation between populations can also be significant, although dental development is considered less variable in this respect (Saunders, 1992).

3.2.2 Adult age estimation

The reliability of age estimation decreases after skeletal maturation is complete. Therefore, without the benefit of developmental age indicators outlined above, markers of skeletal degeneration must be employed.

Pubic symphysis remodelling

Todd (1920) was the first researcher to consider the morphological changes of the pubic symphysis as a means of ageing adult skeletal material. The morphology of the symphysis begins to change from a 'billowy' growth-plate appearance at about 17 years old when the bone tissue is gradually resorbed over time. Therefore, the medial aspect of the pubic symphysis metamorphosises with advancing age and is generally smoother in appearance in older than in younger individuals (Meindl and Lovejoy, 1985). Todd (1920) presented male data which was later extended to use on female samples by Gilbert and McKern (1973). They identified three main components of the pubic symphysis; the dorsal demi-face, the ventral rampart, an epiphysis-like piece of bone, and the symphyseal rim. Each of these components were assigned six metamorphic changes correlated to chronological age (Gilbert and McKern, 1973).

Some variables exist which may skew age estimates based on the morphogenesis of the pubic symphysis. Most notable are the traumatic changes produced by the later stages of parity which can cause a pubis to appear morphologically much older than the actual chronological age (Gilbert and McKern, 1973). Therefore, there are limitations for use in prehistoric samples where many of the female pubes available may have associated parous changes. A difference in the maturation rates of males and females were also noted where the morphological changes in the female pubes were steady until around 40 years old while male maturation appeared to be steep between 17 and 25 years of age (Gilbert and McKern, 1973). The variability between populations of different origin might also be considered as a possible source of error. However, some researchers have tested this possibility and found that variation in the morphology of the os Pubis is not significant between Negroes and whites (Meindl and Lovejoy, 1985), and Negroes, whites and Mexicans (Katz and Suchey, 1989). The general reliability of using the pubis as a method for adult age estimation has been tested by some (Lovejoy, 1985; Meindl and Lovejoy, 1985). Meindl and Lovejoy (1985) concluded this method to be most reliable when used in conjunction with other ageing techniques. It should also be noted that the pubic symphyses do not always survive well in some archaeological contexts.

Other methods of estimating age in adult material are sometimes used when material is available. For example, the progressive ossification of the chosto-chondral junction of the ribs may be used and the age related changes of the auricular surface of the ilium are also useful indicators for age estimation. The progressive development of muscle attachments may also indicate more advanced age and the degeneration of joints has been correlated to advanced age in adults (Buikstra and Ubelaker, 1994).

Despite its early widespread use, the progressive synostosis of the cranial sutures has been considered extremely unreliable and therefore is used only as a last resort when attempting to estimate adult age (Meindl and Lovejoy, 1985). However, Meindl et al. (1985) investigated its reliability, and suggested that many of the difficulties with using suture closure are derived from the methods of the studies. Therefore, they advocate the reliability of suture closure when used with other more 'trusted' anatomical features.

Dental attrition

Dental attrition can be defined as a reduction of enamel and dentine by frictional wear of the teeth and abrasion caused by food material (Hillson, 1986). The analysis of the degree of attrition on adult teeth is a further method of estimating age at death. Dental attrition is useful for archaeological samples because of the high preservation of teeth in archaeological contexts (Saunders and Katzenberg, 1992). It is also a direct record of the chewing activities of the population under consideration (Molnar, 1971).

One method used for ranking the degree of dental attrition in skeletal samples is that of Molnar (1971) who developed a tooth wear evaluation method for all tooth types. This evaluation method also included criteria for recording the direction of wear and the form of the occlusal surface. However, unless the researcher is specifically testing attrition rate and type in a skeletal sample the degree of wear is the most often used aspect of the Molnar (1971) method. From analysis of large skeletal series, Molnar (1971) concluded that the molars are the easiest tooth type to assess. This is because molars are multicuspid in form and therefore the grade will be more precise and less arbitrary than the unicuspid and bicuspid teeth. Others have attempted to develop new methods for evaluating the rate and type of dental attrition (Scott, 1979; Lovejoy, 1985).

The concept of seriation of dental attrition in adult age estimation was first introduced by Miles (1963). He developed a method by which the researcher may utilise young individuals assigned an age estimate based on methods such as dental calcification and eruption. In doing so, the researcher then has a sample of 'known age' by which to seriate other individuals from the sample.

Miles (1963) also initiated the concept of the functional age of teeth as a means of estimating age. This is based on the assumption that a gradient of wear can be developed within a skeletal sample by calculating the degree of attrition to the length of time the tooth has been a functional unit of the mouth. For example, if the first permanent molar erupts at age six, then when the second molar erupts at around 12, the first molar will have been functioning for six years (Miles, 1963). This method could conceivably be employed for any population with a large sample of 'known age' subadults and young adults whose maturation is almost complete.

The use of dental wear is problematic as the degree and type of wear observed in an individual is subject to the type of foods being consumed. The Miles method also assumes that diet is consistent throughout life and does not account for sexual variation in eruption times for the baseline sample (Kieser et al., 1983). The Miles method of seriation has been used extensively by osteologists to estimate adult age and when its reliability has been tested it has been deemed valid and useful (Nowell, 1978; Kieser et al., 1983).

The multifactorial approach

The use of a multifactorial approach to adult age estimation was first proposed by Acsadi and Nemeriski (1970). They suggested the use of four different ageing variables; cranial sutures, dental attrition, pubic symphysis, radiographic analysis of the trabeculae of the proximal femur and humerus (Acsadi and Nemeriski, 1970). Because not all age at death indicators are equally reliable they ranked each indicator according to its reliability.

A later study developed the Acsádi and Nemeskeri (1970) method further (Lovejoy, 1985) and advocated the use of a 'summary age' where two or more age indicators are used. After careful analysis Lovejoy et al. (1985) state that use of dental attrition is the single most reliable age indicator, while current methods for assessing the pubic symphyses create substantial bias. A recent volume that has sought to provide consistency in the methods of recording age estimation indicators is edited by Buikstra and Ubelaker (1994). This work provides a synthesis and critique of age estimation methods most frequently used by researchers of human skeletal materials; including those methods reviewed above.

3.3 Methods

3.3.1 Subadult age estimation

The methods outlined below were employed for the analysis of both Taumako and Tonga. The definition of 'subadult' is variable among researchers, depending on the emphasis of the researcher. For the purposes of this study, subadults are defined as under 20 years of age. The subadult skeletal samples used in this study were highly variable in their state of preservation. The Tongan subadult samples were generally better preserved than the Taumako subadults, as more complete limb bones were present. Where possible all methods outlined in the above section were employed.

The sample of subadults is further divided into subgroups:

1. Foetal = Pre fullterm
2. Infants = 0–0.9 years
3. Young children = 1–5 years
4. Older children = 6–10 years
5. Adolescents = 11–15 years
6. Preadult = 16–20 years

The age categories were selected to provide consistency in the proportions of the mortality sample. These categories were also selected based on the epidemiology of certain diseases in the Pacific Islands covered in chapters two and five to consider whether mortality rates in these samples reflect modern epidemiology of these diseases. The definition of the upper boundary of subadult as 20 years of age is older than many researchers would use. However, the distinction for this study was based on the normal growth and development of the skeleton whereby the age of 20 years a full permanent dentition has normally been attained and all skeletal maturation concerned with linear growth is complete. Therefore, after 20 years of age an individual is fully grown, in height at least. Because this study is attempting to find any differences in growth patterns between the two samples this distinction is considered important to make.

The possible variation in skeletal growth between populations is recognised. At present there are no subadult ageing standards specific to Pacific populations. An accelerated dental development among Polynesian children has been reported (Fry, 1976) and other studies have shown delayed tooth eruption in New Guinean children compared to Europeans (Malcolm and Bue, 1979; Ulijaszek, 1996). This variation can be as much as two years. The problems of using age estimation standards based on populations with varying genetic, environmental and temporal backgrounds is well recognised (Ubelaker, 1989). However, the methods for subadult age estimations have to be based on the standards that are available.

Ubelaker (1989) incorporated the known accelerated dental development of American Indian populations with the data from Moorees et al. (1963a). This may offset some of the accelerated dental development of Polynesians as noted by Fry (1976). Therefore, Ubelaker's adapted dental calcification charts were employed in this study.

Other methods used for age estimation of infants and children in this study were: dental calcification (Moorees et al., 1963a; Moorees et al., 1963b; Ubelaker, 1989); dental eruption (Ubelaker, 1989); diaphyseal lengths (Merchant and Ubelaker, 1977); and epiphyseal union for adolescents (Krogman and Iscan, 1986).

The more reliable method of dental calcification over dental eruption has been stressed by some researchers (Saunders, 1992; Ubelaker, 1989). Therefore, in children under 12 years old the use of dental calcification was given higher priority than dental eruption or diaphyseal lengths. Only teeth that were loose from their crypts were able to be assessed for the calcification stage. As mentioned above, it is recognised that skeletal growth is more susceptible to environmental disruption than dental development (Eveleth and Tanner, 1990). Therefore, when a subadult could be aged only by the use of dental eruption or diaphyseal lengths then the bone size was compared with the limbs of subadults given age estimates based on dental calcification. Some subadults, especially in the Taumako sample, were represented only by fragmented limb bones. Where this was the case, the size of the limb fragments were compared to those subadults with an estimate based on dental evidence and complete limb bones, though this was minimal.

The ages of older children and adolescents were estimated based on the same criteria as the younger children. However, the variation in age estimates based on dental calcification increases with age (Moorees et al., 1963a; Moorees et al., 1963b). Therefore, more emphasis was placed on estimates obtained by epiphyseal union. Sex estimates were not attempted for any individuals under 15 years of age. Therefore, for each available tooth, the mean of the male and female age estimates is presented here.

3.3.2 Adult age estimation

As outlined above, when estimating age in skeletal material a developmental or biological age is yielded that is relative to other individuals within the population. Therefore because the chronological age in years is not known, it was decided to assign adults to an *age group* rather than chronological age. This is still a measure of relative age within a population, but does not have a numerical unit assigned to it. The age groups used roughly correlate to chronological age as follows: Young (21–30 years), Mid (30–40 years), and Old (40+).

Where possible the following age indicators were assessed for adult age estimation: epiphyseal union (Adolescence through to late twenties); dental attrition; and pubic symphysis. For both samples there were very few pubic symphyses that could be used for age estimation. Where a pubic symphysis was available, a comparison with the estimation based on dental attrition was made. Each criterion was assessed separately and a summary of all age estimates was used. Based on the assertion of Lovejoy (1985) that dental attrition is the single most reliable indicator, a seriation method based on Miles (1963) was used. The degree of wear on the permanent molars was recorded and the average of each molar type was used for seriation.

Taumako

Seven individuals from Taumako were of 'known age' and formed the baseline for seriating the progression of dental attrition. Miles (1963) advises that the baseline sample of known age should exceed twenty individuals. However, it was decided to use the material that was available. Few teeth exceeded Molnar's wear grade 5. This minimal progression in attrition from 12 years to 20 years attests to a very slow rate of wear in this sample (Table 3.1 and Table 3.2).

There were 120 adults with age estimations. Of the adults, 105 were able to be aged by tooth wear. Fifteen adults had no dentitions. It was decided that the degree of joint degeneration and the relative size of muscle entheses should be used to estimate the age of these individuals. Therefore, the skeletal material of these individuals was compared to the degree of joint degeneration and muscle development of adults with age estimates based on dental attrition.

Joint degeneration (osteoarthritis, and inflammatory arthropathies) is a pathological process which is not necessarily associated with advancing age. For this reason, it has been emphatically stated that joint degeneration should not

Table 3.1. *Grades of dental attrition in subadults aged by epiphyseal union and/or dental eruption at Taumako.*

Burial	Age	Molar 1	Molar 2	Molar 3
106	12	2	2	-
165	12	2	2	-
192	12	2	2	-
151	15.5	3	2	-
70	17	2	2	0
137	18	2	2	0
139	20.5	3	2	0

Table 3.2. *Wear grade ranges used for relative ageing of the adult dentitions from Taumako.*

Age Group	Years	Molar 1	Molar 2	Molar 3
Young	<30	2-3	2	0-2
Mid	30-39	4-3	3-4	2
Old	40+	5-6	4-5	3-2

be used as an age estimation criteria (Waldron, 1994). Because of this caution, joint degeneration was not considered as an age indicator for the 105 adults with teeth.

Tonga

Adult age estimation was problematic for both the Tongan mounds. Dental attrition was generally low (Table 3.3) and it was difficult to estimate the relative age of the adults. Very few had dental attrition over Molnar's grade 3. The baseline for estimating adult age from dental attrition was based on three individuals aged under 20 years. One 13 year old had wear of Molnar grade 2 on the first molar and no wear on the second, while an individual aged 18 years had grade 3 on the first molar and grade 2 on the second molar. It is recognised that this baseline sample is very small.

From Mound 1, fourteen adults had age estimations and eleven of these were based on dental attrition. At Mound 2, 21 adults were assigned to an age group and ten of these were based on dental attrition. For those adults with no teeth, the same comparative methods as outlined above for Taumako were used.

3.4 Sex Estimation

3.4.1 Theory

A certain degree of sexual dimorphism is present between human males and females. In general, males are taller and more robust than females. In skeletal studies this dimorphism is measurable by observing differences in size and shape between the sexes. Methods used for sex estimation in skeletal material are either based on morphological features, which assess shape differences in certain parts of the skeleton, usually related to different reproductive functions, or metric, which measure size differences which manifest as the dimorphism which exists between males and females (Buikstra and Ubelaker, 1994; White and Folkens, 1991). Because the dimorphic differences between the sexes develop during and after adolescence, it is difficult to estimate sex in juvenile skeletal material (Saunders, 1992). Therefore, the following discussion is relevant only to adult skeletal remains.

Generally, morphological differences in shape are utilised more frequently by osteologists when estimating sex in skeletal studies. Differences in the shape of the pelvis are emphasised as more accurate than the shape of the skull because the anatomical differences in the pelvis are based on biological necessity whereas differences in skull shape are more subject to population variation. The shape of the human pelvis is important for bipedal locomotion and, in females, parturition. Therefore, selective pressures during human evolution have determined the morphological differences in the pelvis between the sexes (White and Folkens, 1991).

Basically the shape differences in the female pelvis have evolved to facilitate the safe birth of a large foetus with a large brain. Therefore, the female pelvic cavity is wider and more open than in males. There have been a number of methods developed which concentrate on specific areas of the pelvis. For example, Washburn (1948) measured the length of the pubis in relation to the length of the ischium using an index which differentiated between male and females *os coxae* (cited in White, 1991).

However, the study which was considered a 'breakthrough' by most researchers was the 1969 paper by T.W. Phenice. This paper emphasises the observation of bony structures in the pubis related to the attachment of the genitalia. There are three aspects of the females ischiopubic ramus which are not usually found in males. In decreasing order of reliability these are: the ventral arc, the subpubic concavity, and a ridge on the medial aspect of the ischiopubic ramus (Phenice, 1969). This method was tested by Phenice (1969)

Table 3.3. *Wear grade ranges for relative ageing of the adult dentitions at both Tongan mounds.*

Age Group	Years	Molar 1	Molar 2	Molar 3
Young	<30	2	2	0–2
Mid	30–39	3–4	3–2	2–0
Old	40+	5–6	4–5	3–2

on a sample of individuals of known sex and found to have an accuracy of 96 per cent. A later study by Kelly (1978) suggested this method was equally powerful when used on prehistoric skeletal remains (cited in Buikstra and Mielke, 1985). While this method is indeed useful it should be noted that it is dependant on a complete and well preserved *os pubis* which may not be present in all skeletal samples.

Furthermore, the hormonal activity and trauma associated with pregnancy can leave markers on the pelvic bones which may help in estimation of sex in females. The resorption of bone on the dorsal aspect of the *os pubis* and in the preauricular sulcus of the ilium have been demonstrated to provide evidence of past pregancy. This bony resorption is a response to oestrogen release in the third trimester of pregnancy which loosens the ligament attachments at these sites (Buikstra and Mielke, 1985; Houghton, 1974). However, it has been noted that some women may bear children without developing these characteristic lesions or bone resorption may occur in women who have not borne a child (Stewart, 1970). Similarly, continual remodelling of bone at these sites may obliterate any markers of past parturition in older women (Kelley, 1979). Despite these caveats, parturition markers in pelvic material are frequently used to estimate female sex in skeletal material.

Because of general greater muscle bulk of males, the muscle and ligament attachment sites are more pronounced and rugose in the male cranium. Therefore, these features of the skull are generally larger and more robust in males than females. Among other characteristics listed below, the male crania usually have larger, more prominent supraorbital 'brow' ridges and heavier, more rugose temporal and nuchal lines. For ease of presentation, Tables 3.4 and 3.5 summarise the features of the pelvis and skull which are commonly used for estimation of sex in skeletal material. While these morphological differences are useful for sex estimation they are sometimes subject to population variation in robusticity. For example, some populations are generally more robust than others and their crania may appear more 'male' in shape. Therefore, it was considered important to familiarise myself with the overall robusticity of the samples before attempting estimations of sex from the cranium (White, 1991).

The use of metric observations as a means of assessing size differences between males and females is another method which is extensively used in sex estimation (Buikstra and Mielke, 1985). While this is a useful and valid method of estimating sex in skeletal material it was not employed in this study primarily because of the lack of complete material.

3.4.2 Methods

Independent assessment of the features of the pelvis and cranium listed in Tables 3.4 and 3.5 was carried out on both samples. Sex estimates were not attempted for individuals less than 15 years of age at death. The priority was given to pelvic features but where possible the assessment of both the cranium and pelvis was carried out. The presence of parturition indicators, as described above, was considered an unequivocal female characteristic. In conjunction with features of the pelvis and/or cranium, a more subjective assessment of overall skeletal robusticity was also made. In the results, the adults with equivocal estimates of sex were merged with those of more positive estimates. This was done in order to increase the sample sizes.

3.4.3 Statistics

The statistical analyses used in this thesis are simple because the use of complex statistics on data derived from small samples can sometimes lead to misleading interpretations. The type of data used in this thesis was nominal scale data which are described by a particular attribute. For example, in the case of dental defects, cribra orbitalia and proliferative lesions, two classes were described with an attribute present (affected) or absent (unaffected). When comparing these proportional data (the number present/total observed), chi-square tests were employed to compare the samples and Fisher's Exact Tests (FET) were employed when sample sizes were small. A critical level of $p = 0.05$ per cent was employed. This means that if a test produced a p-value of less than 0.05, the difference between the data was judged to be statistically significant.

3.5 Results

3.5.1 Age at death and sex estimation

As an indication of the overall impression of the level of sexual dimorphism in all of these samples, the males were generally hyper-robust in muscle attachments of the crania and general skeletal frame, while the females were relatively gracile in overall skeletal frame and cranial features. These impressions of very slight females and robust males was borne out by parturition markers observed on the pelvis of many of these very slight 'female' individuals.

Table 3.4. *Features of the pelvis commonly used to estimate sex.*

Feature	Female	Male
1. Pubic arch angle	wide and open	narrow and steep
2. Ventral arc	present	absent
3. Subpubic concavity	present	absent
4. Sciatic notch angle	wide and open	narrow and closed
5. Ishio-pubic ramus	ridge present	broad
6. Sacrum morphology	wide and shorter	narrower and longer
7. Parturition markers	often present	not present

1,4,6. Buikstra and Mielke (1985); 2,3,5. (Phenice, 1969); 7. Houghton (1974)

Table 3.5. *Features of the cranium commonly used to estimate sex.*

Feature	Female	Male
1. Supraorbital ridge and glabella	not prominent	prominent
2. Orbital rim	sharp	blunt or rounded
3. Nuchal lines and external protuberance of occipital bone	less marked	rugose and prominent
4. Mastoid size	smaller	larger
5. Chin region of anterior mandible	rounded	square
6. Mandibular gonial angle	less marked eversion	everted
7. Orbital shape	round	square
8. Parietal bossing	present	absent

All features were adapted from Buikstra and Mielke (1985).

Age at death: Taumako

The skeletal sample from the Namu burial mound on Taumako was reasonably well preserved and estimations of age at death were possible on all except 16 out of 226 individuals. Table 3.6 provides information of the methods used for estimating subadult ages where the use of dental eruption and calcification comprised the primary means of age estimation. Table 3.7 presents the general age structure of the sample from Taumako. In the subadults, 11 per cent of the total sample died in the first year of life and a total of 41 per cent of the sample died before 20 years of age. Mortality decreased after five years of age when the children from 6 to 20 years of age comprise only half of the total subadult sample. Of the adults the highest mortality was found in the Young group (Figure 3.1).

Sex distribution: Taumako

Sex estimation was possible for 94 per cent of adults (n = 134/143). Overall, the distribution of the sexes is close to unity in the Taumako sample (Table 3.8) with a 1.2:1 ratio of males to females. Different mortality patterns are seen between the sexes. Female mortality was highest in the under 20 year age group, while male mortality was higher among the Mid aged adults. When comparing the mortality rates of age groups between the sexes it is shown that 80 per cent of the individuals with a sex estimate under 20 years of age were female. In all age groups over 20 years, more males were represented than females. In the Mid adults nearly twice as many males as females died (64 per cent and 31 per cent respectively). Chi-square statistical analysis showed no significant differences in mortality within or between the sexes. Of the 134 adults with sex estimates the diagnosis was uncertain in 15 females (F?) and 11 males (M?). Nine adults had no sex estimate.

3.5.2 Tonga

Age at death: Mound 1:

The subadult material from Mound 1 was well preserved and age estimations were predominantly based on dental development and combined dental and diaphyseal data (Table 3.9). A total of 45 burials were present from Mound 1 and all except five adults were assigned age estimations. The burial sample from Mound 1 consisted of 26 subadults and 19 adults. The age structure of this burial sample indicates that more individuals died during infancy and childhood than as adults (Table 3.10). The mortality rate is the same for infants and children under five years, then decreases to under four per cent in the 6–15 year olds. A high proportion of the adults died young while no individuals were classified as Old (Figure 3.2).

Sex distribution: Mound 1

Overall, a difference in sex distribution was found at Mound 1 where more females were represented than males (Table

Table 3.6. *Methods used for subadult age estimation: Taumako.*

Age (yrs)	Dental	Diaphyseal	Epiphyseal	Combined	Comparative	Total
Foetal	-	1	-	-	2	3
0–0.9	10	2	-	0	13	25
1–5	21	1	-	1	9	32
6–10	5	-	-	2	3	10
11–15	1	-	1	5	3	10
16–20	1	-	3	6	0	10
Subadult	-	-	-	-	3	3
Total	38	4	4	14	33	93

Dental = number of individuals with dental material available for estimation of age; Diaphyseal = number of individuals with complete diaphyses; Epiphyses = number of individuals with age estimations based on epiphyseal union; Combined = number of individuals with both dental and diaphyseal, and/or epiphyseal material for age estimation; Comparative= number of individuals with fragmentary skeletal material where age estimation was based on a comparison of skeletal size with individuals aged using dental development

Table 3.7. *Age structure of sample: Taumako.*

Age Group	N	Per Cent Total Sample
Foetal	3	1
0–0.9	25	11
1–5	32	14
Child Subtotal	60	27
6–10	10	4
11–15	10	4
16–20	10	4
Subadult	3	1
Subadult Total	93	41
Young	50	22
Mid	36	16
Old	34	15
Adult	13	6
Adult Total	133	59
Total	226	100

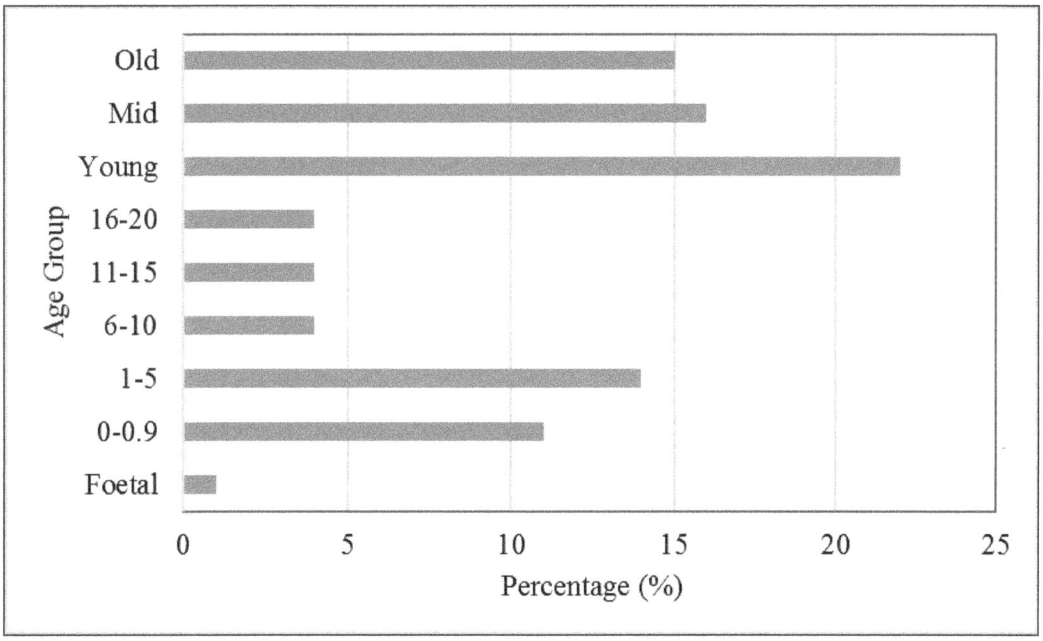

Figure 3.1. *Age at death of total sample: Taumako.*

Table 3.8. *Adult age and sex distribution: Taumako.*

Age	Male	Per cent	Female	Per cent	Subtotal	Unknown	Per cent	Total
<20 yrs	2	20	8	80	10	0	-	10
Young	27	54	23	46	50	0	-	50
Mid	23	64	11	31	34	2	6	36
Old	17	50	15	44	32	2	6	34
Adult	3	24	5	38	8	5	38	13
Total	**72**	**54**	**62**	**46**	**134**	**9**		**143**

Per cent = per cent total for each age group

Table 3.9. *Methods used for subadult age estimation: Mound 1, Tonga.*

Age (yrs)	Dental	Diaphyseal	Epiphyseal	Combined	Comparative	Total
0–0.9	2	1	-	5	1	9
1–5	5	-	-	3	1	9
6–10	2	-	-	-	-	2
11–15	2	-	-	1	-	3
16–20	1	-	-	-	-	1
Subadult	-	-	-	-	2	2
Total	**12**	**1**	**0**	**9**	**4**	**26**

Dental = number of individuals with dental material available for estimation of age; Diaphyseal = number of individuals with complete diaphyses; Epiphyses = number of individuals with age estimations based on epiphyseal union; Combined = number of individuals with both dental and diaphyseal, and/or epiphyseal material for age estimation; Comparative= number of individuals with fragmentary skeletal material where age estimation was based on a comparison of skeletal size with individuals aged using dental development

Table 3.10. *Age structure of sample: Mound 1, Tonga.*

Age (yrs)	N	Per cent total sample
Foetal	0	0
0–0.9	9	20
1–5	9	20
Child Subtotal	**18**	**40**
6–10	2	4
11–15	3	7
16–20	1	2
Subadult	2	4
Subadult Subtotal	**26**	**58**
Young	9	20
Mid	5	11
Old	0	0
Adult	5	11
Adult Subtotal	**19**	**42**
Total	**45**	**100**

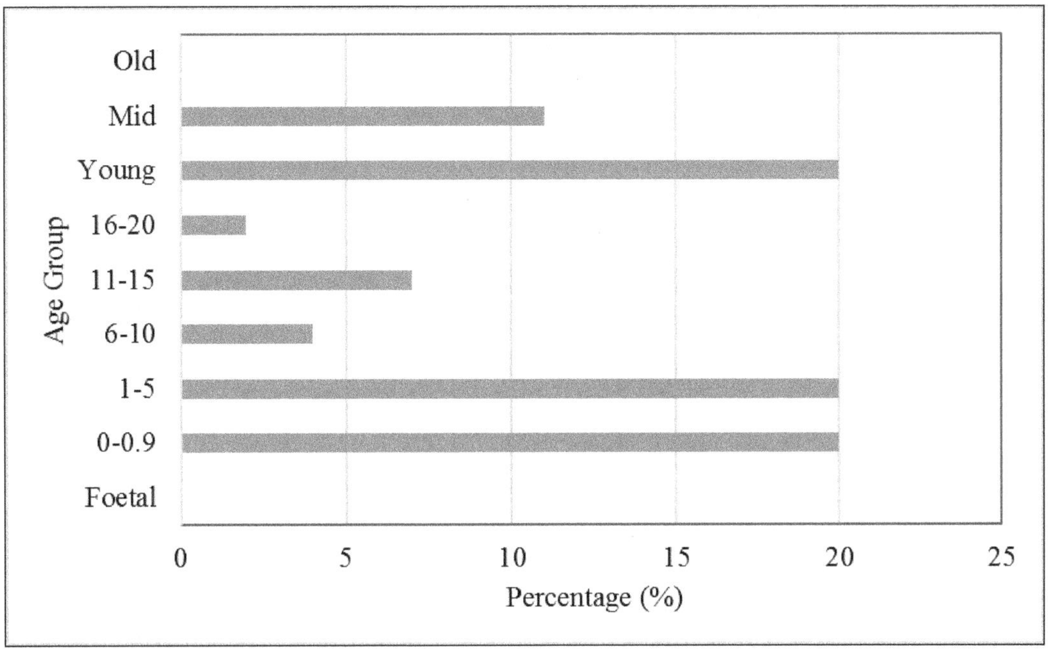

Figure 3.2. *Age at death of sample, Mound 1, Tonga.*

3.11). This is a 0.6:1 ratio of males to females. A difference in mortality is seen between the sexes where more Young females died than older females. Conversely, less Young males were present than Mid aged males. Of the 16 adults with sex estimates the diagnosis was uncertain in two females (F?) and one male (M?).

Age at death: Mound 2

Table 3.12 illustrates the methods used for age estimation of subadults from Mound 2. Of the 54 burials from Mound 2, a greater proportion were adult, with only 28 per cent of the sample under 20 years of age at death (Table 3.13). In subadults, 6 per cent died as infants and a total of 28 per cent of the sample died before 20 years of age. One third (33 per cent) of the adult sample did not have enough material present to provide age estimations. However, of those that were assigned ages, a higher representation of young adults is shown (Figure 3.3).

Sex distribution: Mound 2

The overall distribution of the sexes was comparable with a ratio of 0.9:1 males to females (Table 3.14). It was not possible to assign sex estimates to sixteen adults which constitutes 41 per cent of the adult sample. Overall, female mortality decreased with age while male mortality increased. Of the 25 adults with sex estimates the decision was uncertain in 4 females (F?) and 1 male (M?).

3.5.3 Fertility

An estimate of fertility was made for each sample by observing the number of female pelvic bones with evidence of parity. As described above, the pre-auricular sulcus of the ilium and the dorsal pubis are sites of bone resorption as a response to pregnancy. These characteristic lesions were interpreted as evidence that the female had carried a foetus to the third trimester at least once before death. The absence of these markers was interpreted as evidence of nulliparity. Exceptions to these assumptions regarding evidence of parity in skeletal material have been outlined in the methods. Table 3.15 presents the results of parity in all of the samples. The highest level of fertility was observed in Mound 2. Similar levels were observed between Taumako and Mound 1.

The proportion of females with evidence of parity was not universal among the age groups at Taumako (Table 3.16). In all age groups except the Mid adult females evidence of parity was greater than evidence of non-parity. Nearly half (45 per cent) of the Mid aged females did not show any evidence of having given birth before death. Chi-square and FET statistical analysis showed no significant differences between the age groups of Taumako females and evidence of parity. There were too few data from either of the Tongan mounds to analyse the distribution of age of females with evidence of parity.

3.5.4 Comparison of age at death and sex distribution between Taumako and Tonga

For this section, the proportions of individuals are calculated from within the different age groups rather than the sample as a whole. For example at Taumako, 25 out of 93 subadults died between birth and one year, which constitutes 27 per cent of the subsample of subadults from this site. All raw data are provided in Tables 3.7, 3.10, and 3.13. It is acknowledged that the samples of the Tonga mounds are small relative to Taumako. However, it is possible to

Table 3.11. *Adult age and sex distribution: Mound 1, Tonga.*

Age	Male	Per cent	Female	Per cent	Subtotal	Unknown	Per cent	Total
<20 yrs	-	-	-	-	-	-	-	-
Young	3	30	7	70	10	-		10
Mid	3	60	2	40	5	-		5
Old	-	-	-	-	-	-	-	
Adult	-	-	1	20	1	4	80	5
Total	**6**	**37**	**10**	**63**	**16**	**4**	**25**	**20**

Per cent = per cent of total for each age group.

Table 3.12. *Methods used for subadult age estimation: Mound 2, Tonga.*

Age (yrs)	Dental	Diaphyseal	Epiphyseal	Combined	Comparative	Total
0–0.9	1	1	-	1	-	3
1–5	1	-	-	2	2	5
6–10	1	-	-	1	-	2
11–15	-	-	2	-	-	2
16–20	-	-	4	1	-	5
Subadult	-	-	1	-	-	1
Total	**3**	**1**	**7**	**5**	**2**	**18**

Dental = number of individuals with dental material available for estimation of age; Diaphyseal = number of individuals with complete diaphyses; Epiphyses = number of individuals with age estimations based on epiphyseal union; Combined = number of individuals with both dental and diaphyseal, and/or epiphyseal material for age estimation; Comparative= number of individuals with fragmentary skeletal material where age estimation was based on a comparison of skeletal size with individuals aged using dental development

Table 3.13. *Age structure of sample: Mound 2, Tonga.*

Age	N	Per Cent Total Sample
0–0.9	3	6
1–5	5	9
Child Subtotal	**8**	**15**
6–10	2	4
11–5	2	4
16–20	2	4
Subadult	1	2
Subadult Subtotal	**15**	**28**
Young	10	19
Mid	4	7
Old	7	13
Adult	18	33
Adult Subtotal	**39**	**72**
Total	**54**	**100**

Table 3.14. *Adult females with skeletal evidence of pregnancy: Taumako and Tonga.*

Age	Male	Per cent	Female	Per cent	Subtotal	Unknown	Per cent	Total
<20 years	1	33	2	67	3			3
Young	3	30	6	60	9	1	10	10
Mid	3	50	3	50	6			6
Old	4	80	1	20	5			5
Adult	1		1		2	16	89	18
Total	**12**	**48**	**13**	**52**	**25**	**17**		**42**

Per cent = Per cent of total for each age group

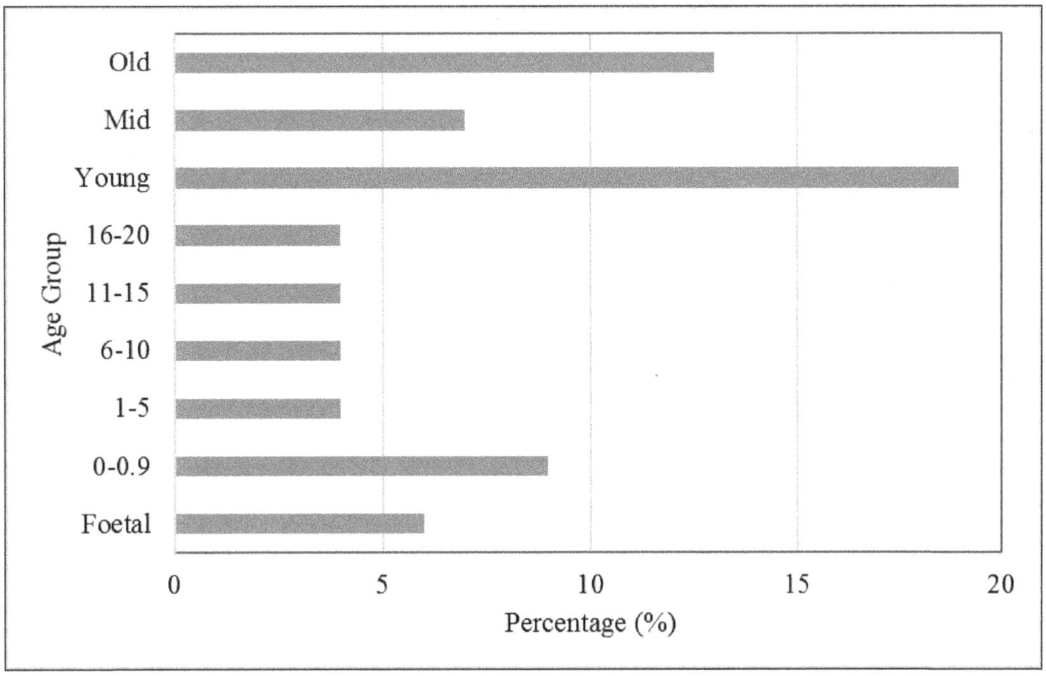

Figure 3.3. *Age at death: Mound 2, Tonga.*

Table 3.15. *Adult females with skeletal evidence of pregnancy: Taumako and Tonga.*

Site	N Pelvis	N Parous females	Per cent	Youngest pregnancy
Taumako	46	31	67	15.5–17.5 yrs
Mound 1	5	3	60	Young adult
Mound 2	4	4	100	Young adult

Table 3.16. *Age distribution of pregnancy markers in females from Taumako.*

Age	N pelves	N parous	Per cent	N nulliparous	Per cent
<20	8	5	62	3	38
Young	13	11	85	2	15
Mid	20	11	55	9	45
Old	5	4	80	1	20

make some general comparisons of mortality between the samples.

3.5.5 *Subadults*

In a comparison of the subadult and adult mortality rates between the three samples, Mound 1 on Tongatapu had the highest proportion (58 per cent) of individuals dying before 20 years of age (Figure 3.4). This difference was significant only between Mound 1 and Mound 2 (chi-square p-value = 0.041).

Similar proportions of infants died in all samples. Within the samples of Taumako and Mound 2, the highest proportions of young children died between one and five years of age, while at Mound 1 the same mortality rate was found between infants and young children (Figure 3.5). In all samples fewer children died after five years of age. However, at Taumako the risk of death was the same for older children and adolescents between six and twenty years of age, while fewer 16–20 years olds died at the Tongan mounds. Despite this difference, the 16–20 year olds have a consistently lower mortality than the younger age groups in all samples. Only at Taumako was there any evidence of foetal mortality which may be an artefact of sampling and ageing methods.

3.5.6 *Adults*

The proportion of Young adults is higher than older adults in all samples. No old adults were identified at Mound 1. It is relevant to note a higher proportion of Adult material that was not able to be assigned an age estimate at both the Tongan mounds, particularly at Mound 2. This is probably the result of sampling issues outlined in chapter two.

Overall, at both the Tongan mounds there was a higher proportion of females than males, while at Taumako the opposite was found (Figure 3.7). Chi-square and FET statistical tests revealed no significant differences in the numbers of males or females between the sites. However, when the proportion of males and females are separated by age group a different pattern emerges. Figure 3.8 illustrates that in the proportion of males to females in the individuals under 20 years of age with a sex estimate, females were at a higher risk of death than males. The proportions are similar between Taumako and Mound 2 while at Mound 1 the one individual between 16 and 20 years of age was not assigned a sex estimate. A similar pattern is seen in the Young adults at the Tongan mounds while more males than females died as Young adults at Taumako (Figure 3.9). However in the Mid adults, more males died at Mound 1 and the same mortality rate is shown between the sexes at Mound 2. Again at Taumako more males died as Mid adults than females (Figure 3.10). This same pattern is seen in the Old adults (Figure 3.11).

Taumako compared to combined Tongan results: Mortality rates and sex distribution

With the data from the Tongan mounds combined the differences in mortality become minimal when compared to Taumako (Table 3.17). The adult age groups have not been compared but results reveal that overall the same proportion of adults died at both Taumako and Tonga. With the males and females from the Tonga mounds combined a similar pattern of male and female mortality was found as when the mounds were dealt with separately (Table 3.18). The ratio of males to females in the combined Tongan mounds is 0.7:1 which is similar the sex ratio with the mounds separate.

3.6 Census Discussion

Before a discussion of the possible causes of differential mortality in age and sex is presented it must first be established whether these samples are representative of the mortuary population.

Firstly the sizes of the samples vary markedly between Taumako and the Tongan mounds. This is largely a reflection of the excavation strategies employed by the archaeologists who recovered the material. Also, Waldron (1994) demonstrates a number of extrinsic factors which can influence whether the skeletal sample is representative and considers the introduction of certain systematic errors in the sample. The extrinsic factors that should be considered are: 1) the proportion of those dying which are buried at the site; 2) the proportion of individuals whose remains are preserved well enough to be discovered; 3) the proportion of skeletons discovered through archaeological investigation and 4) the total number of skeletons recovered (Waldron, 1994). As mentioned above, a further extrinsic factor may be the research objectives of the archaeological investigation.

If we consider these factors in relation to each site, then the reasons for the different sizes of the samples become clearer. Firstly, in relation to all of the samples Waldron's first extrinsic factor is unknown, as it would be for most prehistoric skeletal samples. It cannot be determined whether there was any differential burial practices of certain ages or sexes in any of these sites. However, the preservation of the bone material was generally very good in all of the samples, and the presence of infant material would suggest that differential preservation has not greatly affected the sample.

The Taumako excavation was conducted with the primary purpose of recovering a large burial sample to determine the biological origins of the island inhabitants (Leach and Davidson, 2008). Therefore, a large area (8x8 square meters) was opened up for excavation. As outlined in chapter two, time constraints compelled the archaeologists to concentrate on the north east quadrant of the site. Therefore, we know that not all individuals discovered during excavation were recovered and that the total size of the sample was reduced by this strategy.

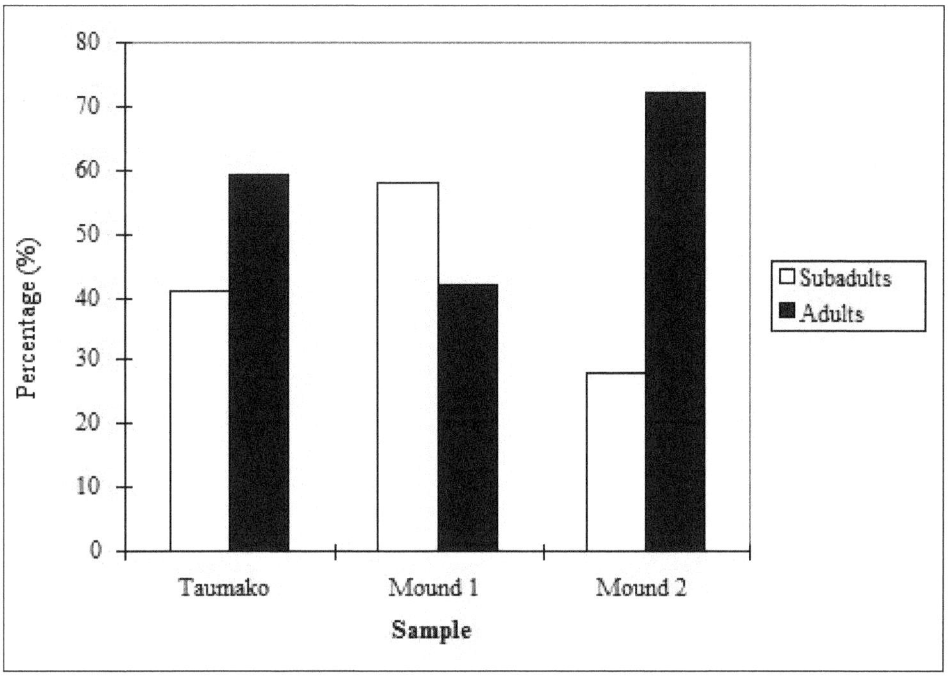

Figure 3.4. *Comparison of subadult and adult mortality rates among Taumako and Tonga.*

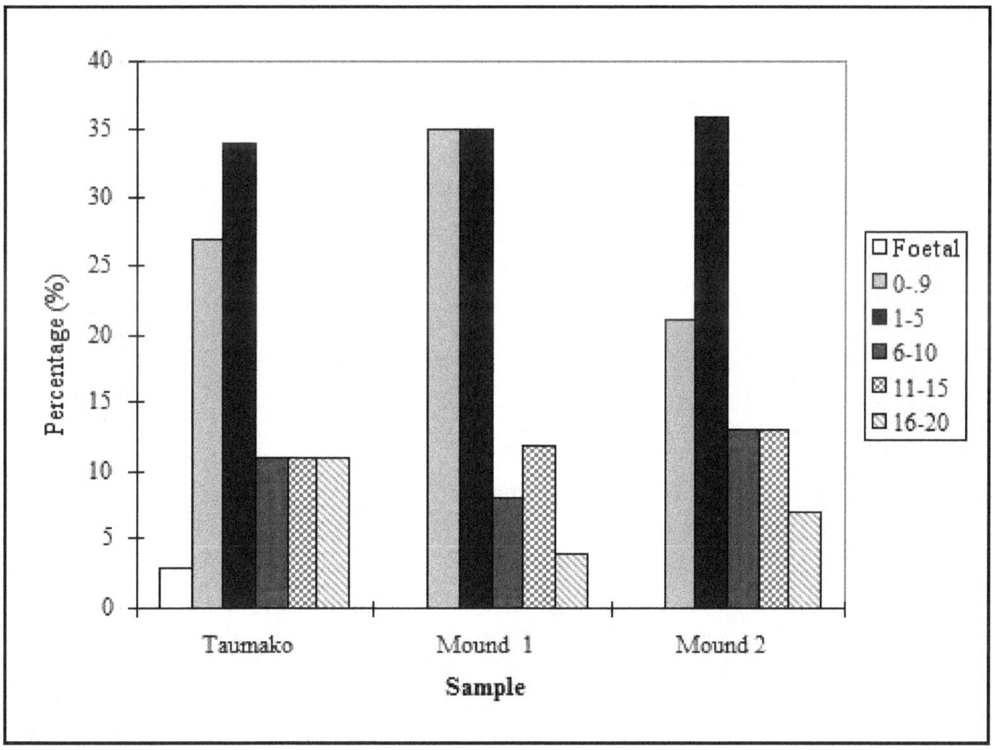

Figure 3.5. *Mortality of subadult age groups at Taumako and Tonga.*

Table 3.17. *Comparison of mortality rates between Taumako and combined Tongan data.*

Age Group	Taumako n/N	Per cent	Tonga Combined n/N	Per cent
0–0.9	25/93	27	12/41	29
1–5	32/93	34	14/41	34
<6	60/93	65	26/41	63
Subadults (<20)	93/226	41	41/99	41
Adults	133/226	59	58/99	59

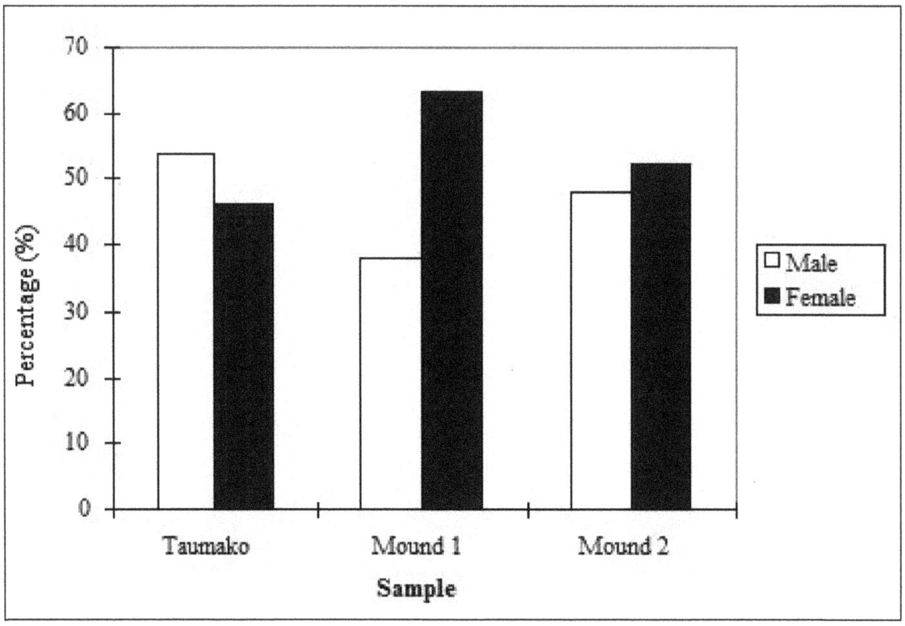

Figure 3.6. *Mortality of adult age groups at Taumako and Tonga.*

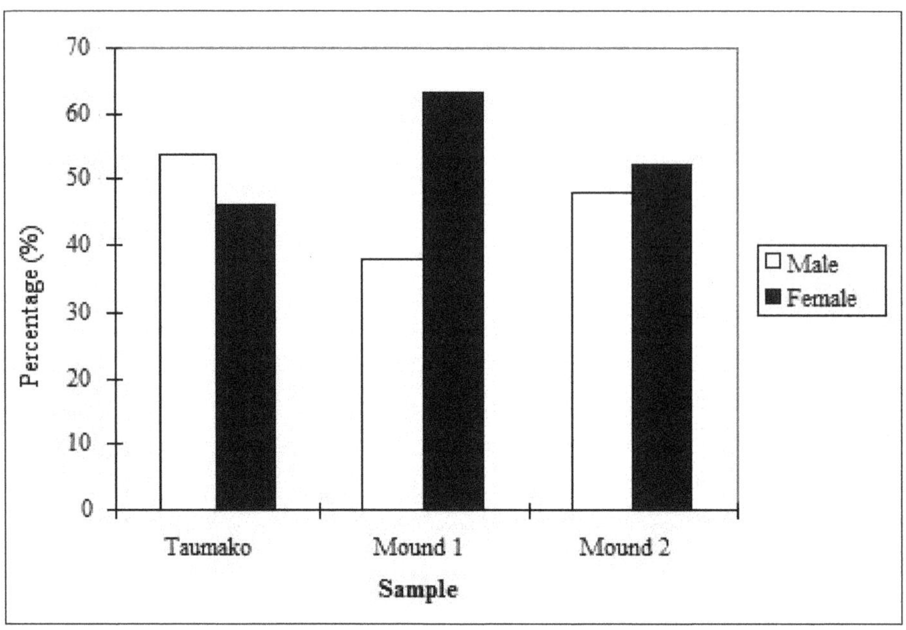

Figure 3.7. *Overall proportion of males and females at Taumako and Tonga.*

Table 3.18. *Comparison of sex distribution between Taumako and combined Tongan data.*

Sex	Taumako n/N	Per cent	Tonga Combined n/N	Per cent
Male	72/134	54	62/134	46
Female	18/41	44	23/41	56

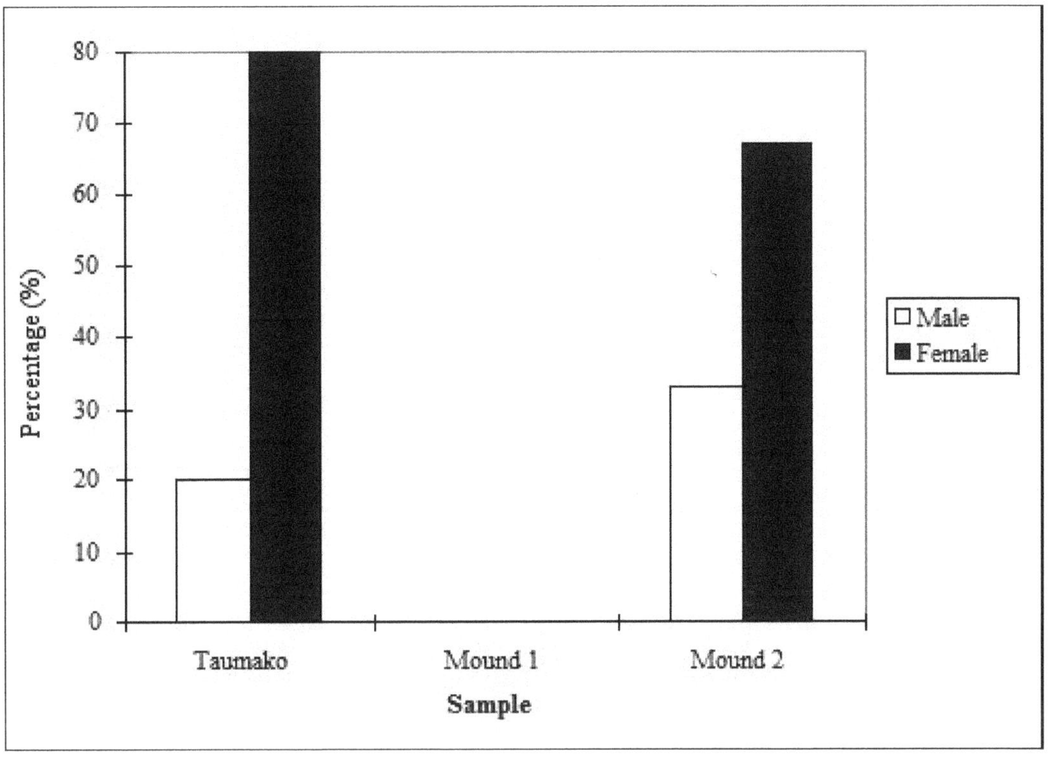

Figure 3.8. *Mortality of males and females under twenty years of age at Taumako and Tonga.*

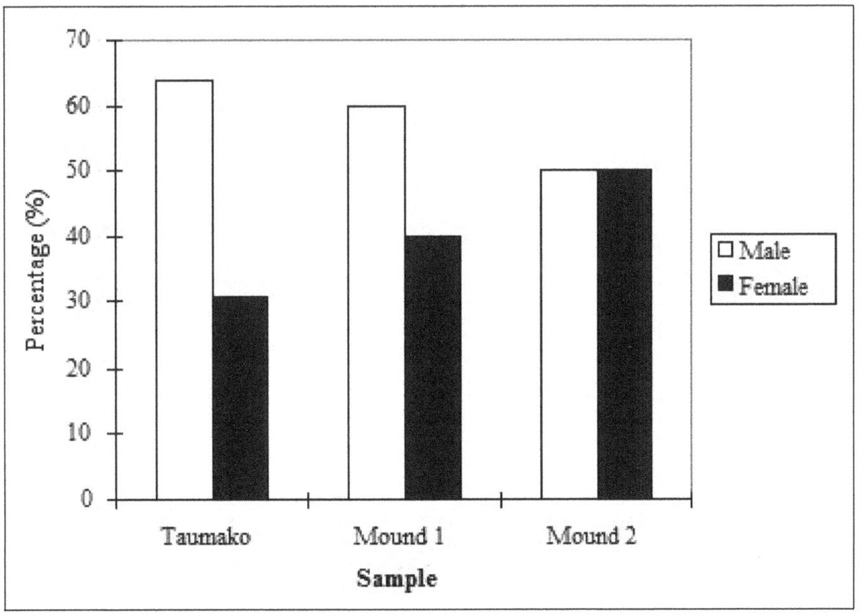

Figure 3.9. *Mortality of males and females under twenty years of age at Taumako and Tonga.*

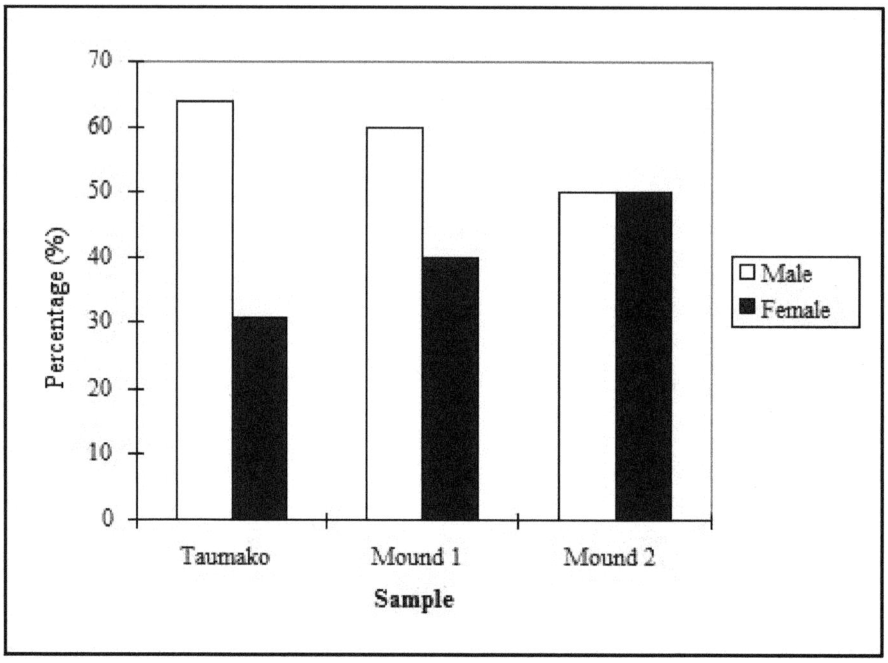

Figure 3.10. *Mortality of males and females under twenty years of age at Taumako and Tonga.*

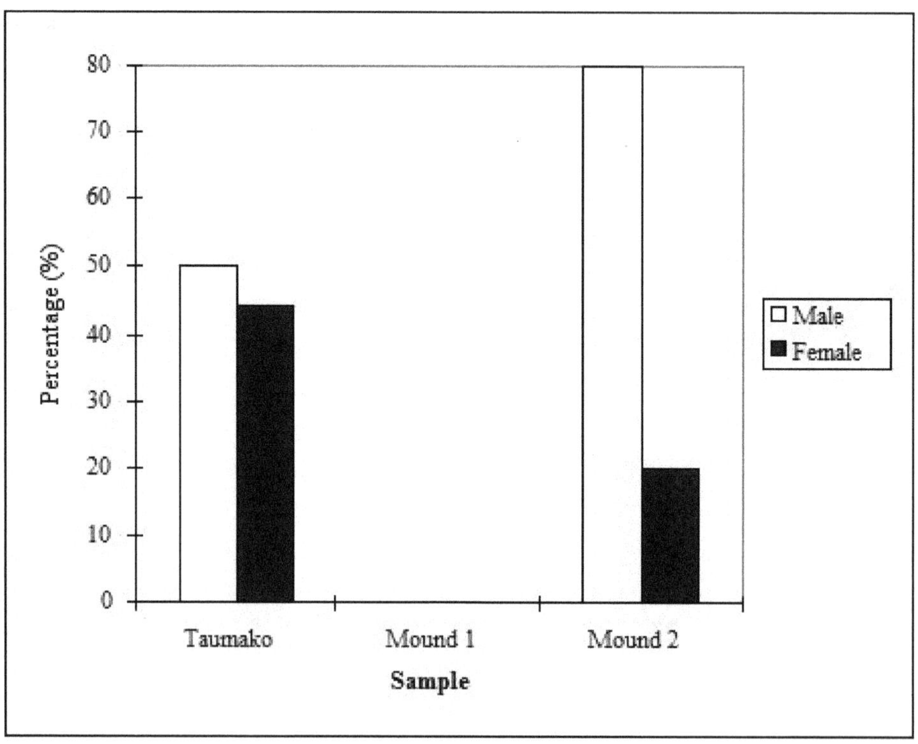

Figure 3.11. *Mortality of Old adult males and females at Taumako and Tonga.*

As discussed in chapter two, the research objectives of the excavations at the Tongan mounds were different from that of the Taumako excavation. The primary objective was to investigate the structure of the mounds themselves (Davidson, 1969) and the discovery of burials was secondary to this objective. The size of the excavated area at Mound 1 was considerably smaller than at Taumako, but all areas that were investigated were excavated to the sterile substrate. At Mound 2, the strategy of excavating a one meter wide trench through the middle of the mound meant that only a small portion of the individuals buried at the site were discovered which may account for the lower number of subadults recovered at this site. This strategy also meant that not all individuals were completely recovered.

As the results section of this chapter has shown, the excavation strategy of Mound 2 introduced considerable problems for determining age and sex in 33 per cent of the total sample recovered. So, based on consideration of Waldron's four extrinsic factors it would seem that Taumako is the most representative of the samples, while Mound 2 from Tonga is the least representative.

Furthermore, a skeletal 'population' represents those individuals which have died and can therefore never be representative of the living population from which it was derived. However, some checks and balances can be made to determine whether the sample fits with expectations of age and sex mortality of a population in an analogous state of technological development. It is generally agreed that most prehistoric populations would fit the demographic profile of a developing country, where mortality is highest at the extremes of the spectrum of life span (Waldron, 1994). Waldron (1994) suggests that most pre-industrial societies will have a substantial number of juveniles within the population, up to 30 per cent. Another author suggests that the proportion of juveniles in a pre-industrial society will range from 19–60 per cent (Weiss, 1973).

Waldron (1994) also states that if the proportion of juveniles in a skeletal sample is less than 30 per cent then extensive epidemiological studies should probably not be carried on the sample. The results of the census for each sample in this study show that the proportions of subadults in the sample fit within the expected range of juvenile mortality for a pre-industrial society. The significantly higher proportion of subadults at Mound 1 compared to Mound 2 is probably more a reflection of the number of subadults recovered during excavation than any higher risks of mortality at Mound 1. Based on Waldron's suggestions then, the parameter of subadult mortality would suggest that further study of the paleoepidemiology of infectious disease between these samples is warranted. It is recognised that the significant difference between Mound 1 and Mound 2 in subadult mortality would preclude any serious study of comparative epidemiology between these two sites. The primary objective of this study is to measure the relative health between Taumako and Tonga and not between the Tongan mounds. When the Tongan data were combined, the mortality rates were similar to Taumako which suggests that further epidemiological analyses between Tonga and Taumako are warranted.

A second check of the representativeness of a sample which can be made is the sex ratio of adults. This ratio should be close to unity (1:1) and a low ratio of one sex to another may indicate serious errors in a sample. The samples of Taumako and Mound 2 were close to unity in sex ratios and no significant differences were found between any of the samples. A ratio of 0.6:1 of males to females in Mound 1 is perplexing. In defence of the sample from Mound 1, it should be noted that Figures 3.9 to 3.11 show that the higher proportion of females to males is only evident in the Young age group. In the older age groups more males than females were represented. Previous research of Pacific Island skeletal material has found a similar under-representation of males which may be explained by factors such as sea burials, or burial on the battleground during inter island warfare (Pietrusewsky, 1994). As mentioned in chapter two, the period when the Tongan mounds were in use probably coincided with political unrest during the struggle for power of the first Paramount chief of the Tongan islands. It is possible that the lack of males at Mound 1 may be a reflection of young males being killed away from home and buried elsewhere. This is offered merely as a suggestion for the under-representation of males at Mound 1. With the Tonga sex data combined, the ratio of males to females was only a fraction higher than at Mound 1 alone. This might be a reflection of the sampling errors inherent in both these sites with regard to sex estimation. With these possible sample errors associated with the Tongan mounds in mind, all samples were still analysed for evidence of infectious disease.

The differential mortality of age and sex in skeletal samples is often used as a measure of population health (Roberts and Manchester, 1995a). In terms of subadult mortality, infants from birth to one year of age are generally considered the most vulnerable individuals in a pre-industrial society. This vulnerability is frequently reflected in high mortality rates among this young section of the population in skeletal samples. A high infant mortality rate is thought to reflect the acute stresses of new life in an environment with high pathogen loads and deficient diets (Roberts and Manchester, 1995a).

Contrary to these usual results in skeletal studies the census of the samples from this study show that a greater proportion of subadults died between one and five years of age and not in the first year of life. This result is similar to a study carried out on a prehistoric Amerindian sample (Powell, 1988) but is markedly dissimilar to the findings of a recent study based on material from prehistoric Thailand (Domett, 1999). In a study of four prehistoric sites from Thailand, Domett (1999) found that the infant category of all the samples was higher than in the children aged between one and five years of age.

There are a number of questions this difference in infant mortality raises. Firstly, it is possible that the infants who died at these Pacific Island sites were not buried in the cemetery with the rest of the population. However, if this were the case then it would be expected that the samples of infants and young children would be smaller than they actually were. Yet, at Mound 1 the proportion of infants and young children is the same and a fair number of infants are represented at each site. Secondly, this higher proportion of young children may be an artefact of the age categories that were chosen for this study, where a category spanning five years would contain more individuals than a span of one year. However, it is doubtful the differing proportion can be explained by this factor because the Domett (1999) study used similar age categories to the present study.

A third factor to consider is the ageing techniques used for this study. The ageing methods were based mainly on data derived from Amerindian sources. These were chosen partly because they recognised the accelerated dental development of Amerindian populations and may have accounted for the supposed advanced development of Polynesian children. This does not account for any differences in skeletal growth. One study that considered the linear growth of Samoan babies found they were considerably advanced in linear growth and weight in the first six months of life, even compared to European Americans (Baker et al., 1986). Conversely, skeletal studies have shown that subadult linear growth in Amerindian populations were retarded compared to modern Europeans (Mensforth, 1985; Wall, 1991). With this in mind, it is possible that some of the subadults from the one to five year age category from Taumako and Tonga have been over-aged due to different growth rates between Polynesian and Amerindian subadults.

Similarly, modern studies have shown that south-east Asian subadult growth rates are markedly retarded when compared to European populations (Chandrapanond et al., 1973; Vathakanon and Chavalittamrong, 1978). Therefore, the very high infant mortality demonstrated in the Domett (1999) study may be an artefact of consistent under-ageing of young children due to different growth rates of Thai samples and reference standards used. The possibility that the present study may have over-aged infants is acknowledged. Tables 3.6, 3.9, and 3.12 illustrate that a considerable proportion of the subadults from each sample were aged using standards based on dental development. As explained above, the dental development of Amerindians and Polynesians is similar. Therefore, the age estimates of this study may be as close an approximation to chronological age as possible and the higher proportion of young children might be real.

A final factor that may have influenced this higher proportion of one to five year olds in the mortality sample is that young children were at a greater risk of mortality in all samples than the infants. With a fair representation of infants in all the samples, it could also be argued that the causes of mortality were more chronic than in the studies where infant mortality alone was high. In tropical environments the risk of death in the first year of life is equally likely as in pre-industrial societies of temperate zones. In temperate countries it is expected that the risk of death will dramatically decrease after one year of age, in the tropics this is not the case. The causes of death in the tropics vary with age in most tropical countries (Jelliffe, 1970). During the first few weeks of life infants are especially vulnerable, depending on maternal health and exposure to various pathogens. After one year of age one of the principal causes of mortality in the tropics, besides malaria, are complications associated with various nutritional disorders (Hendrickse and Brabin, 1996). This may explain the differences in subadult mortality.

However, when viewed in the broader context of Pacific tropical environments the mortality of young children among the samples in this study is not so unusual. Figures 3.12 to 3.14 compare the Taumako and Tongan subadult samples to the two Polynesian skeletal series of Mokapu, Hawaii and Hane Dune, Marquesas (Snow, 1974; Pietrusewsky, 1976). Raw data is extracted from Pietrusewsky (1976) and Snow (1974). It is recognised that the age categories employed by Snow (1974) and Pietrusewsky (1976) are not directly comparable to one another, or to those used in this study. However, the age categories are broadly similar and illustrate a subadult mortality pattern reminiscent of those found in Taumako and Tonga. It would be ideal to compare the subadult mortality of Taumako and Tonga to other large Pacific Island samples. However, much of the comparative data from previous research in the Pacific is trapped within the so-called 'grey literature' of commercial reports and are not accessible to most other researchers. This is particularly relevant to the Marianas Islands in Micronesia where a considerable amount of rescue excavation has been carried out in recent years. Given the comparability with two other skeletal series from the Pacific Islands it is suggested that the subadult mortality pattern of Taumako and Tonga is probably a fair representation of the true death rates of these prehistoric samples. The possible causes for the subadult mortality pattern observed in this study will be discussed in conjunction with multiple indicators of disease and stress in chapter seven.

3.6.1 Adult mortality

As shown in Figure 3.6 the proportion of Young adults to older adults was high in all of the samples. This indicates that some individuals who survived the stresses of early childhood were vulnerable to the risk of death as young adults. After 20 years of age Young females from the Tongan mounds were exposed to higher levels of risk than at Taumako, compared to males. This pattern was the same as at Mokapu, Hawaii where more young females (20–30.5 years) died than males and more males died in the older age categories (Snow, 1974).

Evidence of parturition in individuals as young as 15 years old at Taumako show that females were exposed to the risk of childbearing from an early age. This may also have been the case at Tonga. The presence of foetal remains from Taumako may indicate that some pregnancies were not successful and may have also contributed to maternal mortality.

In developing countries, pregnancy and childbirth is a sig-

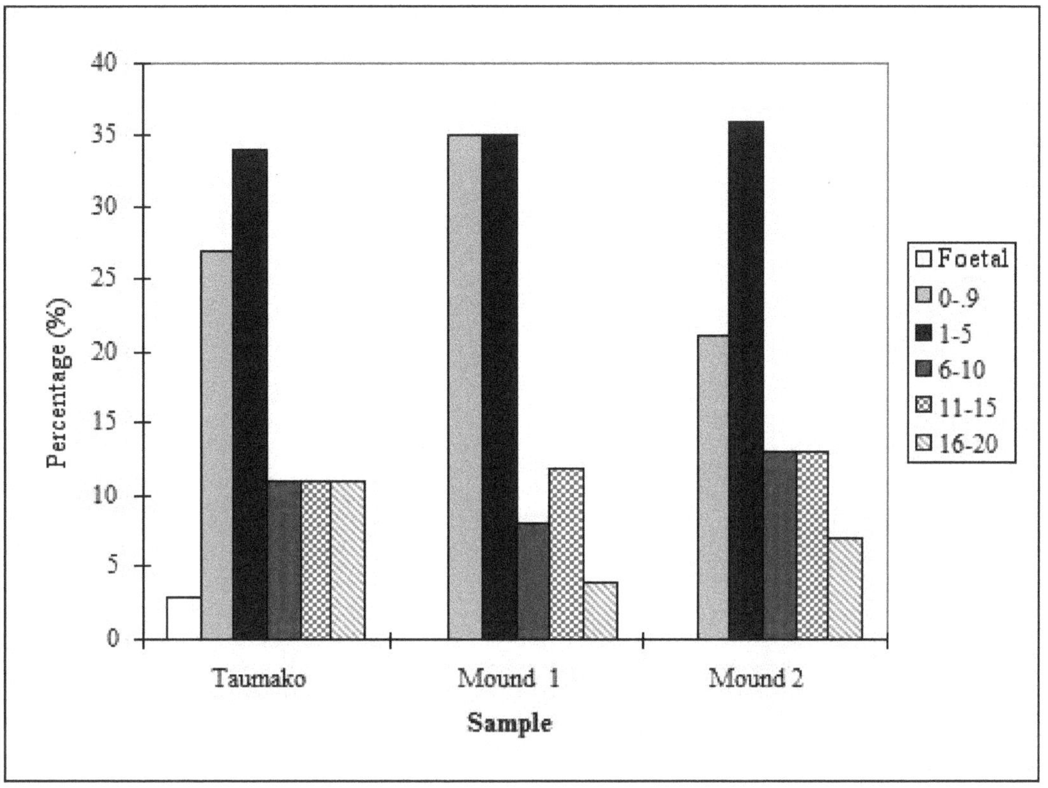

Figure 3.12. *Age at death of subadults from Taumako and Tonga.*

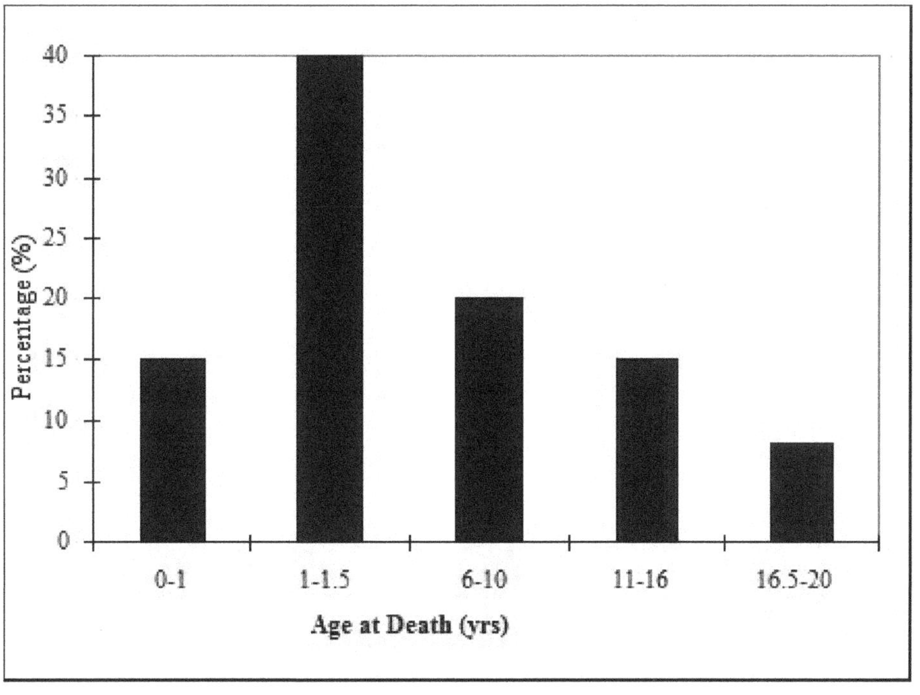

Figure 3.13. *Age at death of subadults from Mokapu, Hawaii (after Snow 1974).*

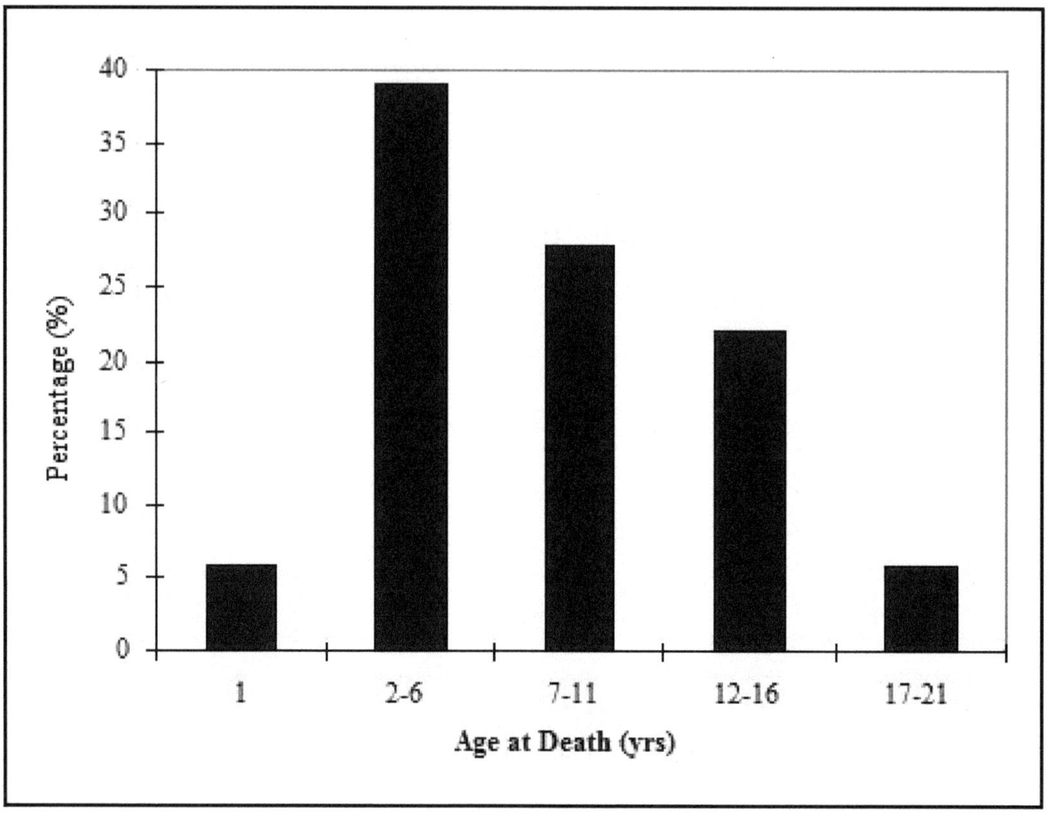

Figure 3.14. *Age at death of subadults from Hane Dune, Marquesas Islands (after Pietrusewsky, 1976).*

nificant cause of death among females (Say et al., 2014). Factors which contribute to deaths associated with pregnancy include direct and indirect causes. Among the direct causes are: infection, haemorrhage, obstetric complications of pregnancy, and childbirth itself. These direct factors contribute to 50–98 per cent of all maternal deaths in developing countries and 50 per cent of these deaths are associated with infection and haemorrhage. Indirect causes include conditions which are present at time of conception. Those that may be archaeologically visible are infectious disease and anaemia, excacerbated by pregnancy (Ortner, 1998).

These factors may have contributed to higher female mortality as young adults from Taumako and Tonga. The possible association between evidence of infectious disease and age and sex mortality rates will be addressed in chapter seven.

3.7 Conclusions

To ensure the validity of the following assessments of health and disease indicators in these samples it is important to consider whether these samples were representative of the cemetery populations. Based on the checks and balances of Waldron (1994) reviewed above it has been argued that the skeletal samples of Taumako and Tonga are more or less representative of the cemetery population. The age at death of subadults and adults is also considered to be representative. Therefore, comparative analyses of indicators of stress and disease in subsequent chapters is considered to be warranted.

The first aim of this thesis is to assess whether the presence of malaria may be reflected in the demographic profile of the Taumako sample. In order to achieve this aim the age and sex structure of the three skeletal sample has been established. The results of these analyses have shown few differences in mortality rates between Taumako and Tonga, particularly when the Tongan data were combined. The implications of the results presented in this chapter with respect to the first aim of this thesis are discussed in chapter seven.

Chapter 4

Growth Disturbance

The second aim of this thesis is to test whether the stature of the Taumako adults was adversely affected by the presence of malaria and to assess whether evidence of non-specific growth disruptions in the form of dental defects was more prevalent at Taumako than Tonga. The objective of this chapter is to address this aim.

Few infectious or nutritional diseases will leave a specific signature on the bones or teeth. For this reason many studies have concentrated on the analysis of so-called non-specific indicators of stress. These may indicate that the growth of an individual has been adversely affected by disease or inadequate nutrition. As outlined in chapter one:

> The popularity of non-specific stress indicators stems from "population thinking" in palaeopathology, applying to the past theories derived from natural or laboratory experiments in contemporary populations....Foremost among human biologists who have influenced this research trajectory is Scrimshaw... who emphasizes the synergism between disease and dietary stress.... (Buikstra, 1980, 444–45)

This chapter examined evidence of non-specific indicators of stress in the skeletal samples used for this study. The most commonly analysed indicators of non-specific stress are; Harris lines, dental defects, and adult stature (Buikstra and Cook, 1980). Only dental defects and adult stature are examined as markers of non-specific stress in this study. A discussion of child growth is also included here, but minimal data precluded any further analysis.

Harris lines are seen radiographically as transverse bands of increased density of bone tissue in the medullary cavity. They indicate a period of resumed bone growth after a period of decreased or ceased growth. The presence of these bands is usually interpreted as an indicator of some stress during the period of linear growth caused either by infection or dietary deficiency (Roberts and Manchester, 1995b).

Theoretically, the age at development of the transverse line should be calculable, and many studies have attempted to demonstrate this. However, these lines are able to remodel and disappear due to continual bone remodelling during the course of an individuals life. For this reason, Lewis and Roberts (1997) suggested that the study of Harris lines should be restricted to the juvenile portion of the sample. Study of this indicator of stress is still fraught with many problems, including notably inter- and intraobserver error (Grolleau-Raoux et al., 1997; Lewis and Roberts, 1997).

4.1 Measures of Growth in Skeletal Samples

The study of growth attainment in past populations is also useful. Each individual is born with a genetic potential for eventual body size and the stature of an adult is a reflection of the amount of growth achieved during childhood. It has long been recognised that linear bone growth is more sensitive to environmental stress than tooth development (Eveleth and Tanner, 1990). Therefore, some studies have concentrated on investigating the growth velocity and rates of growth in subadults by comparing tooth developmental stages and long bone growth (e.g. Wall, 1991). Other studies have concentrated on comparisons of adult stature between social classes in one population (Powell, 1988). An important difference between the study of child growth rates and adult stature, is that the latter tests the ability of *survivors* to adapt to childhood stress while child growth studies are based on *non-survivors*. This study will examine childhood stress by discussing both these indicators.

4.1.1 Child growth

People have been interested in morphological variation in the human form for thousands of years and anthropologists have considered the meaning of growth patterns from early in the discipline (Bogin, 1999). In the early 1900s much of the research concerning growth differences between populations was undertaken using anthropometry, the science of human body measurement. At that time most anthropologists believed that population differences in adult stature could be used as 'racial' markers. This inherent belief bolstered universal ideas of the white, tall peoples of Europe as superior to other short and dark peoples. However, the studies of Franz Boas, around 1910, found that phenotypic change occurred with improved nutrition among European immigrants to America. This study marked the introduction of the idea that environmental influences might affect an inherent genetic potential of growth (Bogin, 1999).

Since then, the plasticity of growth during childhood has been recognised. Developmental plasticity refers to the adaptive advantage of an organism with the potential to respond to changes in the environment by phenotypic change during growth. During childhood an individuals skeletal growth may slow in response to environmental stress. If the stress is relieved, this period of slow growth may be followed by a period of accelerated 'catch-up growth'. Various studies have recorded this phenomenon in living children and skeletal samples (Wall, 1991; Bogin and Loucky, 1997).

Today, scientists and anthropologists examine growth in several ways. The first is the study of growth itself. These studies seek to describe changes in the body that occur with age and data from these studies provide standards of growth in living populations which are used by anthropologists to estimate age in skeletal material, such as those reviewed in chapter three. Another reason for studying human growth is to measure the adequacy of child growth within a population as an index of overall population health (Johnston and Zimmer, 1989).

When attempting to measure skeletal growth in prehistoric populations several methodological biases must be considered. The first is what Saunders and Hoppa (1993) call the *biological mortality bias*. This has been recognised as a potential source of error when examining prehistoric skeletal growth (e.g. Johnston 1962) . The biological mortality bias is described as 'whether or not children who die are more likely to have greater growth deficits in attained growth than their counterparts who survive' (Saunders and Hoppa 199, 136). In an extensive review of this subject, Saunders and Hoppa (1993) concluded that any bias that may exist is minimal compared to the extent of error that can be introduced by methods used to estimate chronological age; the considerable variation in the reliability of age estimation methods. As discussed in chapter three, to reduce the amount of error introduced, age estimation standards should be based on populations genetically related to the skeletal samples (Johnston and Zimmer, 1989).

Factors that influence skeletal growth in children

It is generally accepted that poor childhood growth is an indication of unfavourable or stressful conditions in the community, such as inadequate nutrition and/or infection (Eveleth and Tanner, 1990). Studies of growth in prehistoric skeletal material derive data primarily from measurements of long bones which represent a simple measure of linear growth achieved at time of death (Saunders et al., 1993).

Growth retardation is a widely recognised symptom of a limited nutrient supply at a cellular level (Martorell and Ho, 1983). The most widely stated aetiology of growth faltering in living children of developing countries is malnutrition. Malnutrition does not always relate to a lack of food but is often the result of an inadequate balance of essential nutrients for maintaining growth. The amount of retardation of linear bone growth is related to the severity and duration of the malnutrition (Eveleth and Tanner, 1990). If an adequate nutrient supply is reinstated early, then normal linear bone growth can resume. However, if the period of slowed growth is prolonged enough then a child's stature may be permanently affected by the period of retardation (Bogin, 1999).

With respect to permanent negative affects on stature, it has been argued that the period from birth to three years of age is the most crucial (Martorell and Ho, 1983). Because growth during the first two years of life is particularly rapid and produces increased energy demands, any prolonged nutrient starvation during this time will have the most profound effects on later stature attainment. Therefore, after three years of age a child's growth may be either advanced or retarded throughout the rest of their growth period, depending on the experiences of early childhood (Martorell and Ho, 1983; Bogin, 1999).

As mentioned above, the synergistic relationship between nutrition and infection is also important to consider in any study of child growth. When a child is malnourished it is more susceptible to infections and a disease state may be more prolonged (Scrimshaw, 1981). Also, during periods of infection, nutrients are not as efficiently absorbed and each condition can exacerbate the other, leading to either growth retardation or ultimately, death. Diarrhoea has been shown to be closely associated with growth retardation in living children, which usually coincides with age at weaning (Saunders and Hoppa, 1993).

Growth retardation and weaning

The occurrence of retarded growth in subadults between two and four years of age has been extensively reported in living children and in skeletal samples. The cause of growth retardation has been uniformly interpreted as an indication of nutritional stress or infection, or a combination of both, during weaning. For example, Wall (1991) tested the weaning stress hypothesis by assessing whether there was any deviation in rates of growth between dental development and skeletal growth in a sample of prehistoric Amerindian bones. Wall (1991) found a clear indication of reduced skeletal growth from two to three years of age which correlated with ethnohistoric accounts of weaning age.

He also noted a short period of increased velocity after this age which might indicate catch-up growth after weaning (Wall, 1991). Similarly, Mensforth (1985) found a difference in linear bone growth in children between two prehistoric Amerindian samples. One sample was from a hunter-gatherer population and the other was from a sedentary agricultural population. They found that children from the sedentary population had consistently retarded growth compared to the earlier hunter-gatherer population. This growth retardation was attributed to differences in food availability, disease, and increased population density that occurred with the adoption of a sedentary lifestyle (Mensforth, 1985).

From the studies which have assessed child growth in prehistoric samples several general assumptions have been adopted. Firstly, that in the absence of more specific indicators, child growth can be used to evaluate the overall health of a population. Also, that growth differences in early childhood are the result of environmental factors such as nutrition and infection. Finally, that child growth patterns are indicators of the adequacy of the environment, so that poor growth equals poor environment. However, genetic factors are thought to play a role in any observed differences if the groups being compared are not closely related (Johnston and Zimmer, 1989). Unfortunately, there were not enough data available to perform any analyses of child growth for either sample in this study. However, the theoretical basis of child growth reviewed above is useful to consider when discussing the cause of population differences in adult stature and enamel defects.

4.1.2 Adult stature

> "The child is the father of the man;" (William Wordsworth)

As mentioned above, the eventual height of an adult is the result of a continuous and complex interaction between genetic and environmental influences (Eveleth and Tanner, 1990). At birth, humans are considered to have a predetermined genetic potential for growth, meaning that the genotype of each individual determines the upper limit to adult stature. Whether the person reaches this potential height is influenced by environmentally induced experiences during childhood (Bogin, 1999).

As outlined above, if environmental conditions are inadequate during periods of growth that are particularly sensitive, such as under three years of age, then an individual's adult stature may be significantly affected. The true nature of the interaction between genetic potential for growth and environmental influences is not fully understood. As outlined above, inadequate nutrition during childhood is considered the leading cause of differences in adult stature between genetically similar populations (Eveleth and Tanner, 1990). Therefore, differences in stature of genetically close populations living under different environmental conditions can provide a reflection of overall population health.

Sexual dimorphism

The anatomical basis of sexual dimorphism has been outlined in chapter three. The morphological differences between the sexes can aid in estimating sex in skeletal samples. However, the biological basis of sexual dimorphism is also important when considering sexual differences in response to environmental stress. On average males are larger and heavier than females from birth and throughout the growth period (Bogin, 2001). During the growth period, males begin the adolescent growth spurt later and the spurt lasts longer than in females. This contributes significantly to the differences in size seen in adults, because males have a longer period of linear growth (Bogin and Loucky, 1997). In general females also appear to have a superior immune response which may affect growth. These differences may relate to the greater importance of the female role in reproduction during gestation and lactation (Stini, 1985).

Males are considered more vulnerable to disruptions in growth from environmental stress than females. Therefore, a population reduction in sexual dimorphism may suggest the occurrence of inadequate nutrition or disease during the growth period of these adults (Stini, 1969, 1985; Bogin, 1999). The mean difference in human sexual dimorphism in stature is around five per cent (Eveleth and Tanner, 1990). Therefore, a lower level of dimorphism could indicate that the males of the population were experiencing a greater degree of stress, indicated in their stature. This theory of females as more buffered against environmental stress was reviewed by Stinson (1985). She found that the relationship between reduced sexual dimorphism and environmental stress may not be as highly correlated as some would suggest (Stinson, 1985). However, in the present study it is accepted that if a reduced sexual dimorphism is found, then the concept of increased female adaptability to environmental stress as a cause is a reasonable one to consider.

Recording adult stature from the skeletal record

Because a linear relationship exists between the length of limb bones and adult height, regression equations based on bone length are used to estimate the stature of an individual from skeletal remains (Trotter, 1970). Because of the sexual dimorphism in size, all available equations provide sex-specific equations for each bone.

However, considerable variation exists among populations in body proportions. That is, the ratio of trunk and head length with limb length in contributing to total stature. Therefore, population specific regression equations should be used. Population specific regression equations are available from a wide variety of populations including European Americans, Africans and Asiatics. Therefore, if equations are not available for the skeletal sample under study or the population relationship is unknown, then a variety of equations can be applied. The results which provide the least variation between limb bones probably represents the closest approximation to the actual stature of the individual (Trotter, 1970).

Population specific equations for Polynesians are available. In a review of historic literature and using the anthropometric data from a study of Maori Battalion soldiers by Peter Buck in 1923, Houghton et al. (1975) argue that the Polynesian inhabitants of New Zealand had markedly different body proportions from Europeans. For example, the New Zealand Maori had a longer axial length and shorter leg length than other populations (Houghton et al., 1975). Also, the forearms are generally longer and the tibiae shorter in Polynesians than other in other populations (Houghton, nd). Because of this difference in body proportions none of the previously established regression equations for stature are appropriate for New Zealand Maori, or other Polynesian populations. Therefore, (Houghton et al., 1975) developed stature equations specific to this population.

Some differences in limb proportions were noted between the study group of Houghton et al. (1975) and groups from Western Polynesia. Therefore, they warn that this equation should be used with caution on populations from outside Eastern Polynesia. However, later this equation is used for a large comparative study of body proportions in other Pacific populations (Houghton, 1996).

4.2 Adult Stature

Methods

In his (2008) report on the skeletal biology of the Taumako sample, Houghton argued for the appropriateness of using the Polynesian regression equations on this sample. He found that the low tibio/femoral ratio and long forearms of these people were in accordance with Polynesian body proportions. He also found there was less variation in results when using the Polynesian equations than equations based on other populations. On the basis of Houghton's argument, coupled with the Polynesian linguistic and cultural affinities of the Taumako people outlined in chapter two, it was considered appropriate to use the Polynesian regression equations. The previous study of the Tongan sample by (Pietrusewsky, 1969) was conducted prior to the work of Houghton et al. (1975). Therefore, stature estimations based on European and Mongoloids were used. The statures of Tongan adults have been recalculated using the equations of Houghton et al. (1975) in this study.

The regression equations are provided in Appendix A. The left tibia has the lowest standard error of all bones and the right tibia was the bone with the next lowest standard error. Therefore, the left tibiae was the preferred bone for stature estimation. If the left tibia was not available, the right tibia was used. Because the lower limb bones contribute more to actual stature than the upper limb bones, the femur was used in favour of upper limb bones where no tibiae were available. If no lower limb bones were present then upper limb bones were used.

All long bones with fused epiphyses were included in this study. The maximum lengths of the bones were obtained by the use of an osteometric board based on standard criteria for limb bone measurements as outlined in Houghton et al. (1975). Lengths of the limb bones were recorded in centimetres. Where a firm sex estimate was possible the sex specific equation was used. Where sex was unknown these individuals were excluded from further analysis.

The average adult stature for each sample is presented here and differences between the sexes are considered. A greater range of stature within a sample may reflect a population with more stress during childhood. A large range of short and taller individuals within a sample may indicate greater genetic heterogeneity within the sample. It is more likely to be a reflection of social stratification within a population where there is differential access to resources depending on social status and/or sex. The presence of such a pattern is investigated in the results. Sexual dimorphism was measured as the difference between average statures of males and females as a percentage of male stature.

Finally, the Taumako and Tongan samples are placed within a broader Polynesian context by comparing the stature estimates with other Polynesian groups.

4.3 Stature Results

An estimation of stature was able to be made for 47 per cent (n = 63/133) of adults from Taumako, 42 per cent (n = 8/19) from Tongan Mound 1 and 44 per cent (n = 17/39) from Mound 2. The summary results for all samples are presented in Table 4.1 for females and Table 4.2 for males. A comparison of male and female statures of all samples is presented in Figure 4.1. The sample sizes for the Tongan samples are small relative to the Taumako sample. Therefore, any comparison of stature between the samples is made with caution.

The females from Mound 1 are on average over four cm taller than the Taumako females. The different samples sizes may have contributed to this difference in mean statures. The small sample size of Mound 1 is also reflected in the difference between the range of statures from these samples, with only 3 cm difference between the minimum and maximum, compared to a 16 cm range of difference in the Taumako females. There is very little difference in the stature estimates of the males between samples. The level of sexual dimorphism is presented in Table 4.1. The highest level of sexual dimorphism was found in the Mound 2 sample, while the only sample under five per cent was Mound 1. The small sample size may a factor in this difference.

The range of variation in stature estimates is included in Tables 4.1 and 4.2. Within the samples there is little difference in the variation of stature between the sexes, except at Mound 1. The greatest overall range of stature is found in the Taumako sample.

4.3.1 Comparison with adult stature of other prehistoric Polynesian samples

In a comparison of the mean stature estimates from some other prehistoric Polynesian populations, it is shown that male stature is comparable throughout (Table 4.3 and Figure 4.2). More variation is seen in the females with the Mound 1 females being the tallest of all. All of these data are based on variable sample sizes, some as small as those from the Tongan mounds. Therefore, these comparisons should be viewed with caution. Among all these Polynesian populations, the degree of sexual dimorphism increases with the most easterly island groups of the Marquesas and the Cook Islands.

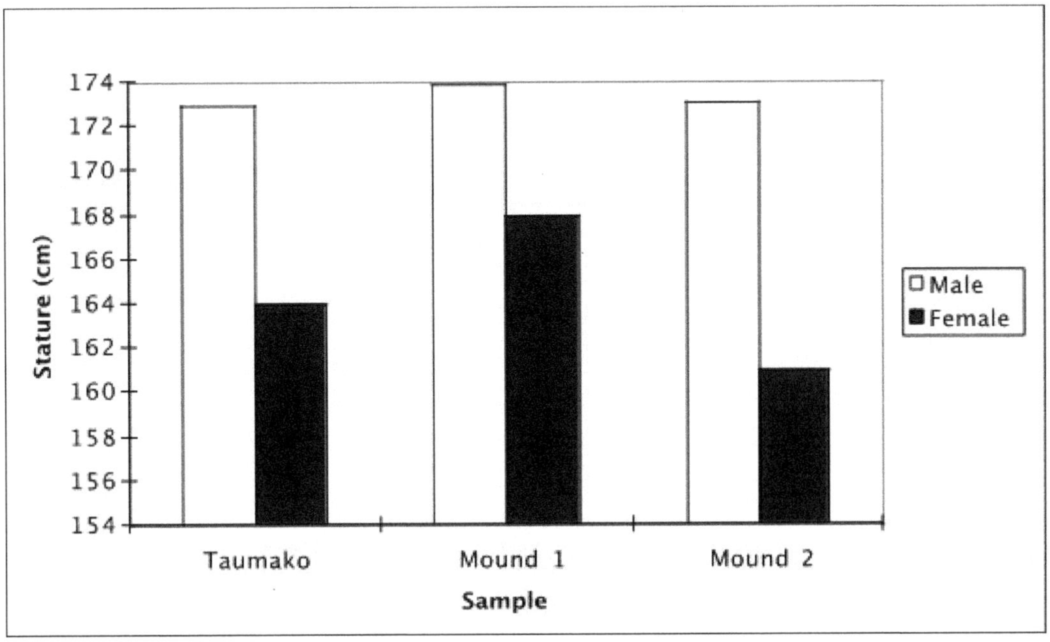

Figure 4.1. *Mean adult stature of males and females from Taumako and Tonga.*

Table 4.1. *Comparison of female stature from the adults of Taumako and Tonga.*

Sample	N	Mean	SD	Range (cm)	Spread (cm)	Sexual Dimorphism (Per cent of male stature)
Taumako	25	164.4	4.3	157-173	16	4.9
Mound 1	3	168.8	1.7	167-170	3	2.9
Mound 2	8	161.6	4.4	152-166	14	6.6

Table 4.2. *Comparison of male stature from the adults of Taumako and Tonga.*

Sample	N	Mean	SD	Range (cm)	Spread (cm)
Taumako	38	172.9	5.0	161-181	20
Mound 1	5	173.9	4.6	165-179	14
Mound 2	9	173.1	3.3	169-179	10

Table 4.3. *Summary of mean estimated statures from some prehistoric Polynesian populations. All estimates are based on the Polynesian stature equations of Houghton et al. (1975).*

Site	Sex	Mean Stature	Sexual Dimorphism (Per cent)
Taumako	Male n = 38	172.9	
	Female n = 25	164.4	4.9
Mound 1 (Tonga)	Male n= 5	173.9	
	Female n = 3	168.8	2.9
Mound 2 (Tonga)	Male n = 9	173.1	
	Female n = 8	161.6	6.6
Marquesas	Male n = 13*	174.1	
	Female n = 6*	156.3	10.2
Hawaii (Mokapu)	Male n = 75**	171.9	
	Female n = 115**	160.3	6.7
Cook Islands	Male n = 38	173.2	
	Female n = 4	160.3	7.4

Source of mean stature for all samples except Taumako and Tonga is Houghton (1996). These mean statures were based on Houghton et al. (1975) stature equations; *individual counts from Pietrusewsky (1976); ** individual counts from Snow (1974).

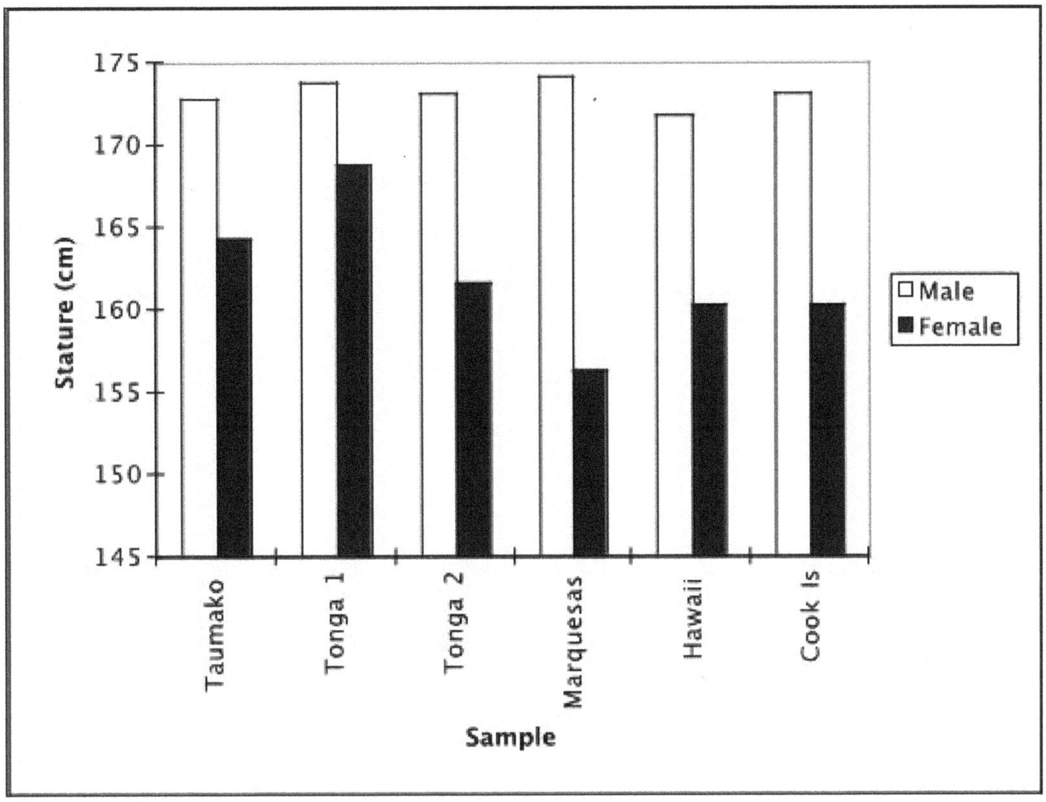

Figure 4.2. *Mean adult stature from comparative prehistoric Polynesian populations including Taumako and Tonga.*

4.4 Dental Developmental Defects

For the reasons stated above, dental defects were chosen as a more reliable indicator of stress than Harris lines. Although the study of dental defects has its own methodological problems, the most notable fact in its favour is the inability of dental enamel to remodel after mineralisation is complete. Evidence of dental defects can be lost through attrition of occlusal surfaces, however on the whole, the researcher is provided with a permanent memory of the health of the individual at the time of tooth crown development. Also, in favour of dental defects, is the fact that teeth are more likely to survive unforgiving taphonomic processes than bone tissue (Goodman et al., 1984). Dental enamel is formed by secretion of enamel matrix and apatite crystallites by ameloblasts (Hillson, 2014). The matrix then undergoes a period of maturation in which the organic component is broken down and crystallites expand to leave a mature matrix almost entirely mineral in content (Hillson, 1996). Dental enamel hypoplasia is a deficiency in enamel thickness caused by a disruption or cessation of amelogenesis during tooth crown formation (Goodman et al., 1980). The cells which form enamel are sensitive to physiological stress such as infection, dietary deficiencies and psychological or physical trauma. Therefore, evidence of a disruption of enamel formation provides a direct non-specific link to a variety of stressors experienced by the individual during the period of enamel formation (Goodman and Song, 1999). The observations of enamel defects can provide a retrospective record of physiological stress in an individual from five months *in utero* in the case of deciduous dentition, and during the first seven years of life in the permanent dentition (Goodman et al., 1980).

Defective enamel formation can be identified on the crown surface in a number of forms, such as transverse grooves, pitting, areas of opacity and localised areas of hypocalcification (Skinner and Goodman, 1992). The term enamel hypoplasia is used for an enamel surface defect including grooves, pits or missing enamel. However, linear enamel hypoplasia (LEH) (Figure 4.3) is the most commonly observed and recorded enamel defect by anthropologists (Goodman and Song, 1999). This is characterised by a linear horizontal groove caused by decreased enamel thickness (Goodman et al., 1980) .

Over the last few decades numerous studies of LEH have been carried out on prehistoric human dental remains. Many have been concerned with recording a shift in health patterns that may correlate with a shift in subsistence or technology (e.g. Cohen, 1984). The prevailing theme has been that a change in economy from hunter-gather to agricultural economy leads to an increase in enamel hypoplasia (Roberts and Manchester, 1995b).

The first systematic study of LEH was performed in 1966 by Swärdstedt who developed a method for estimating the aetiologic age at which an enamel defect has developed (Swärdstedt, 1966). This method followed the tooth development charts of Massler et al. (1941) and was based on the assumption that enamel forms at a continuous rate over the whole development period of a tooth crown. In a series of papers Goodman and various colleagues later developed a method of assessing the age of onset of hypoplastic defects

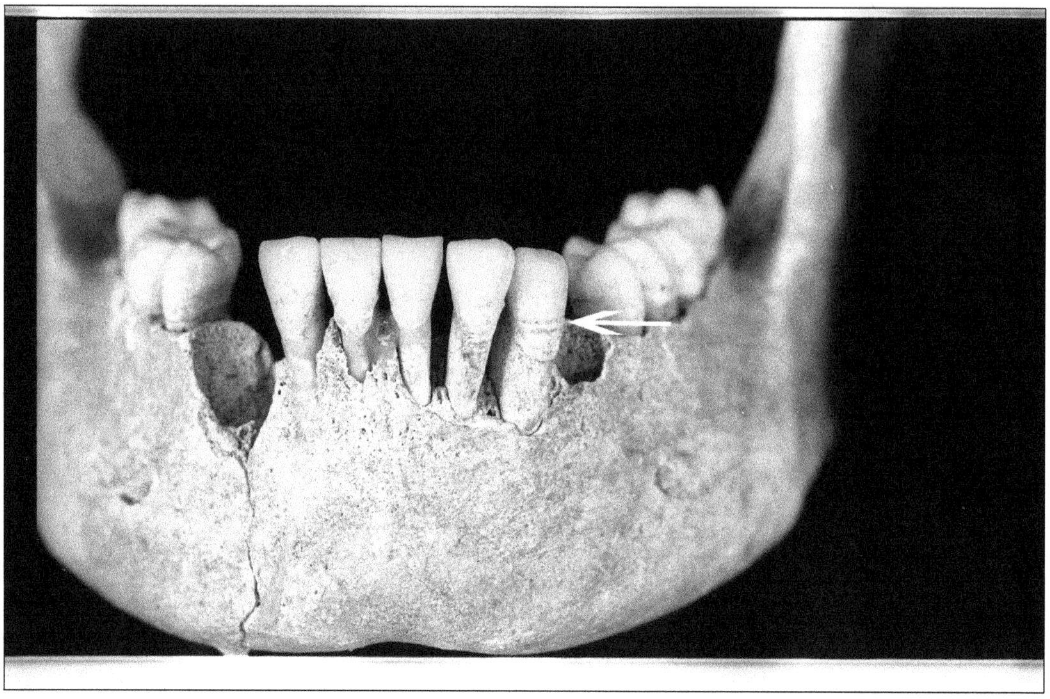

Figure 4.3. *Linear enamel hypoplasia (arrow) of a permanent canine. Photo by author.*

based on a reference graph of tooth development by Massler et al. (1941) and the metric modifications of Swärdstedt (1966) (Goodman et al., 1980; Goodman and Armelagos, 1985, 1988; Goodman and Rose, 1990). The age at which a hypoplasia developed is calculated by measuring the distance of the defect from the cemento-enamel junction and then correlating the position of the defect with a development stage of the crown. This method of assessing the aetiologic age of enamel hypoplasia has been adopted by many anthropologists (Santos and Coimbra, 1999; Wright, 1997; Stodder, 1997).

4.4.1 Methodological issues

Despite the comprehensive adoption of the Goodman et al. method for determining age at defect formation:

> These methods assume growth by accretion where a constant amount is added to the tooth in each time period. In addition to assuming an additive growth process, these methods assume constant timing for crown initiation and termination and in some cases constant finished crown heights (Sciulli, 1992, 31).

The method developed by Goodman and colleagues contains another inherent assumption; that the dental formation in a mortality sample was similar to tooth development in modern healthy children. It also does not account for individual and population variation in tooth formation or variation in tooth size between populations (Lewis and Roberts, 1997). Lewis and Roberts (1997) criticise these assumptions and have suggested that the error they introduce might be reduced by using a standard of dental development that is comparable to, or the same as, the standard that was used for age estimation of the sample.

Other methodological issues have been raised concerning the age of occurrence of LEH. Firstly, an intertooth susceptibility in defect development has been demonstrated (Goodman and Armelagos, 1985), where the anterior dentition have been shown to be more commonly affected than the posterior teeth. Therefore, the maxillary central incisors and the mandibular canines are the teeth most commonly selected in studies of hypoplasia frequencies in prehistoric skeletal samples.

In addition, evidence for intratooth variation in hypoplasia distribution has been found. Goodman and Armelagos (1985) found that most hypoplasias cluster around the middle third of the crown surface with less defects found on the occlusal and cervical portions of the tooth. "These data suggest that there are developmental and/or morphological factors which influence the expression of hypoplasias within teeth (Goodman and Armelagos, 1985, 488)". In other words, more enamel is formed in the middle part of the crown, so more hypoplasia will be expressed there. This structural bias in the formation of enamel has been suggested to possibly influence the reported ages of peak hypoplasia frequencies (Skinner and Goodman, 1992).

The presence of dental defects have also been correlated with an increased risk of mortality in some studies (Goodman and Song, 1999). Depending on the epidemiology of occurrence, these studies have found that the presence of linear dental defects can be associated with reduced age at death. For instance, Stodder (1997) found that subadults and young adults from the Marianas had a greater prevalence of multiple hypoplastic events than those individuals that survived into late and middle adult years. A similar pattern was observed in the Libben sample from Prehistoric Ohio where the peak age at death for individuals with enamel hypoplasia was between 15 and 20 years of age (Duray, 1996).

4.4.2 Dental defects and weaning

The age of occurrence of dental defects in prehistoric remains is largely uniform. Most studies have found a peak occurrence of defects between two and five years of age which is assumed to reflect a correlation with the stress associated with weaning. This possible correlation was first proposed by Cook and Buikstra (1979) in their study of prenatal dental defects in a prehistoric site from the American Midwest. Most subsequent studies have embraced this interpretation when analysing samples from diverse geographical and temporal locations (Goodman et al., 1980; Goodman and Armelagos, 1985; Santos and Coimbra, 1999; Wood, 1996; Stodder, 1997).

However, in a recent critical review of the correlation between weaning and the aetiological age of dental defects, Katzenberg et al. (1996) outline three major assumptions found in the literature. *First, that there is a direct link between weaning and the peak age at occurrence of dental defects.* They point out that many studies have accepted the validity of this link based on earlier statements, even in the absence of any prospective study to support them (Lanphear, 1990; Powell, 1988). Others have attempted to find a link with historical data and age at weaning when interpreting dental defects (Blakey et al., 1994). However Katzenberg et al. (1996) suggest that it is the negative evidence that carries more weight in these studies.

The second assumption is that the estimated age of occurrence in a sample is accurate. As suggested above, this assumption can be argued against on methodological grounds (Skinner and Goodman, 1992). The third assumption "*is that weaning and post weaning periods are a time of increased physiological stress for the infant or child*" (Katzenberg et al 1996, 186). This assumption is complex for a number of reasons, one of which is the variety of ways in which weaning is viewed in the literature. The period of ill-health associated with the weaning period is recognised as the result of contaminated weaning foods leading to diarrhoea (Gordon et al., 1963), or prolonged breast feeding without supplementation leading to growth faltering. However, it should not be automatically assumed that either of these patterns of ill-health occurred in prehistory to the magnitude that some studies would suggest (Katzenberg et al., 1996). In summary:

> Clearly there are limitations to the use of linear enamel hypoplasia data for inferring weaning in past populations. Enamel hypoplasia has a number of causes so it is perhaps best used as non-specific stress indicator along with other markers left on bones and teeth (Katzenberg, 1996, 186)

4.4.3 Dental developmental defects in deciduous teeth

The prevalence of linear enamel hypoplasias of the deciduous dentition are consistently less than those of the permanent dentition (Goodman et al., 1987). A result of this rarity is that most studies concerning dental developmental defects will concentrate on recording evidence from the permanent dentition alone. Observing developmental defects in the deciduous dentition, perturbations of amelogenesis from the third trimester of pregnancy through to one year of age can be inferred (Goodman et al., 1984).

Some studies have demonstrated that non-linear defects can provide an inference of prenatal and perinatal health by recording their presence on deciduous dentition (Cook and Buikstra, 1979; Hanson, 1990; Tayles, 1999). A particular dental pathology that was noted in all these studies is the presence of carious lesions that develop after birth over severe hypoplastic sites. These defects have been termed 'circular caries' (Figure 4.4) and are often overlooked or unreported in studies of prehistoric dental defects (Cook and Buikstra, 1979). Cook and Buikstra (1979) interpreted the occurrence of these lesions in prehistoric American sites as reflecting severe stress associated with malnutrition and infectious disease during the first few weeks of life. These conclusions were based on epidemiological studies of linear enamel hypoplasia in living children from disadvantaged backgrounds (Infante and Gillespie, 1974; Sweeney et al., 1971). These studies found that acute diarrhoea in the first two months of life were positively correlated with the development of hypoplastic defects in the anterior deciduous teeth (Scrimshaw et al., 1968; Sweeney et al., 1971).

Infante and Gillespie (1974) define these severe hypoplastic defects as a horizontal groove, called odontoclasia, on the labial surface of mainly the maxillary incisors which may lead to carious destruction of the enamel. The definition provided by Infante and Gillespie (1974) is useful in light of the variation in descriptive terms used in studies of this defect in Pacific Island populations where it is not always clear what sort of defect the researcher is describing. For the following discussion odontoclasia is considered a hypoplastic defect of the deciduous teeth in which carious destruction of the enamel has probably occurred. Therefore, odontoclasia and circular caries are considered the same dental defect, unless stated otherwise.

Hanson (1990) interpreted the presence of circular caries in a sample of young children from Rota island in the Marianas archipelagos as representing excessive fluoride intake of pregnant and lactating women, which disrupted the normal development of enamel in deciduous teeth. Caries formation over the hypoplastic defect probably resulted from the introduction of cariogenic weaning foods. The introduction of these weaning foods also provided the opportunity for exposure to harmful pathogens, contributing to the development of skeletal subperiosteal lesions in this sample (Hanson, 1990).

A review of the literature concerning dental health in Pacific Island populations supports some of Hanson's (1990)

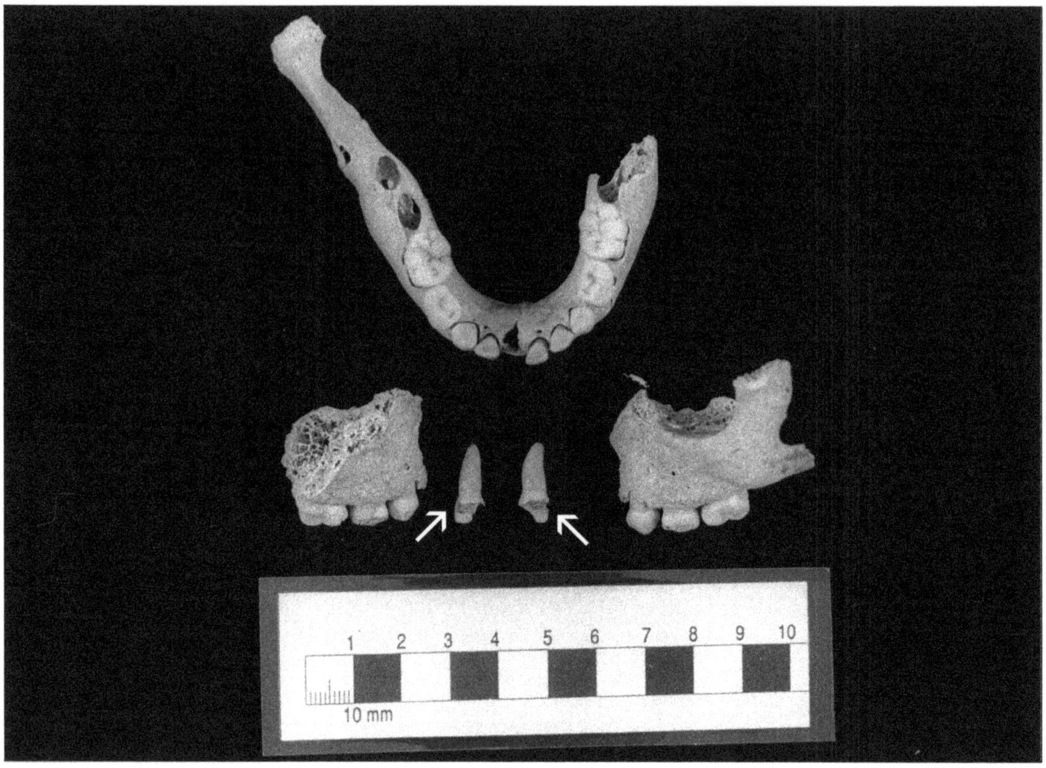

Figure 4.4. *Circular caries (arrow) of the deciduous central incisors. Photo by author.*

conclusions for the aetiology of circular caries. High levels of odontoclasia, with associated carious lesions, have been reported in a number of Pacific Island populations. For example, Kirkpatrick (1935) described odontoclasia of the deciduous teeth in 22 per cent of children under six years of age from Manus island, off the north coast of NG. This was suggested to be the result of inadequate nutrition during foetal growth and was supported by reports of some babies with severe lesions before any weaning foods had been introduced in the diet (Kirkpatrick, 1935). Baume and Meyer (1966) reported high prevalences of a similar condition in children from the islands of French Polynesia. These authors also noted that in children from more remote islands where traditional foods were still consumed, low levels of dental fluorosis, characterised as mottled teeth, imparted a resistance to carious lesions of the deciduous teeth. Odontoclasia was not noted in these children from the more remote islands but in children from more urbanised islands where it was interpreted as a symptom of poor nutrition associated with modernisation (Baume and Meyer, 1966).

An extensive survey of clinical literature from the Pacific Islands reported very high frequencies of circular caries in Hawaiian children in 1930 (Barmes, 1967). The prevalence of these lesions increased with age; a third of infants under one year of age had the lesions while 79 per cent of children aged three to four years old had developed the lesion. This high prevalence of circular caries in older children was attributed to poorly formed enamel predisposing the tooth to caries formation. This study also noted that higher frequencies of odontoclasia were observed in urban Polynesian Hawaiians than those living in rural areas (Jones et al., 1930). This study concluded that "Odontoclasia in breast-fed children indicates a deeply significant prenatal factor, going beyond mere results of poor calcification" (Barmes 1967, 443). Barmes (1967) also noted a high prevalence of odontoclasia in the deciduous teeth of rural Fijian children.

Davies (1958) reported a prevalence of 29.9 per cent of odontoclastic defects in the deciduous teeth of children from Pukapuka, an atoll in the northern Cook Islands. The inhabitants of this island had not experienced any change in their diet since contact with Europeans. The diet of the Pukapuka islanders was very low in protein and based almost entirely on starchy food crops of coconut, and taro, supplemented by fish. Rain water was used for drinking, while food was cooked using sea-water or brackish well-water. The low protein diet of these people was considered to be the cause of poor skeletal and muscle development in adults and minor vitamin deficiencies in children and adults. The presence of structural defects in the deciduous teeth was felt to be a symptom of inadequate nutrition during foetal development Davies (1958).

The prevailing theme of these Pacific Island dental surveys seems to be that the high prevalence of odontoclasia in deciduous teeth of young children is either the result of modernisation or prenatal under-nutrition. Either way, an underlying hypoplastic defect predisposes the tooth to caries. The presence of this defect in the Pukapuka islanders, and rural Fijians would negate an explanation based entirely on the effects of a poor European diet. These defects have been noted in diverse ecological zones in the Pacific Islands, from malarious Manus island off the north coast of NG to the islands and atolls of eastern Polynesia. Therefore, the aetiology of this lesion can probably not be explained by a single factor.

Hypomineralisation of deciduous and permanent teeth of Polynesian peoples has also been proposed as a result of a hereditary form of amelogenesis imperfecta (Smillie et al., 1986). This hereditary dental defect which is particularly prevalent in New Zealand Maori has been described as dark brown stained enamel, sometimes of the whole crown or as a transverse band of pigmented enamel. It is frequently observed in multiple members of a family and affects both anterior and posterior teeth (Smillie et al., 1986). This is a further aetiologic factor that should probably be considered when assessing dental defects in Pacific populations.

A further underlying cause for this dental defect has been proposed. Ortner et al. (1992) argues for a diagnosis of congenital syphilis for a three to four year old child from pre-Columbian North America with dental lesions suggestive of circular caries. Ortner et al. (1992) bases this diagnosis on the occurrence of subperiosteal new bone of the long bones, particularly the tibia, and skeletal lesions pathognomonic of treponematosis in an adult male from the same site. These dental lesions are not usually considered pathognomonic of congenital syphilis (Hillson et al., 1998). As the literature reviewed above suggests, the presence of these dental lesions are probably best considered as a non-specific indicator of prenatal and perinatal stress with superimposed carious infection.

4.5 Dental Defects Methods

Despite the limitations outlined above, the presence of LEH in a skeletal series is still widely used as a measure of comparative health in prehistoric populations.

The methodological problems are recognised in the following analysis of enamel defects in the skeletal samples of Taumako and Tonga. The presence of enamel defects in both the deciduous and permanent teeth were recorded in the present study. It is acknowledged that enamel defects in subadult individuals may introduce a significant mortality bias in subsequent analysis (Katzenberg et al., 1996). However, including subadults is considered useful for providing a continuum of stress events from foetal life through to around seven years of age.

4.5.1 Deciduous teeth

Linear enamel hypoplasia was recorded in the deciduous teeth. Circular caries were identified macroscopically based on the descriptions of Cook and Buikstra (1979). Although pitting and staining of the deciduous crowns were also observed in these samples, only the prevalence of circular caries of the anterior teeth is presented here.

The deciduous maxillary and mandibular incisors and canines were chosen for study. The development of these teeth encompasses a period of prenatal development from 14 weeks after fertilisation in the case of incisors and 18 weeks for canines. The crowns of these teeth are completely developed by three postnatal months in the incisors and nine months in canines (Hillson, 1996). Therefore, observations of defects in these teeth may serve as indicators of stress from early intrauterine life, providing a reflection of maternal as well as foetal health and continue throughout the early postnatal months. As mentioned above, circular caries are believed to develop within pre-existing hypoplastic sites of tooth enamel (Cook and Buikstra, 1979). Therefore, they can provide information on an individual's intrauterine health but may also provide some information on postnatal feeding habits. For instance, the presence of circular caries may indicate the introduction of cariogenic supplementary weaning foods prior to death.

Circular caries were distinguished from other types of carious lesions by their formation on the buccal and/or lingual aspect of the tooth crown. Other caries usually form on the occlusal or root surfaces of the tooth (Hillson, 1996). A true carious lesion was distinguished from enamel discoloration by the outer layer of enamel having been breached by the pathological process (Roberts and Manchester, 1995b). The age at occurrence of either LEH or circular caries was not attempted for the deciduous teeth, although there is a method available for calculating this (Sciulli, 1992). Because the majority of circular caries were observed in the middle third of the crown, it is assumed that most of the underlying enamel defects occurred prenatally (Cook and Buikstra, 1979).

4.5.2 Permanent teeth

The permanent maxillary and mandibular incisors and canines were also selected for study. These teeth were chosen based on the discussion of Goodman et al. (1980) where they argue that these teeth are the most sensitive to enamel defect development. They also argue that because of this sensitivity, minimal information will be lost if not all permanent teeth are included. As outlined in chapter three, the levels of dental attrition were low in both the Taumako and Tongan samples. Therefore, it was not necessary to exclude individuals from analysis on the basis of tooth wear.

As mentioned above, linear enamel hypoplasias (LEH) are the most frequently recorded of all enamel defects. This is mainly because of the ease of recognition and they are also considered the most common of enamel defects Goodman and Rose (1990). Based on these assertions, only the presence of linear enamel hypoplasia was recorded in the present study. Linear enamel hypoplasia was identified as a transverse line of depressed enamel on the labial surface of the tooth crown. The LEH were identified macroscopically.

The units of analyses are both the tooth and the individual. By using the tooth as the unit of analysis, the sample is increased and removes some of the problems of missing data in some individuals. The individuals with one or more of the requisite teeth were then assessed in order to consider the epidemiology of dental defects in the samples. In order to determine any correlation between reduced age at death and enamel growth disruption, the prevalence of LEH among individuals is also presented. Any difference in LEH presence between the sexes is investigated.

The counts described above also provide data on the number of teeth with more than one defect. Multiple defects of the crown indicate repeated episodes of stress rather than a single episode. The relationship between age at death, sex, and the likelihood of having multiple periods of stress is presented in the results.

4.5.3 Age at occurrence of LEH

Following the method of Goodman et al. (1980) described above, the age of defect development was calculated. The total crown height was measured using digital callipers from the cemento-enamel junction (CEJ) to the occlusal surface and the position of the defect was measured from the CEJ to the middle of the linear defect.

The regression equation used here was modified by Wright (1997) and is based on the assumption of constant incremental growth of crown development (Goodman et al., 1980). Skinner and Goodman (1992) argued that the first year of enamel development will be buried under the subsequent cuspal enamel in some teeth. Therefore the first year of tooth development is considered unobservable. However, because the true extent of the buried enamel phenomenon has not been quantified (Wright, 1997), the assumption of constant crown development must be adopted in this study. The data on age at occurrence of enamel defects are presented by tooth and the mean age of occurrence is presented in developmental stages of six months.

Where a tooth had more than one defect, the age of occurrence of the defect furthest away from the CEJ was calculated. This method of dealing with multiple defects is derived from Corruccini et al. (1985) and provides a record of the earliest stress event in the tooth. This is preferred over presenting an average age of all defects in tooth which does not provide any biological correlate, only the age of the individual in the midst of a period of nonspecific stress. By presenting the age at occurrence of the earliest defect, it is possible to estimate at what age non-specific stress begins in a sample and discuss the possible causes.

Where data are available, all the above analyses are carried out on both the Taumako and the Tongan samples separately and then a comparison of prevalence of dental defects is made between the two samples.

4.6 Results

Evidence of linear enamel hypoplasia and circular caries was found in both the Taumako and Tongan samples. Because there were fewer data in the Tongan samples it was decided to carry out the intertooth analysis of age at occurrence of LEH only on the Taumako sample. However, in order to compare the age at occurrence between Taumako and Tonga, an analysis of the mean age at occurrence in each age group in the Tongan mounds is presented.

4.6.1 Taumako

Deciduous teeth

In the deciduous teeth, the data on the maxillary and mandibular LEH were merged. Of 138 deciduous teeth present, only four were affected with LEH (Table 4.4). The sample of teeth with LEH consisted of three individuals, two under two years of age and one 6.5 year old.

The prevalence of circular caries in the Taumako sample is high compared to LEH (Table 4.5). Overall, 22 per cent of all deciduous teeth were affected with circular caries. The sample with circular caries comprises 11 individuals separate from those with LEH. Eighty-two per cent (n = 9/11) of these were between the ages of one and two and a half years at death. Of the Remaining two, one individual was nine months old at death and another was 6.5 years old at death.

Permanent teeth

The prevalence of LEH in the permanent teeth of subadults is shown in Table 4.6. Overall, 29 per cent of all subadult permanent teeth were affected with LEH. Of the adult permanent teeth, 21 per cent were affected (Table 4.6).

Individuals with LEH

Figure 4.5 and Table 4.7 present the prevalence of individuals with LEH by age group. The data from the incisors and canines have been merged. In children younger than six years of age some of the data represent completely developed but unerupted permanent tooth crowns. There was an increase in prevalence with age, which peaks in the adolescents and then declined in old adults. Chi-square statistical tests showed that none of these differences were significant.

The lowest prevalence was seen in the 1–5 year age group at 20 per cent. This low prevalence is to be expected considering that it is not until after around four years of age that permanent incisor crowns would be fully developed. Similarly, the permanent canines do not complete crown development until around six years of age. Therefore, the low prevalence of LEH in the 1–5 year age group was probably a reflection of development. After six years of age the permanent incisors and canines would have developed fully. Therefore, comparisons in the older age groups can be made with more confidence. Nevertheless, the 1–5 year olds are included in this comparison because they provide a greater dimension of time. Overall, of the 106 individuals with the required teeth over half had evidence of disrupted crown development.

Table 4.4. *Prevalence of linear enamel hypoplasia of the deciduous teeth: Taumako.*

Tooth	N	Affected	Per cent
Incisors	51	1	2
Canines	87	3	3
Total	**138**	**4**	**3**

N = Number of observations; Affected = Number with dental defect

Table 4.5. *Prevalence of circular caries of the deciduous teeth: Taumako.*

Tooth	N	Affected	Per cent
Incisors	51	4	8
Canines	87	26	30
Total	**138**	**30**	**22**

N = Number of observations; Affected = Number with dental defect

Table 4.6. *Prevalence of linear enamel hypoplasia in permanent teeth: Taumako.*

Tooth	N	Affected	Per cent
Subadult			
Max Incisors	45	3	7
Max Canine	25	13	52
Mand Incisors	35	7	20
Mand Canine	21	14	67
Subtotal	126	37	29
Adult			
Max Incisors	167	19	11
Max Canine	94	39	41
Mand Incisors	154	5	3
Mand Canine	99	47	47
Subtotal	514	110	21
Total	**640**	**147**	**23**

N = Number of observations; Affected = Number with dental defect; Max = Maxillary teeth; Mand = Mandibular teeth

Table 4.7. *Age distribution of individuals with linear enamel hypoplasia of the permanent teeth: Taumako.*

Age (yrs)	N	Affected	Per cent
Subadults			
Feotal	0	0	0
0–0.9	1	0	0
1–5	10	2	20
6–15	7	3	43
16–20	8	6	75
Subtotal	26	11	42
Adult			
Young	45	29	64
Mid	25	15	60
Old	10	4	40
Subtotal	80	46	58
Total	**106**	**57**	**54**

N = Number of observations; Affected = Number with dental defect.

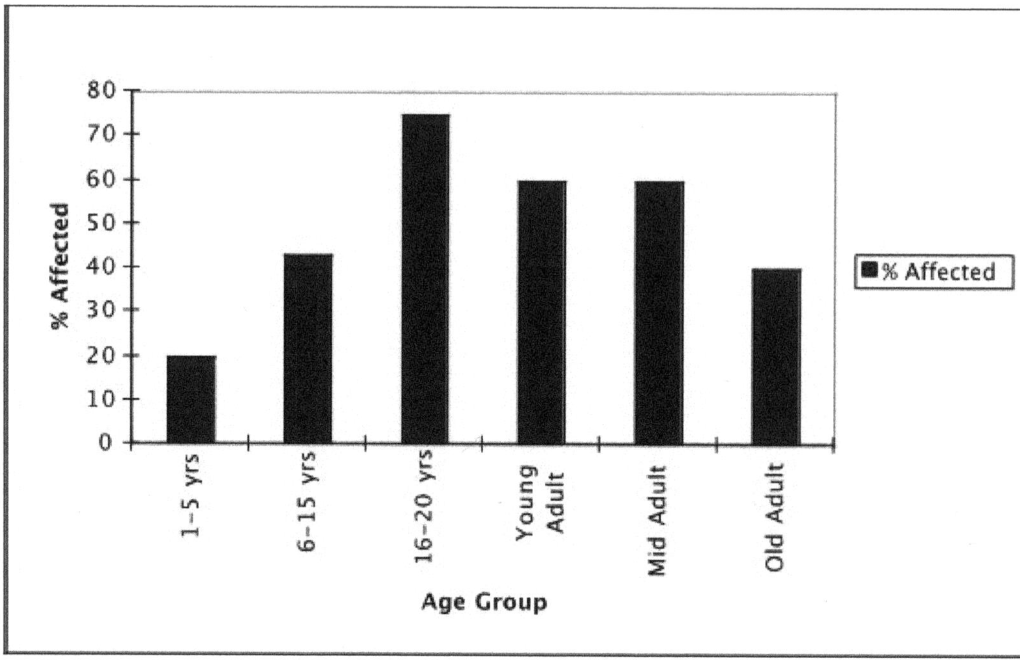

Figure 4.5. *Prevalence of LEH in the permanent teeth of affected individuals: Taumako.*

Number of defects per tooth: Taumako

In the subadult permanent teeth with defects a high proportion had more than one defect per tooth (Table 4.8). However, in the adult permanent teeth a slightly greater number had more single defects than multiple defects (Table 4.9). Chi-square statistical analysis showed that the differences between subadults and adults in the proportion of single versus multiple lesions were not significant.

Of the eleven subadults with LEH of the permanent dentition, six were between 16–20 years of age. Five of these were female and one was male. Twice as many had multiple defects as single defects. Both of those individuals who died before the age of five years had multiple defects (Table 4.10).

In the adults, 46 per cent had single defects of the teeth, while 54 per cent had more than one defect (Table 4.11). The young adults are equally affected by both single and multiple defects. A difference in prevalence was shown between the middle and old age groups where the middle aged adults have a higher number of multiple defects than single and the opposite was shown for the old aged adults. When looking at sex differences between single and multiple defects per tooth, males who survived beyond young adulthod mostly had only one defect. There was no consistent pattern among the females.

4.6.2 Age at occurrence of linear enamel hypoplasia: Taumako

Permanent subadult teeth

In the subadults, LEH in the maxillary central incisors has a higher frequency of age at occurrence between 2 and 2.5 years of age, while no lateral incisors were affected. The peak frequency of maxillary canines is found at 4-4.5 years of age (Figure 4.6). In the mandibular teeth the peak frequency of central incisors and canines is the same at 3–3.5 years (Figure 4.7).

Adult teeth

In the adults, the peak age at occurrence for defects in both of the maxillary and madibular incisors is 3–3.5 years, while the maxillary canines have the peak frequency of LEH developing between 4 and 5.5 years of age (Figures 4.8 and 4.9).

Mean age at occurrence: Individuals

The mean age of occurrence of LEH in both subadults and adults increases with age at death (Figure 4.10). The 1-5 year old age group have the youngest age at occurrence at three years old, while the old adults did not suffer disrupted enamel development until 4.5 yrs of age. The 6-15 year olds are anomalous to this pattern.

When the sex of affected adult individuals is compared, a pattern of increasing age of occurrence with increased age at death is seen among males (Figure 4.11). This is compared to a static age at occurrence among females.

4.6.3 Results Tonga

Of the 81 deciduous teeth from Mound 1, only incisors had evidence of LEH, constituting 2 per cent of the sample (Table 4.12). These two teeth with LEH were from one individual who was nine months old at death. There was

Table 4.8. *Number of defects per tooth in subadult permanent teeth: Taumako.*

Tooth	N	1	Per cent	>1	Per cent
Max Incisor	3	0	0	3	100
Max Canine	13	6	46	7	54
Mand Incisor	7	2	29	5	71
Mand Canine	14	6	43	8	57
Total	**37**	**14**	**38**	**23**	**62**

N = Number of observations; 1 = Single defect; >1 = More than one defect

Table 4.9. *Number of defects per tooth in adult permanent teeth: Taumako.*

Tooth	N	1	Per cent	>1	Per cent
Max Incisor	19	11	58	8	42
Max Canine	39	23	59	16	41
Mand Incisor	5	3	60	2	40
Mand Canine	47	25	53	24	51
Total	**110**	**62**	**55**	**50**	**45**

N = Number of observations; 1 = Single defect; >1 = More than one defect

Table 4.10. *Number of defects in subadult individuals with permanent teeth- teeth combined: Taumako.*

Age	N	1	Per cent	>1	Per cent
1–5	2	0	0	2	100
6–15	3	1	33	2	67
16–20	6	3	50	3	50
Total	**11**	**4**	**36**	**7**	**64**

N = Number of observations; 1 = Single defect; >1 = More than one defect

Table 4.11. *Number of defects in adult individuals: teeth combined: Taumako.*

	N	M	F	M Aff.	Per cent M	F Aff.	Per cent F	Total	Per cent
Single Defect									
Young	29	17	12	7	41	6	50	13	45
Mid	13	5	8	3	60	2	25	5	38
Old	4	2	2	2	100	1	50	3	75
Subtotal	46	24	22	12	50	9	41	21	46
Multiple Defects									
Young	29	17	12	10	59	6	50	16	55
Mid	13	5	8	2	40	6	75	8	62
Old	4	2	2	0	0	1	50	1	25
Subtotal				12	50	13	59	25	54
Total	**46**	**24**	**22**	**24**	**52**	**22**	**48**	**46**	**100**

N = Number of observations; M Aff. = Number of males with defects; F Aff. = Number of females with defects; Per cent M = Percentage of males affected; Per cent F = Percentage of females affected.

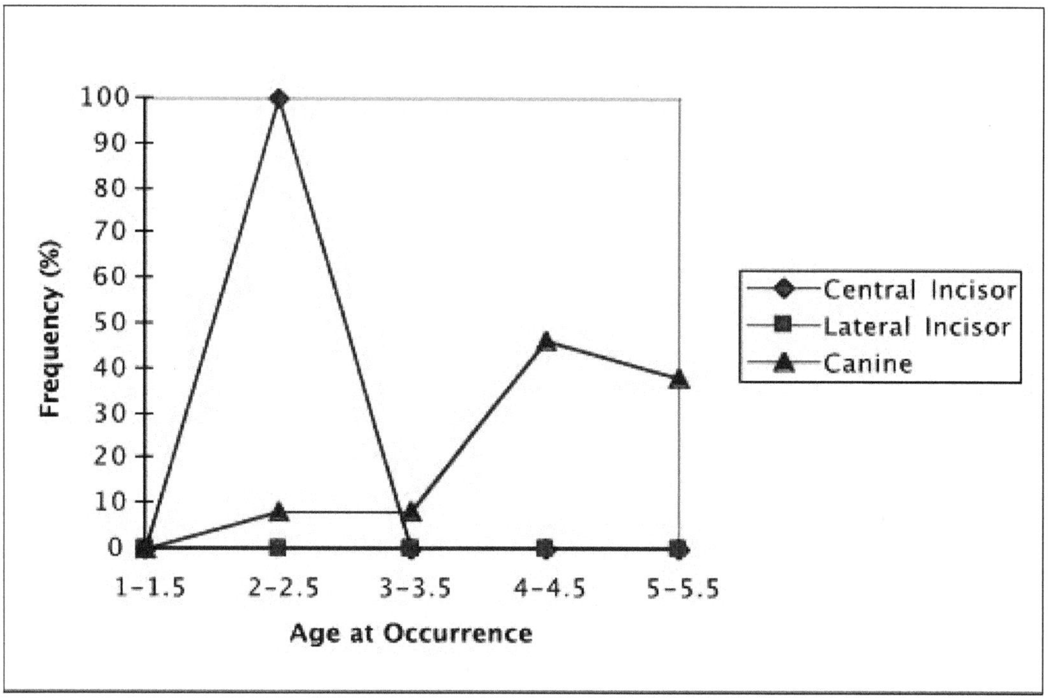

Figure 4.6. *Age at occurrence of LEH in subadult permanent maxillary teeth: Taumako.*

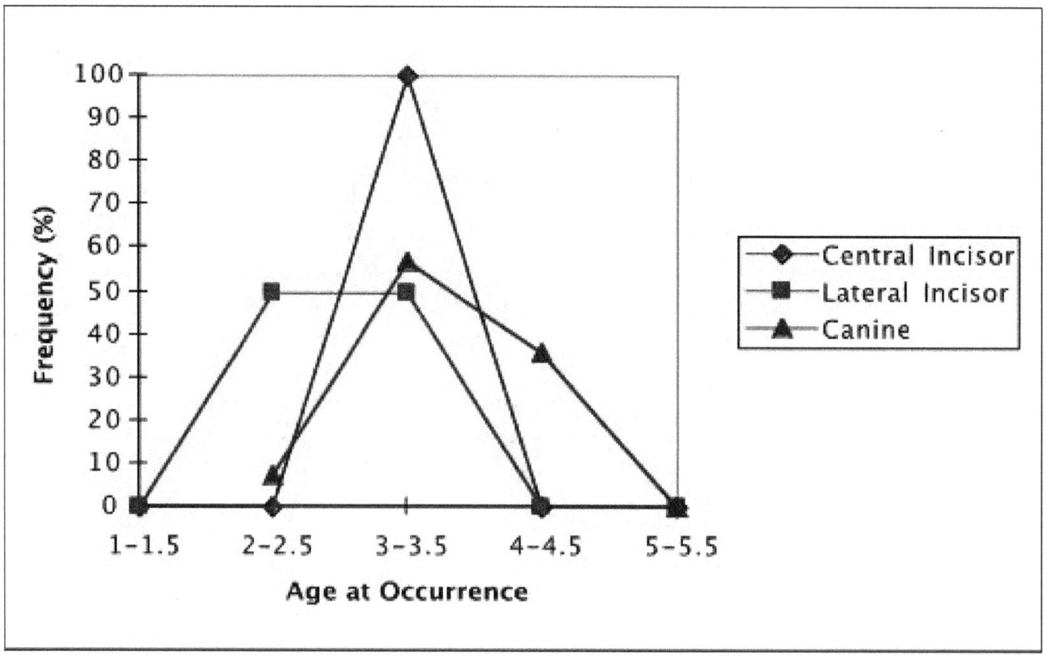

Figure 4.7. *Age at occurrence of LEH in permanent subadult mandibular teeth: Taumako.*

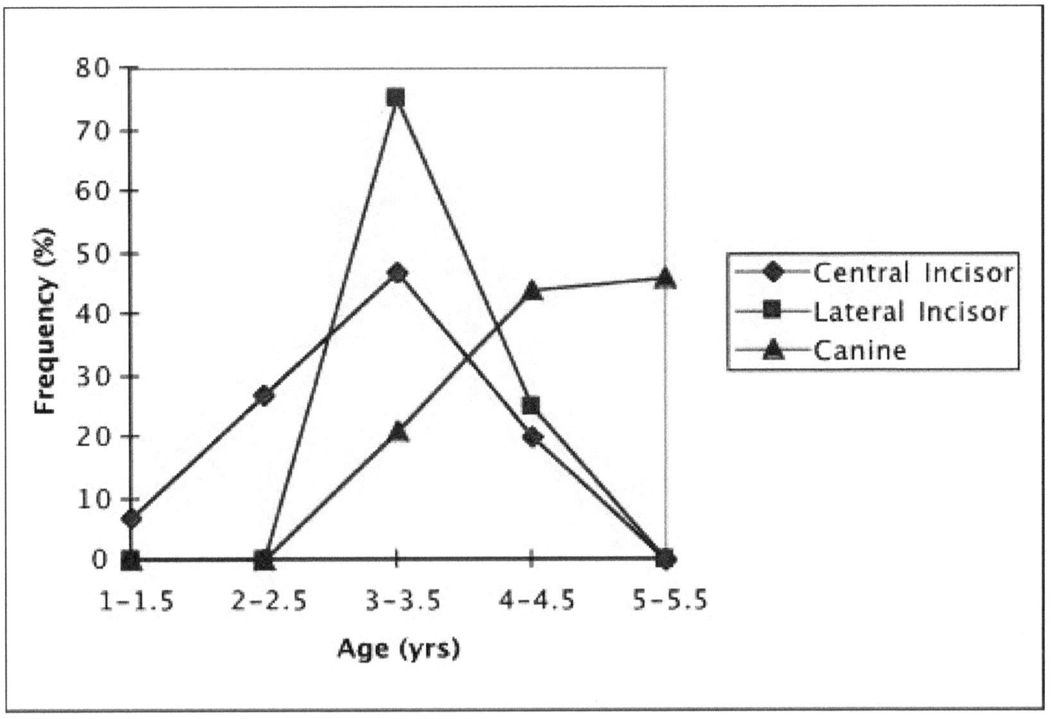

Figure 4.8. *Age at occurrence of LEH in adult maxillary teeth: Taumako.*

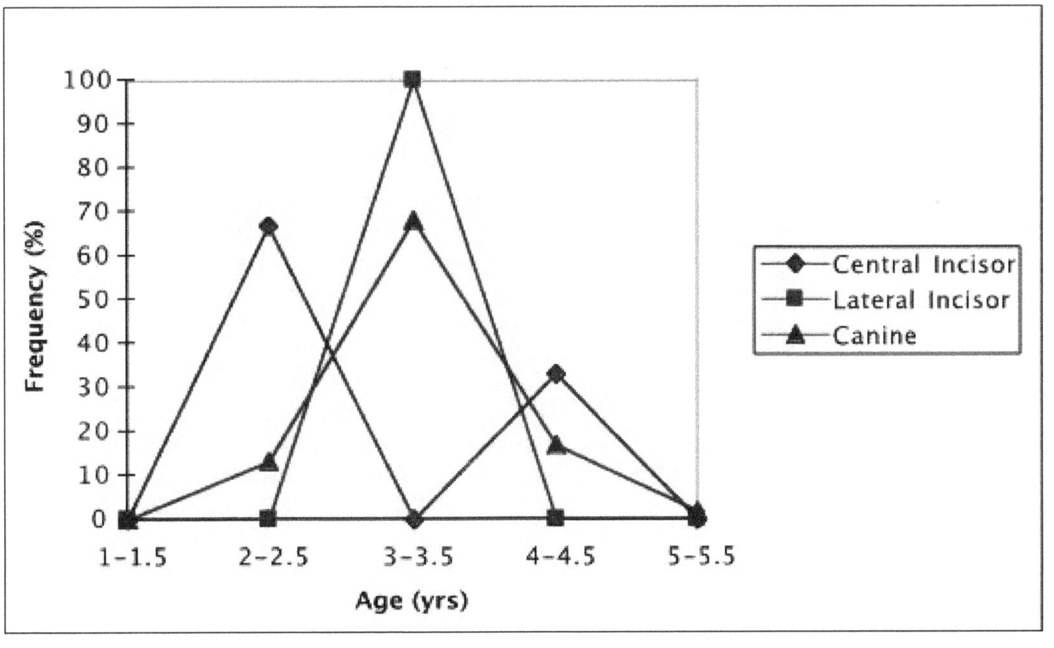

Figure 4.9. *Age at occurrence of LEH in adult mandibular teeth: Taumako.*

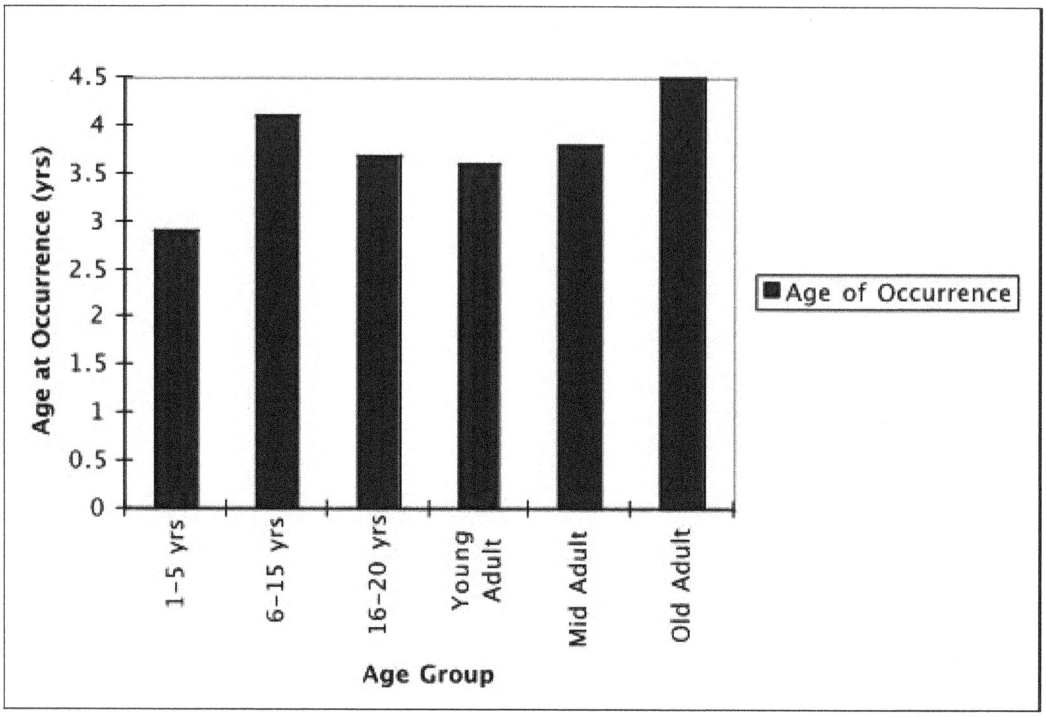

Figure 4.10. *Mean age of occurrence of LEH in individuals with permanent teeth: Taumako.*

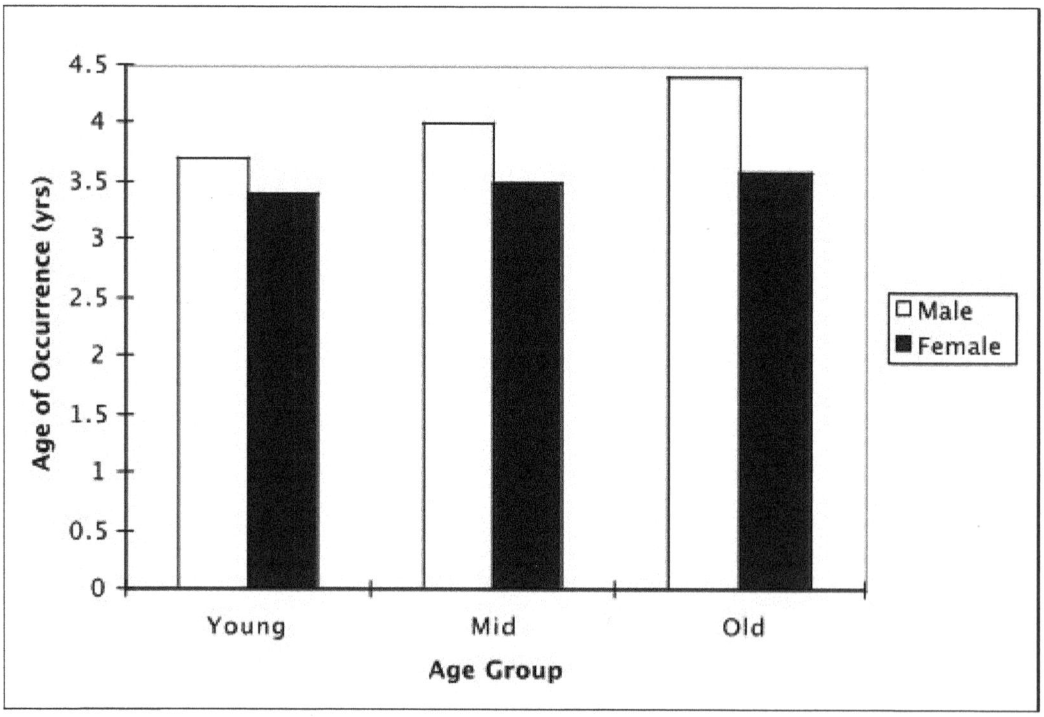

Figure 4.11. *Mean age of occurrence of LEH in adult individuals- by sex: Taumako.*

Table 4.12. *Prevalence of linear enamel hypoplasia of the deciduous teeth: Tonga.*

Tooth	N	Affected	Per cent
Mound 1			
Incisors	31	2	6
Canines	50	0	0
Total	81	2	2
Mound 2			
Incisors	21	0	0
Canines	15	0	0
Total	**36**	**0**	**0**

N = Number of observations; Affected = Number with dental defect

no linear enamel hypoplasia observed in deciduous teeth at Mound 2.

Conversely, no evidence of circular caries was observed at Mound 1, while 8 per cent of the sample had circular caries of the deciduous incisors at Mound 2 (Table 4.13). The teeth with circular caries are from two individuals: a one year old child and a two and a half year old child.

The prevalence of LEH in the permanent teeth is presented in Table 4.14. Overall, 11 per cent of the sample had defects of the enamel. At Mound 2, 9 per cent of the sample had LEH (Table 4.15). The prevalence of LEH was lower in adults than subadults from both mounds (Tables 4.14 and 4.15).

Tables 4.16 and 4.17 present the prevalence of LEH in individuals from Mounds 1 and 2. In the subadults from Mound 1, only two individuals from the 6–15 year age group were affected with LEH. Overall, this prevalence constitutes 29 per cent of the total sample of subadults with teeth. The adults, only one young adult was affected with LEH in the adult sample from Mound 1. At Mound 2, two out three 1–5 year olds were affected and the single 6–15 year old was affected. Overall, 60 per cent of the subadults had evidence of disrupted growth of the enamel. In the adults from Mound 2, the most affected age group is the middle aged adults with 25 per cent of the individuals having enamel defects. It is recognised that these results represent only single individuals in many cases.

By comparing the individual prevalence of LEH between the Tongan mounds, the pattern of affected age groups is seen to be markedly different (Figure 4.12). At Mound 2 all age groups were affected except the 16–20 year old. While at Mound 1 only the 1–5 year olds and one young adult have any evidence of LEH.

At Mound 1, the single adult affected was a young female, while at Mound 2, four of the affected adults were female while the fifth affected individual was of unknown sex.

Number of defects per tooth: Tonga

In the subadults with permanent teeth at both mounds a greater number had single defects of the teeth than multiple defects (Table 4.18). Results for adults are not tabulated as all affected adult teeth from both mounds had single defects only.

Age at occurrence of linear enamel hypoplasia: Tonga

As explained above there were too few data to warrant an intertooth comparison of the age at occurrence of LEH in the Tongan samples. However, an investigation of the mean age at occurrence of enamel defects of both mounds reveal a reasonably uniform distribution (Figure 4.13). The mean age at occurrence for Mid adults is younger than the other age groups at 2.5 years compared to 3.5 for all other age groups. However, this constitutes a single individual which has probably skewed the result.

4.6.4 Results summary

Adult stature

Between the samples the means of females from Mound 1 were taller than the other two groups. The spread of difference in stature ranges was lowest in Mound 1 females and highest in Taumako females. There was little difference in the height of males between Taumako and Tonga. The Taumako males had the highest spread in stature ranges compared to the Tongan males but the spread was higher in males of Mound 1 than in the females from this site. Compared to the other Polynesian samples the females at Mound 1 were the tallest of all and those from the Marquesas were the shortest. The Taumako females were taller than the Marquesan, Hawaiian and Cook Island females. Compared with the other Polynesian samples, only at Taumako and Mound 1 is the range of sexual dimorphism lower than 5 per cent.

Table 4.13. *Prevalence of circular caries of the deciduous teeth: Tonga*

Tooth	N	Affected	Per cent
Mound 1			
Incisors	31	0	0
Canines	50	0	0
Total	81	0	0
Mound 2			
Incisors	21	3	14
Canines	15	0	0
Total	**36**	**3**	**8**

N = Number of observations; Affected = Number with dental defect

Table 4.14. *Prevalence of linear enamel hypoplasia of permanent teeth: Mound 1 Tonga*

Tooth	N	Affected	Per cent
Subadult			
Max Incisors	10	3	30
Max Canines	9	2	22
Mand Incisors	7	0	0
Mand Canines	5	1	20
Subtotal	31	6	19
Adult			
Max Incisors	9	1	11
Max Canines	9	0	0
Mand Incisors	12	0	0
Mand Canines	12	1	8
Subtotal	42	2	5
Total	**73**	**8**	**11**

N = Number of observations; Affected = Number with dental defect; Max = Maxillary teeth; Mand = Mandibular teeth.

Table 4.15. *Prevalence of linear enamel hypoplasia in permanent teeth: Mound 2, Tonga.*

Tooth	N	Affected	Per cent
Subadult			
Max Incisors	12	3	25
Max Canines	6	0	0
Mand Incisors	5	0	0
Mand Canines	3	1	33
Subtotal	26	4	15
Adult			
Max Incisors	35	6	17
Max Canines	19	1	5
Mand Incisors	36	2	6
Mand Canines	23	0	0
Subtotal	113	9	8
Total	**139**	**13**	**9**

N = Number of observations; Affected = Number with dental defect; Max = Maxillary teeth; Mand = Mandibular teeth.

Table 4.16. *Age distribution of individuals with linear enamel hypoplasia of the permanent teeth: Mound 1, Tonga.*

Age (yrs)	N	Affected	Per cent
Subadults			
1–5	2	0	0
6–15	4	2	50
16–20	1	0	0
Subtotal	7	2	29
Adult			
Young	6	1	17
Mid	2	0	0
Old	0	0	0
Subtotal	8	1	13
Total	**15**	**3**	**20**

N = Number of observations; Affected = Number with dental defect.

Table 4.17. *Age distribution of individuals with linear enamel hypoplasia of the permanent teeth: Mound 2, Tonga.*

Age (yrs)	N	Affected	Per cent
Subadults			
1–5	3	2	67
6–15	1	1	100
16–20	1	0	0
Subtotal	5	3	60
Adult			
Young	8	1	13
Mid	4	1	25
Old	5	1	20
Subtotal	17	3	18
Total	**22**	**6**	**27**

N = Number of observations; Affected = Number with dental defect.

Table 4.18. *Number of defects per tooth in subadult permanent teeth: Tonga.*

Age	N	1	Per cent	>1	Per cent
Mound 1					
Max Incisor	3	3	100	0	0
Max Canine	2	2	100	0	
Mand Incisor	0	0	0	0	0
Mand Canine	1	0	0	1	100
Total	**6**	**5**	**83**	**1**	**17**
Mound 2					
Max Incisor	3	2	67	1	33
Max Canine	0	0	0	0	0
Mand Incisor	0	0	0	0	0
Mand Canine	1	1	100	0	0
Total	**4**	**3**	**75**	**1**	**25**

N = Number of observations; 1 = Single defect; >1 = More than one defect

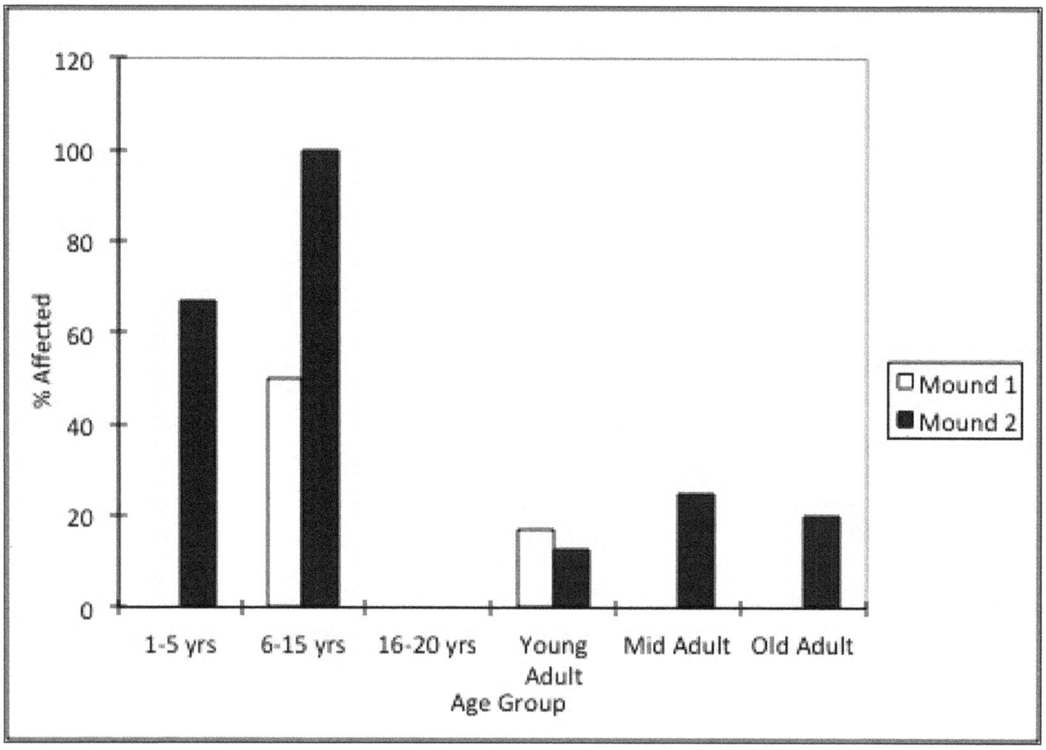

Figure 4.12. *Prevalence of LEH in the permanent teeth of affected individuals: Tonga.*

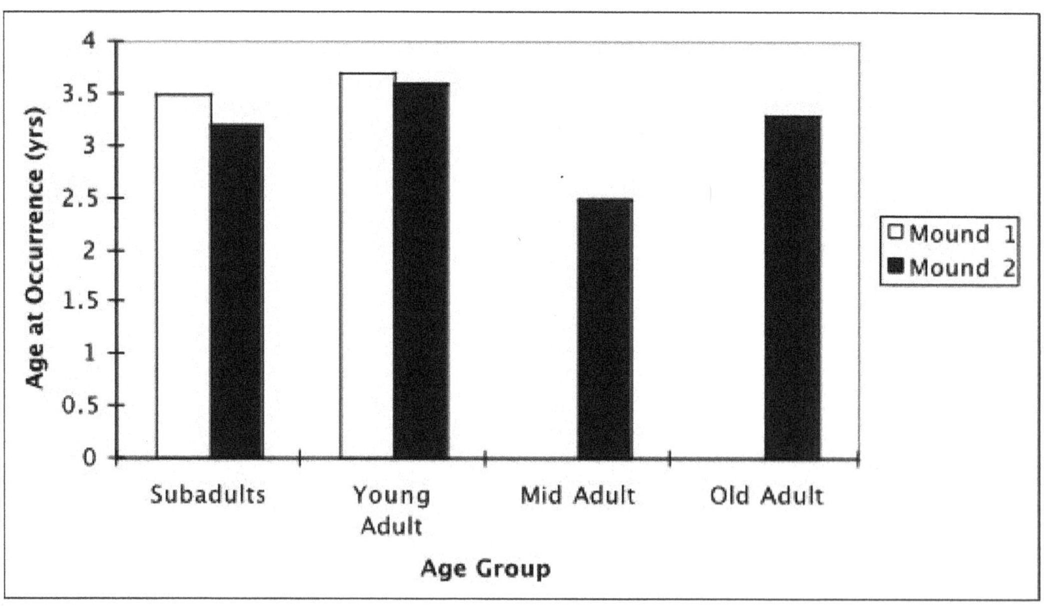

Figure 4.13. *Mean age at occurrence of LEH in the permanent teeth of subadult and adult individuals: Tonga.*

4.6.5 Dental defects

Taumako

The limitations of the sample are recognised. However, some general patterns in the dental defects can be outlined.

Comparison of subadult and adult prevalence of LEH

Evidence of prenatal defects in the deciduous teeth were found in the form of LEH and circular caries. As expected a low prevalence of LEH (3 per cent) was found in the deciduous teeth, however, 22 per cent of deciduous teeth had circular caries.

In total 23 per cent of all teeth available, excluding deciduous teeth, had evidence of LEH. At the individual level, 54 per cent (n = 57/106) of individuals with one or more of the requisite teeth had LEH. At the intertooth level, where the tooth is the unit of analysis, a slightly higher prevalence of LEH was found in the subadult permanent teeth compared to the adults, while the individual prevalence was greater in adults than subadults. As Figure 4.5 showed, an increasing level of LEH prevalence was found throughout childhood and adolescence to reach a peak of 75 per cent in the adolescents. The prevalence of LEH was the same in young and mid adults, while it was lower in old adults.

Number of defects

At the intertooth level, subadults had a higher proportion of teeth affected with multiple defects. While adults had more single defects in affected teeth than multiple. In both the intertooth and inter-individual analyses, subadults had a higher proportion of multiple defects than adults. This difference was greater in the intertooth analyses than in the individual counts.

Within the adult individuals a difference in prevalence is shown between the middle and old adults, where the middle adults have more multiple defects than the older group. A sex difference was seen where more young males are affected with multiple defects more than young females. However, more mid and old females are affected than males.

Age at occurrence

The age at occurrence in the maxillary teeth of subadults and adults showed a difference, where subadults had a consistently younger peak of age at occurrence than adults. In subadults, the peak age at occurrence ranges from 2–2.5 years old, while in the adults the range is from 3-3.5- 5-5 years. No difference in age at occurrence was seen in the mandibular teeth. At the individual level, a general pattern of a younger mean age at occurrence was shown in association with reduced age at death. This may reflect a difference in the development between maxillary and mandibular teeth. In the adults a sex difference was found where an increasing mean age at occurrence was correlated with increased age at death among males only.

Tonga

Compared to the Taumako data, the Tongan sample is poor. It is difficult to discuss the pattern of LEH prevalence when in most cases only a single individual may be affected. Evidence of prenatal defects in the deciduous teeth was found in the Tongan mound samples. However, very low levels of LEH were observed only at Mound 1, while circular caries was found only at Mound 2 (8 per cent). The pattern of prevalence of LEH in the permanent teeth of subadults and adults was similar between the mounds, with more subadults affected than adults.

In the subadults, all age groups except 16–20 year olds were affected at Mound 2 while only one young adult and two 6–15 year olds were affected at Mound 1. When looking at the number of defects per tooth, both mounds reveal a pattern of more single stress events than multiple in the subadults. All adults with LEH had only single defects. The mean age at occurrence at both mounds was 3.5 years.

Comparison between Taumako and Tonga

Overall, the proportion of LEH was higher at Taumako than both the Tongan mounds. This was shown in the intertooth and individual levels with the Tongan data combined (Table 4.19). The intertooth prevalence of Taumako was statistically higher than both of the Tongan mounds and with the Tongan data combined (Table 4.20). At the individual level, no statistically significant differences were found between Taumako or either of the Tongan mounds. When the Tonga data were combined the prevalence of LEH at the individual level was found to be significantly higher at Taumako (Table 4.21).

4.6.6 Prenatal stress

As a general indicator of non-specific prenatal ill-health Figure 4.14 illustrates a higher level of circular caries at Taumako. The low levels of LEH in all samples was expected based on previous studies of this pathology in deciduous teeth. Table 4.22 shows a significant difference in circular caries between Taumako and Mound 2.

4.6.7 Childhood stress

The prevalence of LEH in subadult permanent teeth reveals similar differences to prenatal indicators. Again, Taumako has the highest prevalence of LEH in all the subadult samples with the Tongan mounds separate and when the Tongan data were combined (Figure 4.15 and 4.23). Chi-square statistical analyses showed that these differences were not statistically significant. The difference between the two Tongan mounds was also not significant. While there were

Table 4.19. *Comparison of overall levels of LEH between Taumako and Tonga.*

Unit	Taumako A/n	Per cent	Mound 1 A/n	Per cent	Mound 2 A/n	Per cent	Tonga Combined A/n	Per cent
Tooth	147/640	23	8/73	11	13/139	9	21/212	10
Indiv	57/106	54	3/15	20	6/22	27	9/37	24

Indiv = Individual level of analysis; A = affected number of individuals; *n* = number of observations

Table 4.20. *Chi-square p-values for the significance of the prevalence of LEH between Taumako and Tonga: Intertooth level.*

	Mound 1	Mound 2	Tonga Combined
Taumako	.0489	.0024	.0005
Mound 1		ns	

Table 4.21. *Chi-square p-values for the significance of the prevalence of LEH between Taumako and Tonga: Individual level.*

	Mound 1	Mound 2	Tonga Combined
Taumako	ns	ns	.0472
Mound 1		ns	

Table 4.22. *Fisher's Exact p-values for the significance of the prevalence of circular caries of the deciduous teeth between the samples.*

	Mound 1	Mound 2
Taumako	ns	<0.01

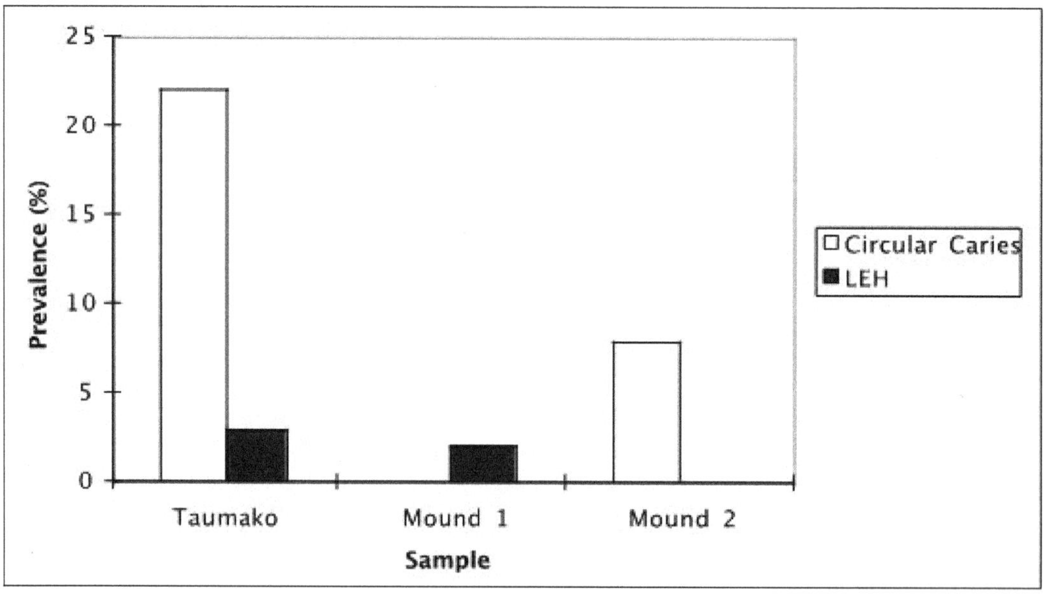

Figure 4.14. *Comparison of the prevalence of circular caries and LEH of the deciduous teeth between Taumako and Tonga.*

Table 4.23. *Comparison of prevalence of LEH in permanent teeth between Taumako and combined Tongan data.*

Age Group	Taumako A/*n*	per cent	Tonga combined A/*n*	per cent
Subadult permanent	37/126	29	10/57	18
Adult permanent	110/126	21	11/155	7

Table 4.24. *Chi-square p-values for significance of the LEH prevalence in adult teeth between samples.*

	Mound 1	Mound 2	Tonga combined
Taumako	0.0248	0.0047	0.0005
Mound 1	—	ns	—

no significant differences between Taumako and combined Tongan data, the prevalence rates between these two samples does suggest the levels of childhood stress were higher at Taumako.

When comparing the levels of LEH in the adult teeth, Taumako stands alone with a higher prevalence than both of the Tongan mounds and when the Tonga data were combined (Figure 4.16 and Table 4.23). The difference in prevalence of LEH at Taumako is significant compared to both the Tongan mounds and when the Tongan data were combined 4.24.

This pattern of greater severity at Taumako is carried through in the comparison of the number of defects per tooth. As shown in Figure 4.17 the prevalence of multiple defects is greater in the permanent teeth of Taumako subadults than both the Tongan mounds. These differences are not statistically significant, but the prevalences support the conclusion that Taumako subadults suffered more episodes of stress than those at Tonga.

No multiple defects were observed in the adult teeth of either of the Tongan mounds, while at Taumako the single and multiple defects were similar in number among the adults (Figure 4.18).

4.7 Conclusions

The second objective of this thesis is to assess any differences in non-specific indicators of growth disruption in dental material and any differences in stature between Taumako and Tongan adults. This objective has been achieved in this chapter. Adult stature was comparable between Taumako and Tonga, but the Tongan females were taller than Taumako females and the range of statures was greater at Taumako.

The analyses of LEH carried out in this chapter found an overall significantly higher prevalence of dental defects in the permanent teeth at Taumako compared to Tonga. Statistically significant higher levels of prenatal stress in the form of circular caries were also recorded at Taumako. The prevalence of adult LEH was significantly higher at Taumako compared to Tonga and higher proportions of subadult permanent teeth affected was also found at Taumako. More evidence of multiple episodes of childhood stress were recorded in Taumako subadults compared to the Tongan mounds and the affected Tongan adults had evidence of single episodes of childhood stress only. These differences in dental defects and stature will be discussed in chapter seven in relation to the second research aim of thesis.

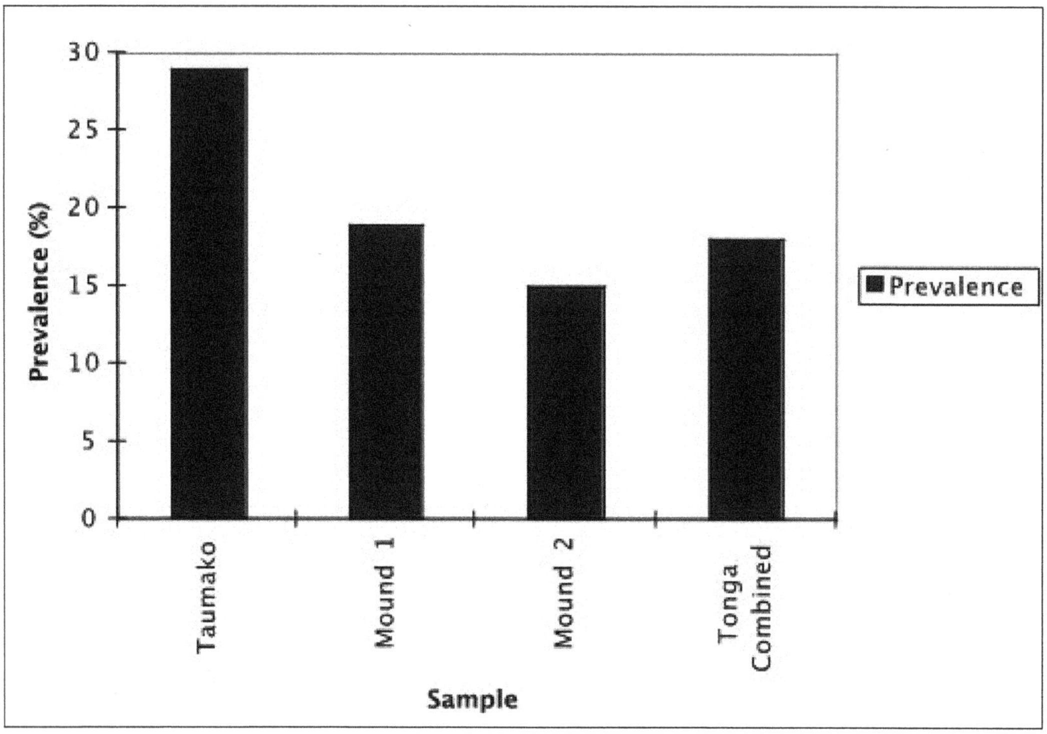

Figure 4.15. *Comparison of prevalence of LEH in the permanent teeth of subadults between Taumako and Tonga.*

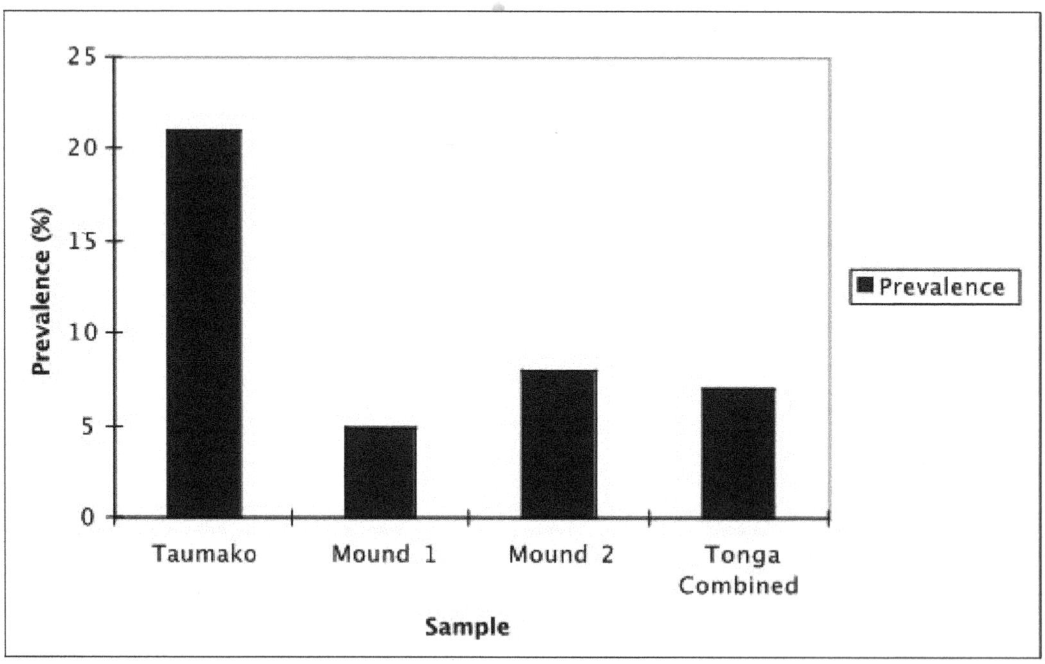

Figure 4.16. *Comparison of prevalence of LEH in adult teeth between Taumako and Tonga.*

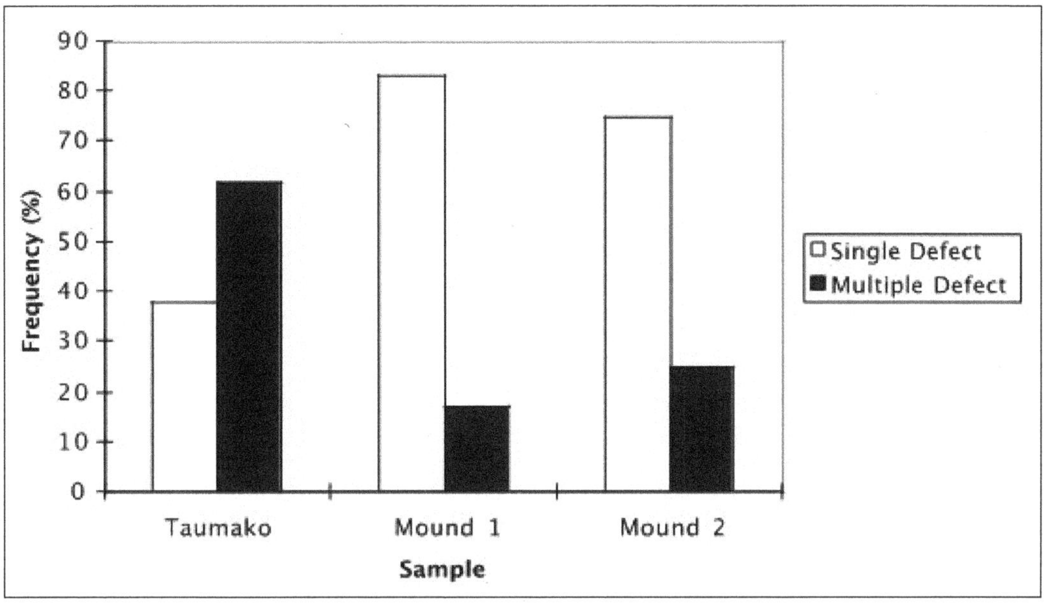

Figure 4.17. *Comparison of number of defects per tooth in the permanent teeth of affected subadults between Taumako and Tonga.*

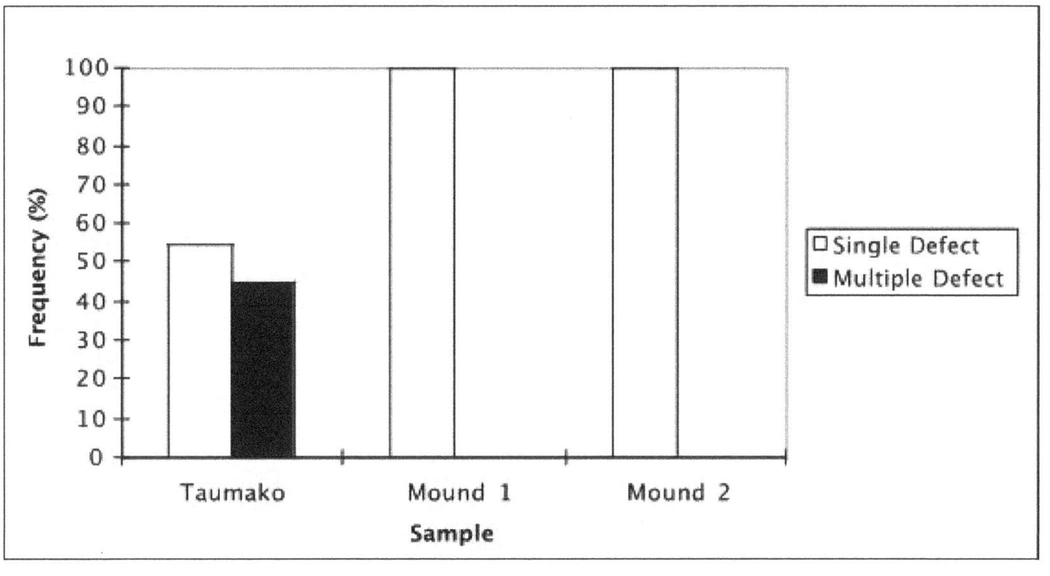

Figure 4.18. *Comparison of number of defects per tooth in teeth of affected adults between Taumako and Tonga.*

Chapter 5

Malaria, Anaemia and Porotic Hyperostosis: A Pacific Island Perspective

Part of the third aim of this thesis is to assess whether anaemia was exacerbated because of the presence of malaria at Taumako compared to Tonga where malaria was absent. This chapter will consider the various causes of anaemia in the Pacific Islands, the most significant of which is malaria. Two causes of iron-deficiency anaemia in the Pacific Islands are outlined here, malaria and hookworm. The epidemiology and illness associated with malaria are addressed in order to understand the possible impact of this disease in the prehistory of the Pacific Islands. While malaria does not directly affect the skeleton, it may cause anaemia as a sequel to infection. The presence of malaria has also been associated with haemoglobin mutations that cause similar skeletal changes to iron-deficiency anaemia. These haemoglobin mutations and the associated skeletal changes will also be considered in this chapter. There are a number of other causes for anaemia, such as hookworm and malnutrition, which will be reviewed in this chapter. A further condition associated with diet and disease is scurvy. This condition will also be considered.

Skeletal evidence of anaemia in the Taumako and Tonga samples will be presented. Comparisons between the two samples in the prevalence of skeletal evidence for anaemia will considered.

5.1 Malaria

Malaria is the most devastating of all parasitic diseases in humans (White, 1996). The name 'malaria' is Italian, meaning 'bad air' which reflects an early belief of a miasmic origin for the seasonal fevers suffered by people in warm climates (Worboys, 1994). There are four plasmodia species that cause disease in humans; *P. malariae*, *P. vivax*, *P. falciparum* and *P. ovale* (Gilles and Warrell, 1993). All of these species are present in the Pacific Islands. The most intensively studied species of the human malarias is the 'malignant' *P. falciparum*. Almost all deaths and severe morbidity of malaria are caused by infection with *P. falciparum*, while infection with the other three species only rarely result in death (White, 1996).

5.1.1 Evolutionary origins of human malaria

Ecological evidence suggests that malaria species which infect humans evolved in Southeast Asia, in particular *P. vivax* and *P. ovale*. A later parallel evolution of the *Lavernian* group, including *P. falciparum*, probably occurred in Africa around 10,000 years ago (White, 1996), with the subsequent introduction of this group into Asia (Poolsuwan, 1995). The life cycles and morphology of primate malaria resemble each other enough to support a hypothesis of a common evolutionary origin (Waters et al., 1991; Poolsuwan, 1995). Excepting *P. falciparum*, they are all 'benign' long-term infections which might indicate a long co-evolution between humans and parasite (Bruce-Chwatt, 1965).

It is traditionally believed that diseases of greater virulence to the human host have had insufficient time to evolve towards a benign relationship with the human host (Cockburn, 1963). However, an evolutionary theorist has recently suggested that 10,000 years should have been enough time for a species with such a rapid reproductive rate to co-evolve towards benignity if the '*insufficient time hypothesis*' is to hold true (Ewald, 1994). Ewald (1994) suggests a 'virulence niche hypothesis' better suits *P. falciparum* because it is more successful than other species only when a large population of both vectors and humans are present throughout the year.

However, the evolutionary origin of *P. falciparum* is equivocal. It is differentiated from the other malaria species by its extreme virulence and rapid pathogenicity. Bruce-Chwatt (1965), among others, has argued for an African origin of the parasite. Experimental attempts to infect both vector and vertebrate host with this parasite have been more successful with Asian species, suggesting an Asian origin of *P. falciparum* (Poolsuwan, 1995). A recent phylogenetic study of all primate plasmodium species discovered a relationship between avian malaria and *P. falciparum*. Through analysis of an asexually expressed small subunit of ribosomal RNA, the authors concluded that *P. falciparum* is directly related to an avian malaria. The avian transmission of the parasite is suggested to have occurred by lateral transfer and not co-evolution with a primate malaria (Waters et al., 1991). Although knowledge of zoonotic transmission by an avian parasite is lacking (Poolsuwan, 1995), Waters et al. (1991) argue that lateral transfer of this type with other animal plasmodia is not uncommon. The avian malarias are unique, because they are capable of invading most tissues of the body, which may have enabled cross-species infection (Waters et al., 1991).

5.1.2 Malaria in the Pacific: When did it arrive?

As discussed in chapter 1, Groube (1993) presented a theory of the role of *P. vivax* in Pacific prehistory based on epidemiological, entomological and archaeological evidence. He asks why the economic development of Australia and Melanesia was retarded after 30,000 years of colonisation compared to their Asian contemporaries:

> ...I would suggest...the role in Sahuland prehistory of the 'predators within', the parasites of infectious diseases, particularly the presence in Melanesia, and probably northern Australia as well, of that most ancient of human-specific diseases, malaria (Groube, 1993, 166).

He argues that the presence of malaria in northern Australia and island Melanesia 'marginalised' populations and restricted population growth and economic development. Based on the conclusions of Waters et al. (1991), that *P. falciparum* has evolved in the last 10,000 years, Groube (1993) argued that:

> The date of the arrival of *falciparum* in Melanesia is unlikely to have been much before a thousand years ago. It is very important to note that this most lethal of malaria parasites had no role in the early prehistory of Sahuland (Groube, 1993, 168).

As will be discussed further, *P. vivax* is a species which causes cyclic disease and can remain dormant within the hepatic cells for very long periods. This ability of *P. vivax* to remain dormant probably supports Groube's (1993) hypothesis that it was carried into the Pacific Islands with the first human settlers. In order for *P. falciparum* to be transported into the Pacific region, human settlement would have had to be rapid. If settlement was not rapid, then death of the carrier would probably have occurred before the parasite was able to be transmitted to the vector. The technology for a rapid settlement of these islands was not seen until the appearance of the Lapita Culture in the Pacific Islands.

The life cycle of all human malaria parasites are essentially the same across species. It comprises of two phases of multiplication; one in the mosquito, acting as a vector, and one in the human host (Gilles and Warrell, 1993). Infection is transmitted when a female *Anopheline* mosquito inoculates the human host with plasmodial sporozoites during feeding. The clinical disease associated with malaria is characterised by a debilitating fever. This fever is the result of destruction of the red blood cells when they are invaded by the parasite (Wyler, 1982). Infection with *P. falciparum* is continuous and more severe than with the other species because the parasite is able to multiply in blood cells of all ages. *P. vivax* is restricted to the invasion of young blood cells (Wyler, 1982). This species can remain dormant in the liver cells for months. The tendency for a long dormant period in *P. vivax* is the reason for the intermittent fevers that characterise it (White, 1996). *P. vivax* is a therefore a more chronic form of malaria than *P. falciparum*.

5.1.3 Epidemiology and disease

The haemolytic anaemia of malaria is multifactorial and complex. With parasitaemia there is an obligatory destruction of red blood cells containing merozoites. There is also precocious destruction of non-parasitised red cells yet the mechanism for this is unclear (Wyler, 1982; White, 1996). Haemolytic anaemia is compounded by a dysfunction of the bone marrow, in which erythropoiesis is defective for some time during and after infection (White, 1996). Hyperplasia of the normoblastic bone marrow has been shown to be an important factor in the pathogenesis of malarial anaemia (Hendrickse, 1987).

The threat of severe illness and death from infection with malaria varies with its endemicity. The degree of endemicity of malaria determines whether infection rates are stable or unstable. In areas where it is unstable then the risk of death is greater than in stable areas. Traditionally, the degree of endemicity of malaria is defined in terms of rates of parasites in the spleens of children between two and nine years of age.

- Hypoendemic: Parasite rate 0–10 per cent. This stage of endemicity is considered unstable where premunition is often not attained.
- Mesoendemic: Parasite rate 10–50 per cent– Stable
- Hyperendemic: Parasite rate 50–75 per cent– Stable
- Holoendemic: Parasite rate over 75 per cent

In stable, or holoendemic and hyperendemic areas of malaria, infection occurs repeatedly from the first year of life and this constant exposure aids in achieving a state of premunition or partial immunity. If this partial immunity develops, then clinical symptoms are minimal or completely absent. Malaria can also be expressed as an epidemic disease. Epidemics are caused by migrations where new susceptible hosts move into the area and disrupt the fine balance between the parasite and human host. Mortality during epidemics is generally high compared to holo- and hyperendemic areas (White, 1996).

Within a population where malaria is present, certain individuals are more at risk of severe disease than others. These individuals are: pregnant women, the foetus carried by an infected woman, and young children between the age of one and five years. As stated in chapter one, these subgroups of the population would be expected to exhibit a higher degree of anaemia in Taumako than in Tonga.

5.1.4 Pregnancy and foetal malaria

> 'The risks and manifestations of malaria in pregnancy vary considerably in relation to maternal immune status and parity, but it is universally recognised that pregnancy, per se, increases the risks of malaria in *all women*' (Hendrickse, 1987, 499).

A greater susceptibility to malaria is especially found in women who are pregnant with their first child. The risk of death to the woman and foetus are increased by severe anaemia and depression of blood cell production. Where malaria is unstable, the risks of maternal and foetal mortality are great. However, in areas where exposure to the parasite is constant from birth, a partial immunity to severe disease has usually developed in all women by the time of child bearing age (Hendrickse, 1987). If infection does occur in areas of stable malaria, then maternal illness, particularly anaemia, may significantly affect foetal growth and development and may also induce spontaneous abortion, still birth, or prematurity (Lee, 1988).

Another clinical aspect of malaria and pregnancy is the parasitisation of the placenta which may compromise foetal health (Lee, 1988).The incidence of placental infection is greatest among semi-immune primigravidae (Hendrickse, 1987; Lee, 1988). The ability of the parasite to transfer across the placenta has been found in all plasmodium species (Lee, 1988). As discussed in chapter two, *P. vivax* was probably the main malaria species in the Pacific Islands in prehistory. Therefore, if *P. vivax* is able to cause congenital disease it should be considered as a cause of prenatal ill-health in the prehistoric Pacific Islands.

5.1.5 Infants and children

The epidemiology of infant and childhood malaria is complex. In infants under three months of age parasitism does occur but the clinical symptoms of malaria are rare (White, 1996). This is probably due to maternal immunity and also to the presence of foetal haemoglobin (Hbf) which restricts the multiplication of parasites in the young infant (Pasvol and Weatherall, 1976). After early infancy, where exposure to the parasite is constant, parasite loads with associated clinical symptoms and risk of mortality increase steadily until five years of age when the incidence and severity of clinical malaria usually wanes (Hendrickse, 1987; Greenwood, 1997).

The behaviour of the human host plays an integral role in the epidemiology of malaria. For the parasite to transmit efficiently there must be a large reservoir of young parasites in the human population at all times. As stated above, children under five years are more susceptible to malaria than older children and adults. Therefore the younger children are probably the reservoir from which malaria can maintain the level of endemicity. If a child survives the vulnerable period of under five years a state of 'premunition' is achieved where little or no morbidity is suffered by the human host (White, 1996). While it seems that children under the age of five years bear the brunt of malaria related morbidity in endemic areas, their role in maintaining the stability of endemicity is significant. It demonstrates the efficacy of an acquired partial immunity to the parasite. It is notable that where malaria endemicity is unstable the impact of malaria associated morbidity and mortality on the whole population is considerably greater than in stable malarial areas (Hendrickse, 1987).

5.1.6 Malaria in the Pacific Islands

Vector distribution and ecology

There are three primary vectors for malaria in the Pacific Islands; *A. lungae*, *A. punctulatus* and *A. farauti*. The distribution of these vectors in the Pacific is patchy and probably influenced by the varied ecological environments of island groups. It has been suggested that this patchy distribution of vectors is an artefact of the general zoological impoverishment of species in the Pacific Islands from west to east, as explained in chapter two (Laird, 1956). However, entomological studies have shown that mosquito distribution in the Pacific was probably independent of this recognised biological phenomenon (Belkin, 1962).

Belkin (1962) composed a sequence of dispersals of mosquito species in the Pacific that may have begun in the Cretaceous period. The Lungae Complex of mosquitoes may have moved from the west into the Solomon Islands early in the sequence, probably via NG, where *A. lungae* is now isolated. While the Solomon Islands were still attached to Vanuatu via a land bridge, *A. farauti* was able to disperse east into Vanuatu in the late tertiary period. By this time, volcanic upheaval had broken the land link into the Fiji/Tonga/Samoa region so *A. farauti* was halted in its eastern dispersal. Some time after the isolation of the eastern Solomon Islands by rise in sea levels, *A. punctulatus* moved into the greater NG region where it became firmly established (Figure5.1).

If this dispersal sequence of Belkin's (1962) is to be accepted, it would seem a vector able to transmit malaria was firmly established in the Western Pacific well before human colonisation. It is interesting that *A. farauti* is not solely reliant on human blood for survival, and may feed on birds or other animals available. This is attested to by large populations found in uninhabited areas of NG (Spencer et al., 1974). Therefore, these species could have survived in the Western Pacific Islands before human colonisation. As discussed in chapter two, *P. vivax* was probably introduced into the region with the first human colonisers.

Prior to eradication in NG and related archipelagos, *A. punctulatus* was patchy in its distribution. *Anopheles punctulatus* was generally restricted to inland regions away from river valleys. However, it is capable of rapid spread into coastal zones. This species also has the ability to dramatically increase its density by using opportunities of environmental fluctuations such as river valley flooding. Therefore, it is a

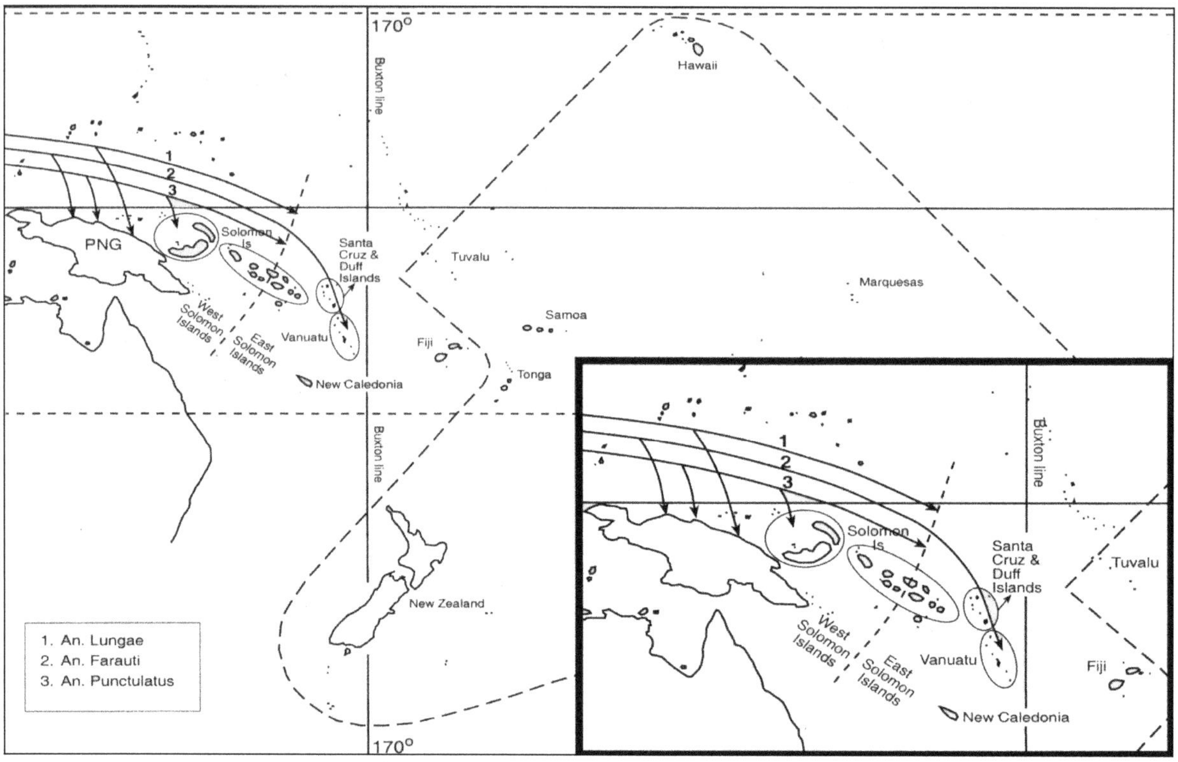

Figure 5.1. *The distribution of malaria vectors in the Pacific Islands. No Anopheline mosquitos are found east of the Buxton Line. (Sources: Belkin, 1969 and Laird, 1956. Published in Buckley, 2006.)*

source of epidemic malaria in marginal zones. In contrast, *A. farauti* is universal in its distribution, but particularly dominates coastal zones (Laird, 1956).

Anopheles farauti is widespread throughout the Solomon Islands chain while *A. punctulatus* and *A. Lungae* are restricted to the western islands (Belkin, 1962; Avery, 1974). In the Santa Cruz group, Taumako and Vanuatu, *A. farauti* is the sole malaria vector. This is the case in all island groups east of the mid Solomons Islands (Belkin, 1962). East of Buxton's Line (170 east longitude) neither the vector nor the parasite are present (Laird, 1956). However, some islands in the malarious west are also free of malaria; Bellona in the Solomons, Belep and Ovuea to the south and west of Vanuatu, and Futuna in the Vanuatu Chain (although Futuna is east of Buxton's Line) (Laird, 1956).

Endemicity

The endemicity of malaria is variable throughout the Pacific island groups and is probably dictated by the distribution of the vector. Generally, the prevalence of malaria is greater in the western island groups, with a patchy distribution east of the mid Solomons Group (Figure 5.2). In NG, malaria is holoendemic on the Northern coasts with lower levels on the Southern coasts (Flint et al., 1986). The most highly endemic areas are coastal, where because of the natural immunity discussed above, malaria poses little threat to the lives of adult individuals. Inland, the endemicity of malaria in NG is largely dominated by altitude. However, between about 600 meters and 1,300 metres above sea level, epidemics of malaria are a threat to survival (Peters and Christian, 1960). Regions above 1,300 meters are entirely free of malaria (Parkinson, 1974). This relative freedom from malaria is solely the result of vector ecology as the mosquitoes are unable to survive the colder temperatures of higher altitudes (Spencer, 1971).

In recent decades the predominant malaria species in NG has been *P. falciparum*, however, this was not always the case (Genton et al., 1995). Since 1950, there has been a shift in species predominance from *P. vivax* to *P. falciparum* (Peters and Christian, 1960; van Dijk and Parkinson, 1974). Presently in the Northern region of Wosera, East Sepik the species predominance is *P. falciparum*, *P. vivax* and *P. malariae* respectively. It is thought this shift in species predominance was brought about by spraying programmes which have changed the distribution of vectors in these areas (Genton et al., 1995).

Prior to eradication programmes in the 1960s the Solomon Islands were generally mesoendemic while some of the outer islands and the Central and Eastern Districts (i.e., Northern Gaudalcanal, Northern San Cristobel and Neggella) were hyperendemic. Some of the outer islands such as Tikopia, Anuta, Santa Ana and Santa Catalina experienced only hypoendemicity (Avery, 1974). Similarly, Rennell Island and most of the Reef Islands were mesoendemic with the vector only found in very low frequencies. Most relevant to this study is that the Duff Group of islands were known to be hyperendemic prior to spraying programmes (Avery, 1974).

In the Solomon Islands malaria severity is in part due to

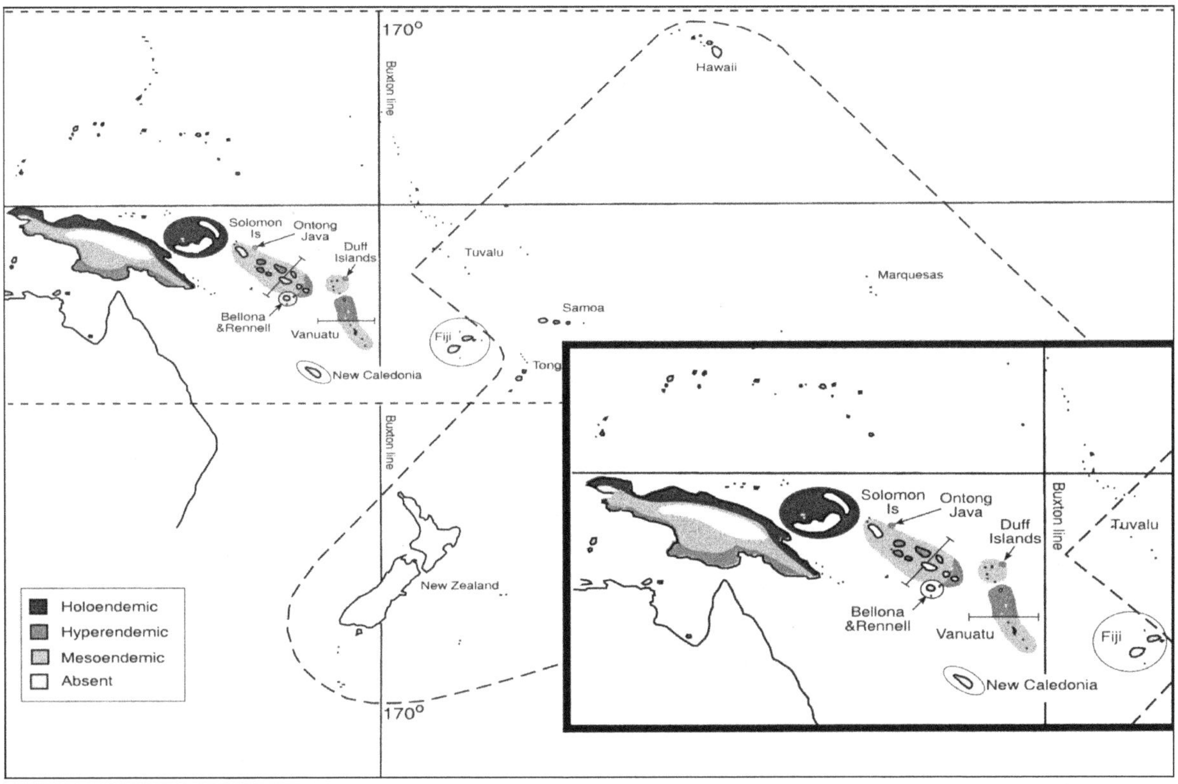

Figure 5.2. *The variable endemicity of malaria in the Pacific Islands. The endemicity is probably related to the distribution of the vector, as shown in Figure 5.1 (sources of information from studies cited in text).*

seasonality. The number of malaria cases increases dramatically in the rainy season from November to April. All species except for *P. ovale* are found in the Western chain and *P. falciparum* is the most predominant species. However, in 1943 *P. vivax* accounted for more infections in the first year of life (Sayers, 1943).

According to Maitland (1996), malaria transmission in the Vanuatu chain of islands is constant throughout the year. However, other studies have suggested a seasonal fluctuation in endemicity (Flint et al., 1986). Malaria endemicity is more intense in the Northern islands than the Southern Islands, while it is completely absent on Futuna (Flint et al., 1986; Williams et al., 1996, 1997). In Vanuatu, malaria endemicity also varies within islands. For example, on the island of Espiritu Santo endemicity ranges from hypoendemic to hyperendemic (Ganczakowski et al., 1995). Pre-eradication surveys in Vanuatu recorded a stable coexistence of *P. falciparum* and *P. vivax*, but *P. vivax* was uniformly recorded as the dominant species (Maitland et al., 1996). A seasonal variation in species prevalence was found, where *P. falciparum* predominated in the wet season and *P. vivax* prevailed in the dry season. This pattern is thought to be a general feature of malaria throughout Vanuatu and not restricted to Espiritu Santo (Maitland et al., 1996).

From the discussion above, it is evident that the endemicity of malaria in the Pacific Islands is highly variable. This probably relates to the distribution of the vector. As Figure 5.1 illustrates, east of the mid Solomon islands only *A. farauti* is found, whereas in NG and the Bismarck Archipelago all three vectors are present. This corresponds to holoendemic malaria in the western island groups and a more variable endemicity in the eastern groups. As mentioned above the different mosquito vectors prefer certain ecological niches for breeding, which in the western islands means both coastal and inland zones would harbour these vectors. However, *A. farauti* is capable of colonising any ecological zone, but it prefers coastal zones. This mosquito is also more sensitive to environmental changes than *A. punctulatus* and more finicky in its choice of breeding place (Laird, 1956). Therefore, it could be argued that where malaria is hyperendemic east of the mid Solomons these ecological zones must favour the survival of *A. farauti*. This pattern of malaria endemicity has probably changed through time and it cannot be assumed that the disease affected people in prehistory in the same manner as that observed by early Europeans. The question of whether malaria was present on Taumako during the period in which the burial mound was in use will be addressed in chapter seven.

5.1.7 Epidemiology of malaria in the Pacific Islands

The epidemiology of malaria in the Pacific is similar to other malarial areas of the world. The population that suffers most morbidity and risk of mortality are young women and children between the ages of one and five years (Sayers, 1943; Avery, 1974; Parkinson, 1974; van Dijk and Parkinson, 1974; McMahon, 1974; Zigas and Morea, 1974; Genton et al., 1995; Maitland et al., 1996).

In the early 1990s in the Wosera District, east Sepik, on the southern coast of NG, the highest mortality rate due to malaria was in the 1–4 year age group. Infants under one year old did die as a result of malaria infection but the mortality rate was relatively lower than in the older children. By five years of age premunition seems to have been achieved in this population. *Plasmodium vivax* was the dominant parasite responsible for febrile episodes in infants, but *P. falciparum* predominated in all other age groups. Generally, after five years of age the risk of death and illness associated with malaria is minimal in NG (Genton et al., 1995).

In the New Georgia region in the western part of the Solomon Islands a similar pattern to NG has been recorded:

> ...during his life-time the Melanesian develops a very effective immunity to his own local strains of malarial parasites. The infant and child suffer severely. The adult does not have many attacks of malaria. Such attacks are not usually very severe....(Sayers, 1943, 5).

Some differences between NG and the Solomons can be seen. The level of parasitaemia in the Solomon Islands rises steadily from birth up to the age of ten years, whereas in NG parasite loads are relatively low in infants under six months old. The peak prevalence of malaria in both island groups is at five years of age with a steady drop into adulthood (Sayers, 1943; Genton et al., 1995). However, in the Solomon Islands illness associated with malaria continued until ten years of age. The prolonged morbidity of malaria reported in the Sayers (1943) study may be a reflection of the earlier time in which the study was carried out. For example, the Genton et al. (1995) study was carried out in the 1990s which is after fifty more years of spraying and prophylactic treatment than when the Sayers (1943) study was carried out. It is possible that prior to widespread eradication programmes, the epidemiology of malaria in NG, and elsewhere, was similar to the pattern reported by Sayers (1943) in the Solomon Islands. In the Solomon Islands, infant cases of malaria were dominated by *P. vivax*. In the second year of life *P. vivax* still predominates but *P. falciparum* and *P. malariae* begin to rise in frequency. Between the ages of five and 10 years *P. malariae* is still clinically important, but thereafter *P. falciparum* clearly dominates (Sayers, 1943).

On Espiritu Santo in Vanuatu, the incidence of malaria episodes was greater in children under five years old than in children between five and nine years old. Infants under six months did suffer from clinical malaria, but this was caused solely by *P. vivax*. *Plasmodium vivax* remained the dominant species until the age of three years and was the cause of little morbidity after the age of six years. The role of *P. falciparum* as a cause of morbidity was less significant and slower to rise than *P. vivax*, until it peaked in importance in the three to four year olds. This species was still a substantial cause of illness in children over five years old (Maitland et al., 1996).

Maitland et al. (1996) found a strong correlation with higher *P. falciparum* densities in children in the 1–4 year age group in the wet season, but *P. vivax* showed little seasonal variation. They suggested that a reciprocal relationship exists between the two malaria species which may indicate that infection with *P. falciparum* inhibits relapse and/or the erythrocytic stage of infection with *P. vivax*. Conversely, early infection with *P. vivax* may also ameliorate a later infection with *P. falciparum*. It is interesting that while severe anaemia was rare in this population, haemoglobin levels were more negatively affected by *P. vivax* than *P. falciparum* (Maitland et al., 1996). This suggests that *P. vivax* is a more likely cause of anaemia than *P. falciparum* which probably relates to the differing pathogenesis of the disease caused by each parasite.

Some intriguing differences between malaria infection in Vanuatu and Africa have been documented. In the Gambia, Africa, severe malaria is categorised as malaria affecting the brain and severe anaemia with a fatality rate as high as 10–20 per cent (Weatherall, 1997). In Vanuatu a recent study of the mortality rates of malaria in children found that during the period of research no deaths could be attributed solely to malaria (Maitland et al., 1997).

In Melanesia it has been recognised that illness associated with endemic *P. falciparum* is less severe than in comparable areas of Africa (Weatherall, 1997). The reasons for the lower impact of *P. falciparum* malaria in Melanesia cannot be explained by a genetic factor similar to the sickle-cell trait in Africa because the thalassaemias that have some protective advantage over malaria are found only in very low frequencies in this Vanuatu population (Ganczakowski et al., 1995). Probably the most likely explanation for the lower mortality rates in Melanesia is a possible cross-species immunity to *P. falciparum* imparted by infection with *P. vivax* during the first two years of life (Williams et al., 1996; Maitland et al., 1997).

Although this phenomenon is not directly related to prehistory, assuming *P. falciparum* malaria was not present until more recently, the Vanuatu studies have provided information on the clinical importance of *P. vivax* in young children in this part of the world. These studies would suggest that *P. vivax* is a significant cause of ill-health in children up to five years of age which may have contributed to the development of iron-deficiency anaemia and early death before the introduction of *P. falciparum*. When *P. vivax* is considered alongside its more lethal cousin, it is be easy to conclude it is an essentially 'benign' disease. However, while it may not have swiftly resulted in death, its chronicity is more relevant to the study of prehistoric illness, simply because of its more chronic and benign character. As explained earlier palaeoepidemiology is equally concerned with markers of prehistoric illness than speedy death because a chronic disease is more likely to leave behind traces of its existence.

Malaria in the context of this thesis

From the clinical reports reviewed above, it would be reasonable to assume that in prehistoric Melanesia pregnant women, especially young women, were more at risk of illness and death due to malaria than adult men. Similarly, these studies have shown that children between one and five years of age were also at greater risk of death than older children. One study (Sayers, 1943) has also provided evidence suggesting children as old as ten years may still have suffered malaria related illness in prehistory. Because of acquired immunity to malaria, in areas where it is stable, adolescents and adults were less likely to suffer from malaria related illness than young children. Infants from birth to around six months of age were probably spared the rigours of illness from malaria, due to maternal immunity and high levels of foetal haemoglobin. Any evidence of prenatal illness may be interpreted as partly due to maternal illness from malaria. As discussed above, while *P. vivax* is considered a benign disease, it has been shown that it is a cause of iron-deficiency anaemia in young children and has also been implicated as a cause of malnutrition in Vanuatu (Williams et al., 1997). These factors support my assertion that the presence of malaria should be considered as a significant cause of iron-deficiency anaemia and its skeletal manifestations in the prehistory of the Pacific Islands.

5.1.8 Thalassaemia, malaria, and the Pacific

The haemoglobinopathies may cause anaemia. A group of haemoglobinopathies, thalassaemias, are inherited disorders of the blood which depress the normal synthesis of haemoglobin, leading to an imbalance in globin-chain production, defective erythropoiesis and haemolysis (Weatherall and Clegg, 1981; Resnick, 1995b).

In normal adults, 97 per cent of the red blood cell is made up of polypeptide chains of globin, and three per cent of haeme groups. The two pairs of polypeptide chains consist of paired alpha chains (HbA) and paired β-chains (Hbβ). In the foetus another type of haemoglobin is present, HbF. This is found in varying degrees in infants at birth and is usually replaced by HbA by four months of age. In adults one percent of HbF remains (Resnick, 1995b; Fleming, 1996). There are two main types of thalassaemia, those which produce a deficiency in alpha globin-chain synthesis and those which produce a deficiency in beta globin-chain synthesis. Clinical severity of the conditions is highly variable and is dependant on the type of molecular abnormality (Huntsman and Lehman, 1984).

β-thalassaemia

β-thalassaemia is the most clinically important type of thalassaemia, in which depression of globin-chain synthesis is confined to the beta chain of haemoglobin. Symptoms of β-thalassaemia become apparent after four months of age when HbA would normally replace HbF. At this age an abnormally high production of HbA is detected which indicates a depression in the beta globin-chain production and compensatory action of the alpha globin-chain (Resnick, 1995b).

The gene variants of β-thalassaemia are either classified as β°-thalassaemia, where suppression of the β-globin synthesis is complete, or β^{+}-thalassaemia, where suppression is incomplete and variable amounts of β-globin and HbA are able to form. The clinical effects of thalassaemia are dependant on the type of mutation and are highly variable. Clinically, the β-thalassaemias are classified as 1) *thalassaemia major* (homozygous alleles), a condition that involves chronic illness, anaemia, and survival is solely dependant on blood transfusions after early childhood; 2) *thalassaemia intermedia*, when blood transfusions are necessary only at intermittent times of crisis, and anaemia is moderate; 3) *thalassaemia minor* (heterozygous alleles) is characterised by little or no clinical indications of the mutation and the anaemia will be mild (Fleming, 1996). Significant skeletal and dental abnormalities can develop in those suffering from β-thalassaemia major and intermedia (Laor et al., 1982; Resnick, 1995b; Ortner, 2003).

Based on these clinical criteria, children in prehistory with β-Thalassaemia Major would not have survived past the first year of life, while β-thalassaemia Intermedia would have left the child vulnerable to infectious disease which may have been exacerbated by any related anaemia of parasitic infestations.

Geographical distribution of β-thalassaemia and the role of malaria

A geographical correlation between β-thalassaemia and malaria has been recognised and malaria is considered the primary mechanism for the globin genes' maintenance in human populations (Fleming, 1996). In its heterozygous form, β-thalassaemia is thought to confer a relative advantage for survival of infection with *Plasmodium falciparum* (Kariks and Woodfield, 1972; Yenchitsomanus et al., 1985, 1986; Fleming, 1996).

β-thalassaemia was first reported in Melanesia in the 1960s and subsequent studies found heterozygote carriers of the disease (Weatherall and Clegg, 1981). Further studies were conducted in numerous coastal and highland populations of New Guinea (NG) revealing carrier rates in coastal populations as high as 5 per cent but these are rare in the highlands (Hill et al., 1989). A patchy frequency of heterozygote β-thalassaemia was recorded in Vanuatu. However, on the island of Maewo in Vanuatu the highest frequency yet was recorded in infants (20 per cent) (Bowden et al., 1985). A more recent study of the carrier gene frequencies in Vanuatu, but using a different methodology, yielded a lower frequency on Maewo (12.6 per cent) in the adult population (Ganczakowski et al., 1995). Studies elsewhere in Island Melanesia (Solomon Islands, Fiji, and New Caledonia) have failed to find β-thalassaemia carriers (Figure 5.3).

Surveys in Polynesia, although uncommon, also found no heterozygotes (Hill et al., 1989). The reasons for this apparent confinement and patchy distribution of β-thalassaemia in Melanesia is unclear. A likely factor restricting the distribution of β-thalassaemia is its clinical severity, because the selective advantage of a heterozygote against malaria is outweighed by the inevitable fatality of a homozygote (Hill et al., 1989). In other words, once the selection *for* the trait is relaxed, selection *against* the trait is strong. Similarly, the previous review of the patchy geographic distribution of malarial vectors may also shed some light on the question. For instance, if the vector for malaria was not present in numbers sufficient for continuous transmission, then the selection of the heterozygote form of β-thalassaemia would not be so intense as in areas where all malaria vectors were present. The patchy distribution of this genetic anomaly may also be a reflection of recording methods and lack of surveys.

5.1.9 α-thalassaemia

α-thalassaemia is an inherited reduction of α-globin synthesis that is the outcome of various deletions at chromosome 16 on the α-globin gene. Deletion of either of the α-globin genes will result in α^+-thalassaemias, while deletion of both α-globin genes causes α°-thalassaemia, and a consequent complete absence of α-globin synthesis. The molecular mutation is similar to β-thalassaemia in that there is a retardation of globin chain production. However, in β-thalassaemia there is compensatory production of haemoglobins F and A2, which is not the case with α-thalassaemia. Because haemoglobins A, A2, and F all contain α-chains, a reduction in α-chain production will result in a reduced production of all three haemoglobins (Huntsman and Lehman, 1984).

The mutations of the α-globin gene which result in α°-thalassaemia are more clinically important than α^+-thalassaemia which is usually asymptomatic. Two clinical forms of α°-thalassaemia have been identified in Southeast Asia and other populations, haemoglobin Bart's hydrops fetalis syndrome and haemoglobin H disease. Haemoglobin Bart's hydrops fetalis syndrome causes death *in utero* or soon after birth. It is not found in Oceania, however haemoglobin H disease (HbH) has been recorded in NG and its offshore islands. It has not been recorded elsewhere in island Melanesia. The clinical manifestations of HbH are similar to those of β-thalassaemia minor (Weatherall and Clegg, 1981), so distinguishing between these two globin traits would not be possible in dry bone. Therefore, HbH is not considered further as cause of anaemia in prehistory although it is acknowledged as a possibility.

Three clinically asymptomatic α^+-thalassaemia genotypes have been identified (Fleming, 1996). The recorded frequencies of these genotypes are utilised for mapping the migrations of past peoples as so-called 'population markers' (Hill et al., 1989). However, these clinically unimportant genotypes do produce mild anaemia that may be exacerbated by malaria or other infections (Ganczakowski et al., 1995; Fleming, 1996). Therefore, the population frequencies of these traits may be relevant to this study.

α-thalassaemia as a selective advantage against malaria in the Pacific

The α^+-thalassaemias are among the most common genetic mutations in human populations, found in up to 70 per cent of the population in some regions (Hill et al., 1989). They are the most widely distributed genetic mutations in tropical and sub-tropical regions, with α^+-thalassaemia deletions reaching some of their highest frequencies in the Pacific (Hill et al., 1989).

While the clinical correlation between sickle-cell anaemia malaria has been established since the 1950s, the genetic complexities of α-thalassaemia make drawing a similar correlation difficult. It appears the strongest evidence that α-thalassaemia has been selected for by malaria is epidemiological (Fleming, 1996). For example it has been demonstrated quite clearly that α-thalassaemia frequencies are positively correlated with the endemicity of *falciparum* malaria in NG and island Melanesia (Flint et al., 1986). Malaria endemicity varies with altitude in NG and latitude in Island Melanesia and the frequencies of globin gene mutations follow the varying endemicity of malaria. Correspondingly, the frequency of α-thalassaemia drops from 70 per cent on the North coast of NG to less than 10 per cent in New Caledonia where malaria is absent (Flint et al., 1986). A type of α-thalassaemia has been recorded in very high frequencies in eastern Polynesia. These frequencies are believed to have arisen as the result of genetic founder effect rather than selection by malaria. This trait has no clinical impact on the population and is therefore not subject to selection (Trent, 1991).

A recent study in Vanuatu has proposed that α-thalassaemia does not impart a protective advantage over malaria but in fact increases susceptibility to *P. vivax* earlier in childhood (Williams et al., 1996). In order to test the hypothesis that α^+-thalassaemia imparts a protective immunity to malaria, Williams et al. (1996) correlated the prevalence of malaria infection with α^+-thalassaemia. Surprisingly, cross-sectional results revealed that all homozygote children under ten years of age had a higher prevalence of parasitaemia, and higher splenomegaly (enlarged spleen) than normals. On examination of the *Plasmodium* species present, it was found that the prevalence of *P. vivax* was higher in younger children than *P. falciparum*. They cite other unpublished studies that state this is a recurrent finding in regions where the two species co-exist. This study demonstrates that early infection with *P. vivax* may provide some immunity against subsequent attacks with *P. falciparum* later in childhood (Williams et al., 1996). Therefore, homozygote α^+-thalassaemia may actually predispose the child to early infection with *P. vivax* (Williams et al., 1996). In other words, it seems that *P. vivax* acts a type of vaccine against the malignant *P. falciparum* (Weatherall, 1997). Therefore, the mechanism by which the α-thalassaemias were selected for may be due to a selective advantage against *P. falciparum*, but the pathway is more circuitous than in β-thalassaemia, and involving *P. vivax*.

While the Williams et al. (1996) study highlights a possible influence of *P. vivax* in the development of α^+-thalassaemia frequencies in the Pacific Islands, they do not consider the role of co-infection with *P. vivax* and *P. falciparum*. A study in the east Sepik region of NG found that over 21 per cent of infected individuals suffered multiple parasitaemia

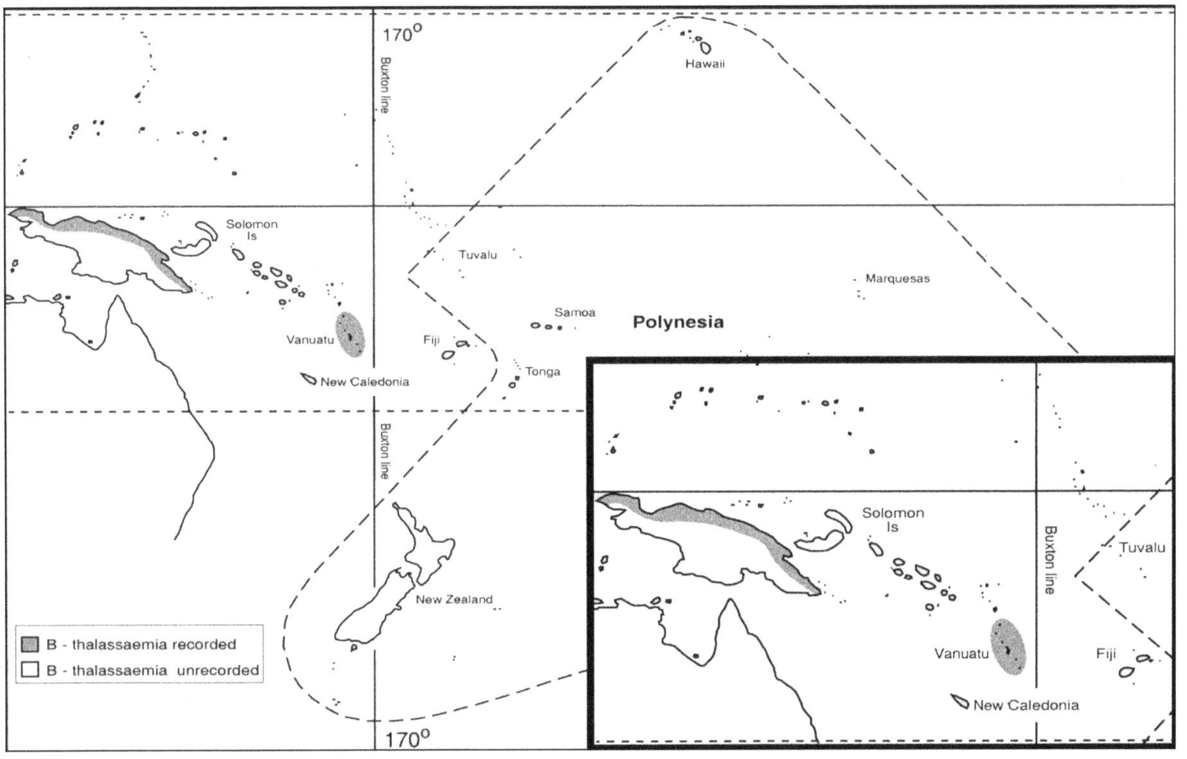

Figure 5.3. *The distribution of β-thallassaemia in the Pacific Islands (sources of information derived from the studies cited in text).*

(Genton et al., 1995). It is interesting that Northern NG is where the highest frequencies of α-thalassaemia have been recorded (Flint et al., 1986). This raises the question of whether malaria endemicity and α-thalassaemia frequencies are correlated with the number of species present; or is it merely coincidental?

Another aspect to consider is the historical shift in the prevalence of *Plasmodium*. As discussed above, a switch from *P. vivax* as the dominant species to *P. falciparum* occurred subsequent to initial eradication attempts (Genton et al., 1995). Therefore, the higher prevalence of *P. falciparum* in recent populations of the Pacific may be an artefact of drug controls that have allowed the more malignant and drug resistant *P. falciparum* to overtake *P. vivax*. This implies that *P. vivax* held *P. falciparum* at bay until drug-therapy tipped the balance and allowed *P. falciparum* to take over. This may also be the case in Vanuatu, and if this prevalence switch has occurred, *P. falciparum* is unlikely to have had much influence on the frequency of α^+-thalassaemia. Therefore it may be useful for future researchers to consider the role of *P. vivax* in the evolution of α-thalassaemia in the Pacific Islands.

5.1.10 Hookworm

Another common aetiologic agent of anaemia in the Pacific is infection with the hookworm species *Necator americanus*. This species is the predominant hookworm of most of Africa, southern Asia and the Pacific Islands (Gilles, 1996). Hookworm was noted by early clinicians as a leading cause of anaemia in the Pacific Islands (Lambert, 1941). Therefore, it is likely to have had a long history in this region and was probably introduced with the first human inhabitants from southern Asia.

Hookworm is extremely prevalent in the tropics and can reach frequencies of 80–90 per cent (Gilles, 1996). The life cycle of the hookworm requires several stages of development both in the human host and in soil. Eggs are excreted with faeces into the soil where larvae hatch and feed. These larvae can survive for up to two years in moist soil waiting for contact with the sole of a foot or other region of uncovered skin. Once in the host they travel through the bloodstream to the lungs, up the trachea where they are ingested through the oesophagus to the stomach and small intestine. The larvae then mature into adult form within the intestine (Gilles, 1996). The adult worms attach themselves to the intestinal mucosa of the host and cause bleeding and anaemia (Fleming, 1996).

The relative severity of anaemia due to hookworm infection depends on; the daily iron absorption from the diet, the size of the body's iron stores and the intensity of the parasitisation. For example, an individual with a poor diet and with low iron stores needs only a mild hookworm infection to cause anaemia. In addition, due to the added pressures of menstruation on iron stores, women have lower thresholds than men. Children usually expose a greater surface of their bodies to infection when playing in the dirt therefore they have heavier infections with hookworm in relation to their body size (Fleming, 1996). The anaemia of hookworm is more severe in young children because of their small blood volume and greater demands for iron and protein during growth. Therefore, a relatively low parasite load can produce a more severe anaemic episode than it would in adults (Jelliffe, 1968).

The transmission of hookworm usually starts when a child begins to move around in the community independently and is exposed to contaminated soils. However, infection can occur early in infancy if exposure to the larvae occurs (Jelliffe, 1970). Several factors favour the transmission of hookworm. For example, livestock such as pigs and dogs which may eat contaminated human faeces will act as a reservoir for the worm. The eggs are not destroyed as they pass through dogs and pigs, which means these species are effective vehicles for dissemination of the worms throughout the community. The most important cultural factors are the defecation habits of people in a community and the use of manure in agriculture. The habit of defecation around houses or close to the communal living area will act as an effective means of continual infection. Similarly, the habit of using human manure as a fertiliser in agriculture will aid in transmission (May, 1958). The latter factor probably does not occur in most of the Pacific Islands where lack of fertilisation of gardens necessitates long fallow periods (Barrau, 1958).

In relation to defecation habits, from descriptions of the ablution exploits of the archaeologists during the excavation at Taumako we know that the latrine was situated over the high wall of the artificial island and the tide would wash away the waste (Leach and Davidson, 2008). However, when gardens were tended on the main island of Taumako, contact with the faeces of humans and pigs would no doubt have occurred. The practices of faeces disposal on the Tongan islands is not known.

This habit of using the sea as a latrine is widespread in the Pacific and would decrease the chances of transmission. However, hookworm infection was highly endemic throughout the Pacific Islands early last century (Lambert, 1941) and there is no reason to believe it was not present prior to European contact.

5.1.11 Iron deficiency anaemia

The previous section covered the complex issues of three causes of anaemia in the Pacific Islands. This section will discuss the equally complex subject of the aetiology of iron-deficiency anaemia in the Pacific Islands. Firstly, the metabolism of iron and iron-deficiency will be reviewed.

Iron is a constituent of haemoglobin, myoglobin (the red pigment in muscle) and enzymes that are responsible for the transportation, storage, and utilisation of oxygen. As an integral part of haemoglobin, iron is essential for the transportation of oxygen from the lungs to the rest of the body's tissues. Similarly, as a component of myoglobin, iron is essential for the storage of oxygen used during muscle contraction (Resnick, 1995b). In addition, 5–30 per cent of iron is found in storage proteins such as haemosiderin and ferritin which are found mainly in the spleen, liver, and bone marrow. These proteins are involved in iron homeostasis by becoming mobilised when the supply of dietary iron is diminished. These storage proteins are transported throughout the body by binding with the protein transferrin (Ryan, 1997). Transferrin is responsible for transporting iron to the body tissues, but can also withhold iron from microbial invaders (Weinberg, 1992).

The amount of iron in the body is partly regulated by the amount of iron absorbed through the intestinal mucosa. However, in adults, approximately 95 per cent of the iron required for haemoglobin is recycled from the products of dying red blood cells and the remainder is absorbed from dietary sources. Because of the rapid production of haemoglobin and myoglobin during the first year of life, young children are dependant on dietary sources for up to 30 per cent of their iron requirements. Usually, the amount of iron absorption is related to the volume of iron stores, the type and amount of iron in the diet and interactions between nutrients and/or foods that either inhibit or maximise the amount of iron available for absorption. When iron is deficient in the diet the rate of absorption increases to compensate for the deficiency. However, the increase in iron absorption or the supply of iron may not be sufficient to prevent anaemia (Ryan, 1997).

Dietary iron consists mostly of iron salts known as non-haeme or inorganic iron. This form of iron is obtained primarily from vegetables. A smaller amount of iron is obtained from haeme proteins, haemoglobin and myoglobin, which are found in meats. In adults, haeme iron is more easily absorbed than non-haeme iron. Furthermore, it is known that absorbic acid (Vitamin C) enhances the absorption of non-haeme iron, while calcium, and phytates (found in cereals such as wheat, oats, and maize) may inhibit iron absorption (Ryan, 1997). In fact, Vitamin C is probably the single most important dietary factor that may increase dietary iron absorption (Wadsworth, 1992).

Early in infant life the sole dietary source is from breast milk. The iron found in human milk is found in fat globule membranes and in lactoferrin (Ryan, 1997). Lactoferrin is a class of transferrin which plays an integral part in the physiological inflammatory response to infection. Because of the high concentration of lactoferrin in human milk, infants that are solely breast-fed are at less risk of infection than those who are not (Weinberg, 1992). Breast-fed infants can maintain adequate iron stores for up to six months of age, however after this time supplementation of iron from other sources is required to maintain growth. It has been found that infants solely breast-fed for up to nine months were associated with a higher prevalence of iron-deficiency anaemia than those fed iron fortified formulas. Therefore, if breastfeeding is the exclusive nourishment, after six months of age additional bioavailable sources of iron should be introduced into the diet. After six months of age, infants must obtain all iron required for growth from dietary sources (Ryan, 1997).

5.1.12 The aetiology and epidemiology of iron-deficiency anaemia

Iron-deficiency anaemia is considered to be primarily caused by a low intake of bioavailable iron in the diet which does not meet physiological requirements (Garn, 1992; Fleming, 1996). Those individuals with the greater physiological demands for iron are at a higher risk of developing iron-deficiency anaemia. These are: infants over six months old, children up to five years old, adolescents, especially menstruating girls, and pregnant women, particularly those who have many closely spaced pregnancies (Fleming, 1996). Others have suggested that the primary aetiology of iron-deficiency anaemia is the interaction between infection and iron-withholding (Stuart-Macadam, 1992).

The pathological bony change of iron-deficiency anaemia and/or any haemolytic anaemia is produced by an increase of haemopoietic marrow activity, due to low haemoglobin levels, that results in a thinning of the cortex and a reduction in trabeculae. This is most readily apparent on the orbital roof where the outer plate of bone is resorbed and the internal spongiosa is exposed. This change at the orbital roof is termed cribra orbitalia in the palaeopathological literature (Hengen, 1971). Slight changes of the orbital roof are seen as an increased porosity and scattered fine foramina. The pathogenesis of cribra orbitalia and associated porotic hyperostosis (changes to the cranial vault) will be discussed in detail below. Some researchers will define porotic hyperostosis as lesions of only the cranial vault, but this definition can also include cribra orbitalia (Ortner, 2003).

There are two main hypotheses concerning the aetiology of iron-deficiency anaemia in prehistory. The first is the dietary model where the skeletal changes of iron-deficiency anaemia are caused by a lack of iron in the diet or a staple diet that inhibits iron absorption. The second is the parasite model proposed by Stuart-Macadam (1992) in which cribra orbitalia and porotic hyperostosis in prehistoric populations is viewed as reflecting:

> ...the attempts of that population to cope with and adapt to its environment. It suggests that there was a primary response of iron withholding as an adaptation to disease and/or pathogen load.... (Stuart-Macadam, 1992, 44)

In the following section both of these models are briefly reviewed and the arguments are put into a context that is relevant to the Pacific Islands.

5.1.13 The dietary model

El-Najjar and colleagues (1976) conducted a study on the prevalence of porotic hyperostosis in two different ecological zones from prehistoric Arizona and New Mexico. The first was the base of the canyons of north-western New Mexico where the predominant source of food is believed to have been cultivated maize. The second ecological zone were the sage plains where maize was not cultivated as successfully and wild plants and animal foods were the primary subsistence base. It was found that the children of the canyon floor were significantly more affected with porotic hyperostosis than populations from the sage plains. Adults from the canyon sites also had higher rates of porotic hyperostosis than the sage plain adults.

This difference in porotic hyperostosis prevalence was interpreted by the authors as influenced by the dependence on maize as a staple food source. Maize, like other cereals such as rice, contain phytic acid which is known to inhibit the absorption of dietary iron (El-Najjar et al., 1976). Cereal based diets also irritate the fragile gastrointestinal systems of young children contributing to the morbidity of weanling diarrhoea (Palkovich, 1987). Also, less animal protein and haeme iron in the canyon sites would have placed children at further risk of developing iron-deficiency anaemia (El-Najjar et al., 1976). Based on this evidence El-Najjar et al. (1976) concluded that maize-dependant populations will have more evidence of porotic hyperostosis than populations with sufficient haeme iron in their diets.

After the publication of the El-Najjar et al paper, the role of dietary iron deficiency or iron inhibitors as the staple diet in prehistory had largely been accepted as the cause of iron-deficiency anaemia and its skeletal manifestations (e.g. Lallo, 1977). However, some authors also suggested a synergistic relationship between nutrition and infection in contributing to the frequency of porotic hyperostosis in some prehistoric populations (Hengen, 1971; Lallo et al., 1977; Mensforth et al., 1978; Walker, 1986; Palkovich, 1987; Powell, 1988; Mittler and Van Gerven, 1994). These authors saw the role of infection and growth demands on children of weaning age as contributing more to the aetiology of iron-deficiency anaemia than the inhibitory properties of maize. Despite this integrated approach to the interpretation of porotic hyperostosis in prehistoric samples, porotic hyperostosis is still considered a useful indicator of nutritional stress (Goodman et al., 1988).

5.1.14 Is iron-deficiency anaemia a disorder or a defence?

'Determining the cause of anemia, i.e., whether it is a disorder or a defence, is critical before establishing programs to eradicate it' (Kent, 1992, 13). The controversy over whether iron-deficiency anaemia is caused by an inadequate diet or as a defence against microbial invasion (Kent, 1992) has serious implications for the interpretation of anaemia in prehistoric populations. One end of the spectrum implies ill-health and the other a successful adaptation to disease.

The inflammatory response to any microbial infection is known to predispose an individual to or even cause anaemia (Ryan, 1997). In this case, anaemia is expressed by lower haemoglobin concentration and saturation of transferrin, and an increase in iron stores. These changes occur because the reticuloendothelial system retains an increased proportion of iron from ageing blood cells (Ryan, 1997). The actual mechanism for this is that macrophages scavenge iron from defence cells at septic sites (Weinberg, 1992).

The lactoferrin in human milk also serves to inhibit the absorption of both dietary and bile-derived iron in infants. It has been found that a ferritin saturation level below 30 per cent is advantageous for inhibiting microbial multiplication (Weinberg, 1984). By the age of six months plasma saturation of transferrin has decreased from 68 per cent to around 25 per cent. The inflammatory response to infection also causes a decrease of general nutrient absorption, including ascorbic acid which would affect iron absorption (Scrimshaw et al., 1968).

The relationship between iron-deficiency anaemia and immunity is fraught with controversy (Ryan, 1997). It is well established that iron is important for bacterial metabolism and survival and if the host is deficient in iron then microbial multiplication may be adversely affected (Weinberg, 1974, 1984). Some studies have suggested that iron fortification regimes in developing countries are detrimental to population health. For example, Murray et al. (1978b) demonstrated an increase in certain infections after iron supplementation in Somali nomads in Africa. The increase in infectious episodes actually reactivated pre-existing conditions of malaria, brucellosis, and tuberculosis. This may suggest that the host immune response is best able to inhibit the multiplication of an intracellular protozoan such as *falciparum* malaria when the infecting organism is already weakened by an iron-deficient cellular environment. The source of the iron-deficiency among the nomads is purely dietary and not from intestinal parasites. They conclude that a balance is struck whereby the nomads are able to coexist with micropredators; but the price is iron-deficiency combined with mild infection (Murray et al., 1978a).

A randomised study was carried out in New Guinea which tested the hypothesis that iron deficiency increased the susceptibility to infectious disease and that supplementation in an iron deficient infant population would decrease mortality (Oppenheimer et al., 1986). Contrary to expectations, Oppenheimer et al. (1986) found that most infectious disease episodes were increased in the group given iron supplements. In particular, respiratory infections and malaria actually increased in prevalence. A study carried out in Africa reported similar findings (Masawe et al., 1974).

Other studies have been carried out which were specifically designed to test the affect of iron supplementation on malaria infection. It was generally found that iron supplementation increased the susceptibility to malaria and also seemed to induce a recrudescence of existing infection (Nurse, 1979; Oppenheimer et al., 1986; Thurnham, 1986). However a recent study testing the association between iron supplementation of pregnant women and susceptibility to malaria found no positive correlation. It was concluded that iron supplementation can be given without the risk of subsequent malaria infection and that birth weights of infants were increased considerably (Menendez et al., 1994). This supports previously held beliefs that low iron levels *in utero* can be detrimental to foetal growth and health (Garn, 1992).

In a comprehensive review of this subject, Ryan (1997) concluded that the apparent benefits of iron-deficiency have been shown only in relation to malaria. Others have suggested that the data from these studies have been overinterpreted (Holland and O'Brien, 1997). The seeming paradox of iron-deficiency as an adaptive response to microbial infection is further confused when the body of literature demonstrating the detrimental effects of any nutritional deficiency on the immune response would suggest otherwise (Scrimshaw et al., 1968; Scrimshaw, 1981).

5.1.15 The 'parasite model': a new perspective: Stuart-Macadam, 1992

In 1992, Stuart-Macadam composed an argument she felt was a new perspective on the interpretation of porotic hyperostosis in prehistory. She argues that previous researchers have stressed the role of diet too strongly, and that new data and her perspective will shed new light on this subject. The premise of this new perspective is based on two main points: 1) Only in cases of severe malnutrition does diet play a role in developing iron-deficiency anaemia and 2) that mild iron deficiency is actually an adaptive response to infection.

The mechanism of iron withholding during microbial infection was outlined above. However, Stuart-Macadam has taken this concept one step further and argues that 'iron withholding [should be viewed] as a positive adaptive response to invading microorganisms' (Stuart-Macadam, 1992, 41). The available data on porotic hyperostosis was interpreted by her as demonstrating an increase in prevalence the closer the sample is to the equator. This correlation is suggested to indicate that 'pathogen load' is more critical to the development of anaemia than diet. This is suggested because tropical climates favour microbial transmission more than temperate climes.

Stuart-Macadam (1992) states that the development of iron deficiency anaemia should be viewed as caused either by direct or indirect agents. The direct causes include malaria and hookworm, and conceivably any other conditions that cause blood loss. The indirect cause of iron-deficiency anaemia is the iron withholding system associated with chronic diseases such as tuberculosis, osteomyelitis, and the mycoses. Stuart-Macadam goes on to stress the role of hookworm infections in her discussion of pathogen loads and their association to iron-deficiency anaemia.

This hypothesis was expanded where Stuart-Macadam and Kent illustrate a geographical cline of hookworm incidence and prehistoric prevalence of porotic hyperostosis (Stuart-Macadam and Kent, 1992). It is interesting that a direct relationship between modern pathogen loads and prehistoric patterns of disease is assumed. It would be reasonable to postulate that the ecology of these diseases, including hookworm, would have been altered over the millennia by environmental changes and population density. Certainly, the distribution and endemicity of malaria is extremely sensitive to environmental change (De Zulueta, 1994) as shown by the historical increase in falciparum malaria in the Pacific and would surely have been markedly different from modern patterns of this disease. Similarly, it should not be automatically assumed that parasites such as hookworm were always present in areas where they are endemic today.

A further point that needs addressing is the confused logic of the basic premise of the new perspective hypothesis. This relates to the fact that Stuart-Macadam (1992) argues that hypoferremia will develop in response to pathogen loads and stresses hookworm as contributing to this. However, no studies examining the protective advantage of iron deficiency found a correlation between hookworm and hypoferremia. As mentioned above, if any studies have demonstrated this then it was only clear in relation to malaria (Ryan, 1997). The most serious morbid sequel of hookworm infection is iron-deficiency anaemia (Gilles, 1996). However, this is caused by loss of blood, leading to haemolytic anaemia, and not iron withholding as Stuart-Macadam (1992) seems to be suggesting. Therefore, the iron deficiency associated with hookworm is a morbid sequel of the disease and not a physiological response to the presence of disease. The same argument could be used for any disease causing haemolytic anaemia, notably malaria.

In a vitriolic critique of the new perspective hypothesis, Goodman (1994, 166) asks 'What is actually new about this new perspective, and more importantly, what is correct and useful?' As outlined above, the idea that diet is not central to the aetiology of iron deficiency in prehistory is not new. Several authors listed above investigated a multiple cause for porotic hyperostosis, including pathogen loads. However, Goodman's main criticism of Stuart-Macadam's (1992) hypothesis is the narrow and confused way in which the concept of adaptation has been used. He argues that any change in the homeostasis of an organism that causes disease is not without its functional costs. For example, iron deficiency will adversely affect an individual's work capacity, ability to resist disease, and cognition and behaviour (Goodman, 1994). More pertinent to the concept of selection however is the fact that iron-deficiency anaemia is associated with greater risk of foetal and neonatal death. Therefore, in terms of Darwinian selection, iron deficiency cannot be viewed as advantageous (Goodman, 1994). Furthermore in terms of successful adaptation, premature death is the ultimate maladaptation and whether it is caused by chronic anaemia or hookworm infection is irrelevant (Holland and O'Brien, 1997).

Goodman (1994) also cautions against interpreting any evidence of disease in prehistory as adaptive:

> Relying on the hope that the human body can homeostatically adapt to insults and deprivations, the model suggests a disease process in which an insult invades the body and the body homeostatically adjusts. All other systems are disconnected in Cartesian terms from the readjusting subsystem. In the most extreme form, that thinking suggests a vulgarization of the concept of adaptation: signs of stress are seen as adaptations for no other reason than that they exist in stressed but surviving organisms (Goodman, 1994, 164).

In a later review of the dietary and parasite models of interpreting porotic hyperostosis in prehistory, Holland and O'Brien (1997) argue that any model which attempts to explain its occurrence should be holistic and include both diet and disease as causal factors.

5.2 The Aetiology of Iron-deficiency Anaemia in the Pacific Islands

A consideration of the aetiology of iron-deficiency anaemia and the occurrence of porotic hyperostosis in prehistoric Pacific populations cannot be strictly defined based on research carried out in Europe and the Americas. All of the debate surrounding the dietary model is based on whether maize and other cereals staples may inhibit iron absorption enough to cause iron-deficiency anaemia. However, Pacific Island subsistence economy is based on starchy root crops and not cereals (Barrau, 1969). Therefore, the questions must be asked if the staple foods in the Pacific Islands are adequate in iron and whether there are any foods that might inhibit iron absorption.

As reviewed in chapter two, the Pacific Island diet was low in haeme irons because red meat was consumed only as a supplement to vegetable foods. In Tonga, the only sources of haeme iron were pigs and birds, which were consumed mainly by the chiefly classes (Pollock, 1992). Marine resources also supplemented the diets of adult Pacific Islanders considerably (Kirch, 2000a). However, most fish species are very low in iron (Dignan, 1994). Shellfish can provide an adequate supply of iron, but these are needed in large quantities to fulfil the requirements of the average daily intake of an adult. Therefore, these resources were not as valuable a source of iron as red meat (Walker, 1986).

The main source of dietary iron for the majority of the people in the Pacific Islands was from the root crop staples (Pollock, 1992). Some plants have an adequate iron content but these are also required in large quantities to fulfil the body's requirements. These plant foods provide only non-haeme iron and are variable in their content. Yams were the staple food of Tongans, supplemented by other root crops. These vegetables provide mainly starch to the diet with little protein or vitamins (Pollock, 1992). They are also very low in iron (Dignan, 1994). While young taro leaves have appreciable levels of iron, the more mature leaves have a high tannin content which can inhibit iron absorption and are not usually eaten (Stahl 1984 cited in Green 1999). Similarly, coconut milk was consumed in large quantities by some Pacific Island children (Davies, 1958), providing an alternative source of calories and fat (Pollock, 1992). However, experimental studies have shown that coconut milk can also inhibit the absorption of iron (Than et al., 1975).

5.3 Expected Skeletal Changes in Anaemia

Essentially, the skeletal changes of different kinds of anaemia are the same, as the cause of the anaemia is similar. The following will outline the biological basis of these changes and suggest ways of differentiating between iron-deficiency anaemia and the bone changes of thalassaemia.

5.3.1 The pathogenesis of skeletal changes in anaemia

As outlined above neonates derive the iron needed for post-natal growth from stores obtained from the mother during foetal life. After six months of age, without dietary supplementation, the demands of the growing body cannot be met by an iron deficient haemoglobin. The marrow of individuals with iron deficiency produce red blood cells that are deficient in haemoglobin content. Therefore, the oxygen demands of the body are not being met. The physiological response to this state is to increase red blood cell production in an attempt to meet the demands of growth in children and tissue renewal in adults. Eventually, the marrow space will expand to accommodate the increase in volume (Aufderheide and Rodriguez-Martin, 1998). It has been demonstrated (and generally accepted), that lesions of the skeletal tissue develop during infancy and early childhood when the blood producing marrow is so highly active during growth (Ortner, 2003). Cranial lesions have been recorded in adults but these are thought to be representative of childhood illness that have not completely remodelled (Stuart-Macadam, 1985). The pathogenesis of skeletal changes is essentially the same in genetic anaemia (Weatherall and Clegg, 1981). Clinically β-thalassaemia major is characterised by severe anaemia and death in early childhood. The anaemia develops shortly after birth and presents as pallor, fatigability, jaundice, deficient skeletal growth and facial abnormalities (Resnick, 1995b). Individuals with β-thalassaemia intermedia may develop skeletal changes, while β-thalassaemia minor is characterised by only mild anaemia and no bone changes (Ortner, 2003).

5.3.2 Porotic hyperostosis

Porotic hyperostosis is the palaeopathological term for the cranial changes that can develop as a result of marrow hyperplasia in severe iron-deficiency anaemia. The lesions of the cranial vault are usually symmetrical and involve the outer table of the frontal and parietals and less frequently the occipital (Aufderheide and Rodriguez-Martin, 1998). The lesions of the vault and orbits are predominantly found in association with each other and are considered to be of the same aetiology (Stuart-Macadam, 1989a). Cribra orbitalia is found more frequently than the vault lesions and is therefore considered a more sensitive marker of iron-deficiency anaemia than cranial porotic hyperostosis (Aufderheide and Rodriguez-Martin, 1998). The lesions of porotic hyperostosis, including cribra orbitalia, are characterised by an expanded diploic space due to overproduction of haemopoietic tissue and resorption of the outer table of compact bone. In extreme cranial cases subperiosteal new bone forms in a series of radiating striations, seen radiographically and giving the cranium a 'hair-on-end' appearance (Weatherall and Clegg, 1981; Resnick, 1995b; Ortner, 2003).

In genetic, haemolytic and iron-deficiency anaemia, the severity of skeletal changes varies with the age of the sufferer (Caffey, 1957). This age variation is because in children the medullary cavities of all bones contain haemopoietic marrow. Because of this, in early childhood both the axial and appendicular skeleton may be affected by hyperplastic activity, but as the individual reaches puberty, haemopoietic activity recedes from the peripheral bones and involvement of the appendicular skeleton diminishes (Resnick, 1995b). Therefore, in children, all bones may be affected by hyperplastic activity but in adults only the bones retaining haemopoietic activity (primarily the skull, pelvis and vertebrae) may be affected (Caffey, 1957; Ortner, 2003).

In thalassaemia, marrow hyperplasia may also affect the facial region of the skull. During infancy and early childhood the paranasal sinuses may be obliterated by osseous expansion of the nasal, maxillary, frontal and temporal bones (Resnick, 1995b). Suppression of pneumatization of the sinuses can occur and this is most severe in the maxillary sinuses (Caffey, 1957). As a result, recurrent sinusitis is a typical complication of inadequate sinus drainage (Weatherall and Clegg, 1981). The external dimensions of the zygomae may also expand lending the individual a 'mongoloid' or swollen cheek appearance of the face (Ortner, 2003). The posterior aspects of the ribs can develop multiple medullary expansions, and thinned cortices and osteoporosis that may persist into adulthood. Furthermore, changes to the ribs can exhibit localised radiolucent lesions, a 'rib-within-a-rib' appearance on radiographs, and osteomas.

Marrow hyperplasia can affect any bone of the body, depending on the age of the individual. In thalassaemia, the effects are most commonly seen in the hands and feet (Middlemiss and Raper, 1966; Resnick, 1995b). In the hands of subadults an interference with the normal moulding process of bone growth can lend the individual bones a 'swollen' appearance. However, with the normal regression of haemopoietic marrow from the peripheral skeleton, these changes may disappear by puberty (Caffey, 1957). The nutrient foramina of the hand phalanges will also become enlarged. This enlargement of the nutrient foramina may be due to the general hypervascularity of the thalassaemic condition (Middlemiss and Raper, 1966; Resnick, 1995b). Once the nutrient foramina have become enlarged, they do not remodel. Therefore, this may be an indicator in adults of a childhood crisis period with thalassaemia. Widened vascular channels of the cranium have also been observed in thalassaemic individuals and are usually associated with changes to the nutrient foramina and porotic hyperostosis of the calvariae (Resnick, 1995a).

Also in thalassaemia, affected appendicular long bones often reveal a widening of the medullary cavity with a resultant loss of normal diaphyseal concavity. The cortex becomes thinned and radiographically the trabeculae appear thickened. An apparent widening of the epiphyseal and metaphyseal regions of the affected long bone has been likened to an Erlenmeyer flask (Resnick, 1995a). The loss of diaphyseal concavity that inspired such a name is the result of inhibited remodelling of the distal metaphyses and is most common in the distal femur (Ortner, 2003).

5.3.3 Growth disturbances in thalassaemia

As well as the 'Erlenmeyer flask' appearance of long bones, Harris lines have been observed (Resnick, 1995b). Premature fusion of the epiphyses can also occur in thalassaemia major. It is most common in the proximal extremities of long bones yet the underlying cause is unknown. The bone most commonly affected by premature fusion is the humerus and is characterised by a medio-inferior tilting of the head (Ortner, 2003). The epiphysis will tilt towards the fusion site with advancing age (Exarchou et al., 1984). The consequence of premature epiphyseal fusion is shortening of the affected bone and structural deformity (Resnick, 1995a).

Skeletal growth can also be locally accelerated. Most typically, hyperplasia of the maxillary region can cause malocclusion of the dentition and lateral displacement of the orbits (Resnick, 1995a). Normal dental calcification and eruption can also be disrupted (Laor et al., 1982; Weatherall and Clegg, 1981; Ortner, 2003). Generally skeletal growth will be retarded and result in a short stature and an infantilistic dwarfed skeleton (Caffey, 1957; Ortner, 2003). Retarded skeletal growth is most apparent between the ages of 8 and 10 years. Sluggish growth or complete growth failure during the adolescent growth spurt is also common. The reason for the apparent retarded growth in thalassaemic children is not known, yet it is most likely a symptom of general anaemia (Weatherall and Clegg, 1981). A recent study has found that the short stature of thalassaemia sufferers is the result of a shorter truncal region with normal limb growth. This is probably due to the delayed puberty associated with homozygous β-thalassaemia where normal endocrine function during puberty is reflected in an increase in trunk length (Rodda et al., 1995).

5.3.4 Fractures and joint abnormalities

Pathological fractures and joint abnormalities are not uncommon in thalassaemic individuals. It seems the osteoporosis associated with thalassaemia may predispose the bone tissue to spontaneous fracture, particularly of the weight bearing limb bones (Exarchou et al., 1984). Fractures can also occur in the forearm and the vertebrae. Healing of the fractures will be slow and usually result in angulation and shortening of the extremities (Resnick, 1995b). Gouty arthritis has been recorded in thalassaemic patients, however, the development of this may be related to prolonged blood transfusions and therefore may not be relevant when exploring prehistoric evidence of thalassaemia. Interestingly, a relationship between thalassaemia minor and articular abnormalities have been reported. These include septic arthritis associated with osteomyelitis and rheumatoid arthritis. This association may prove useful for exploring the aetiology of high frequencies of joint abnormalities in prehistoric populations where thalassaemia was known to occur. The reasons for why these conditions might occur more frequently in thalassaemic individuals is not clear (Resnick, 1995b).

5.3.5 The palaeopathology of thalassaemia

The problems with differentiating between the skeletal changes of the different anaemias will be discussed below. However, some researchers have presented accounts of skeletal changes they believe to be the result of specifically thalassaemia.

Hershkovitz et al. (1991) presents a convincing argument for skeletal changes consistent with β-thalassaemia major in an adolescent male from a Holocene village site off the coast of Israel. The primary evidence for thalassaemic changes in this individual is a left humerus exhibiting changes consistent with published accounts of premature fusion of the proximal epiphysis (Resnick, 1995b). Premature fusion of the proximal epiphysis has caused the shaft to be significantly shortened in relation to the right humerus. The affected humeral head is tilted medially and inferiorally inclined. There are other individuals from this site that present anaemic-type changes to the cranium but with less severe growth disturbances of the long bones. However, these other individuals do lend support to a diagnosis of thalassaemia (Hershkovitz et al., 1991).

Angel (1964) also suggests an aetiology of thalassaemia for anaemic changes to the crania and long bones of infants and children from various Bronze Age Greek sites. Among the pathologies observed are '...a peculiar inner shell, like a sequestrum, surrounded by trabeculae, and representing the non-resorbed shaft at an earlier age (Angel, 1964, 369)'. This description is reminiscent of the 'bone-within-a-bone' of sickle-cell anaemia and thalassaemia major (Resnick, 1995b). The craniofacial changes described in this report, suggest the anaemia of thalassaemia rather than other haemolytic anaemias.

One study of 4,000 year old skeletal material from the site of Khok Phanom Di (KPD) on the central coast of Thailand, presents a convincing case for thalassaemia major as the cause of severe anaemia in several subadult individuals (Tayles, 1996). Of the 13 children over one year old at KPD, 77 per cent ($n = 10$) were found to have the classic thickening and porosity of the orbital roof. Also, three infants with observable craniofacial bones exhibit hypertrophy relative to other infants of the sample. These same infants have thinned cortices of the long bones, and hypertrophy of the hands and feet. A young adult male has one humerus shorter than the other, with a medio-inferior angle to the head. These changes are suggestive of premature fusion of the proximal epiphysis. Although not as severe, this specimen is strikingly similar to the individual described by Hershkovitz (1991). It was concluded that given the bountiful food resources and high malaria endemicity of this region that the anaemic changes were the result of thalassaemia and not iron-deficiency anaemia (Tayles, 1996).

Although thalassaemia major is a significant health problem

in some western Pacific islands there have been few published reports that present convincing evidence of skeletal changes due to the anaemia of thalassaemia. The skeletal sample of Nebira from the south coast of NG exhibit a high prevalence of porotic hyperostosis and cribra orbitalia. This high prevalence was suggested to be the result of thalassaemic anaemia (Pietrusewsky, 1976).

However, no evidence of craniofacial growth disturbances or premature epiphyseal fusion were reported in this sample. It is evident that the paucity of palaeopathological reports on skeletal changes of thalassaemia do not aid the researcher attempting to differentiate between anaemic changes in skeletal material.

5.3.6 Differential diagnosis of genetic and iron-deficiency anaemia

As mentioned above, the underlying mechanisms for marrow hyperplasia in thalassaemia and iron-deficiency anaemia may be different, however the skeletal response is similar. The bony changes of thalassaemia have been reviewed in detail above. It is generally agreed that the vault and orbital changes associated with thalassaemia are also found in iron-deficiency anaemia (Ortner, 2003). However, the growth disturbances and hypertrophy of the facial bones are considered specific to genetic anaemia. Similarly, the postcranial changes described in relation to thalassaemia are not seen in iron-deficiency anaemia (Ortner, 2003).

It should be noted that there is no association between the severity of anaemia and the skeletal changes. Some individuals may have severe anaemia but no skeletal response, while others with seemingly mild anaemia may have dramatic bony changes (Stuart-Macadam, 1987).

In summary, it may prove difficult to differentiate the anaemias based on dry bone alone. It has been shown that while β-thalassaemia is present in the Pacific Islands, the clinical forms of α-thalassaemia are rare, and that a number of factors may predispose young children to the development of iron-deficiency anaemia. Therefore, any differential diagnosis of a skeletal response to anaemia in these samples must consider both conditions.

5.3.7 Metabolic diseases other than iron-deficiency anaemia that may produce skeletal changes in the crania.

It has been argued that scurvy may have contributed to skeletal pathology in subadults from Tonga (Buckley, 2000b). Therefore, the pathogenesis of skeletal involvement in scurvy is considered as a possible contributor to the non-diploic hyperplastic orbital lesions and skeletal pathology observed in both the samples used for this thesis. The presence of scurvy is more likely to be due to an association with other infectious and nutritional diseases. It is therefore included in this chapter rather than included in the chapter on infectious disease.

Scurvy is a condition caused by prolonged Vitamin C deficiency. Scurvy usually develops between the ages of six months and 1.5 years of age (Jelliffe, 1970). This age range is affected because neonates are born with sufficient ascorbic acid stores to protect them from developing scurvy for up to six months. Therefore, it is very unusual to find cases of infants under six months with this disease (Jelliffe, 1970; Krugman and Katz, 1981). Vitamin C is essential for maintenance of an adequate immune response to infection, the absorption of calcium, and also for the production of collagen. A severe deficiency of this vitamin is usually the result of a poor diet, or from a reliance on a limited range of resources (Stuart-Macadam, 1989b).

The bone lesions of scurvy are the result of reduced collagen formation. Vitamin C is also important in the formation of the cement substance which binds the endothelium of blood vessels. A deficiency in this material can cause increased susceptibility to haemorrhage resulting from normal actions such as chewing or movement of the eyes. Mild trauma in a scorbutic individual can therefore induce relatively severe haemorrhage. In particular, a subperiosteal haemorrhage can lift the periosteum away from the bone and new bone formation can be stimulated (Ortner and Ericksen, 1997). Another effect of haemorrhage is the spread of blood into surrounding tissues (extravasation) following trauma. The vascular response to this abnormal occurrence is a proliferation of capillaries at the site. The surrounding bone tissue will become increasingly porous in response to the increased vascularity Haemorrhage can occur in any region where mechanical stress may induce bleeding. Common sites are the alveolar processes of the maxillae, the cranial bosses, the orbital roof, and the temporalis muscle attachments on the vault (Stuart-Macadam, 1989a; Roberts and Manchester, 1995b; Ortner and Ericksen, 1997). Recently, the occurrence of bilateral porous lesions on the outer surface of the greater wing of the sphenoid and adjacent bone has been described as pathognomonic of scurvy (Ortner and Ericksen, 1997). Other bone changes are atrophy of the spongiosa, diminution of the cortex, and 'shell like' periosteal new bone subsequent to subperiosteal haemorrhage of the limbs. The shell-like periosteal new bone can envelop most of the diaphysis and is characteristic of the healing phase of scurvy (Resnick, 1995a). Bone changes are usually bi-lateral and involve multiple bones (Ortner, 2003).

5.3.8 Methods of recording skeletal lesions associated with anaemia

All available cranial material was examined macroscopically for evidence of marrow hyperplasia. The bones of the cranial vault were examined as a unit separate from the orbital plate of the frontal. For the purposes of this study, porotic hyperostosis is defined as lesions of the cranial vault and cribra orbitalia as lesions of the orbits. The severity and type of orbital lesions was recorded based on the descriptions of Stuart-Macadam (1985) as follows:

1. Light: scattered fine foramina.

2. Medium: large and small isolated foramina that have linked to form a trabecular structure.
3. Severe: outgrowth in trabecular structure from the normal contour of the outer bone table

(Stuart-Macadam, 1985, 392)

The criteria for determining whether orbital lesions were active or remodelled was based on the descriptions of the Mensforth et al (1978):

1. Active lesions- these usually exhibit sharp and clearly defined margins in the hyperplastic bone. Active lesions usually exhibit many tiny pores that are able to be observed macroscopically.
2. Remodelled lesions -these typically exhibit a smooth lamellar texture with bone filling in the peripheral pores. There is less diffuse and tiny pitting of the outer table associated with remodelled lesions (Modified from Mensforth, 1978, 23).

Figures 5.4–5.5 illustrate the grading system used in this study. Porotic hyperostosis was recorded on the basis of expanded diploic bone as opposed to new bone deposition on the outer cranial cortex (Ortner, 1992). This was only identifiable if the cranial material was fragmented (Figure 5.6). Porotic hyperostosis was recorded as present or absent and was not recorded as active or remodelled.

It is recognised that recent work suggests that slight porotic lesions of the orbital roof may be associated with scurvy (Ortner et al., 1999). However, lesions resulting from subperiosteal haemorrhage of the orbital roof are able to be differentiated from the hyperplastic lesions of anaemia because new bone is formed over the outer table with a distinct margin, while in anaemia the outer table is penetrated due to diploic expansion. Therefore, the following interpretation of 'slight' lesions will consider both scurvy and anaemia in the differential diagnosis. In this section, the individual is the unit of analysis. Prevalences were calculated only using those individuals with one or both orbits present.

5.4 Results: Porotic Hyperostosis and Cribra Orbitalia

No evidence of appendicular changes due to marrow hyperplasia were observed in either of the samples and no growth disturbances specific to thalassaemia were observed. The only lesions indicative of anaemia were cribra orbitalia and porotic hyperostosis. This was the case in both samples.

5.4.1 Taumako

Porotic hyperostosis and cribra orbitalia were recorded in high frequencies in the Taumako sample. Overall, 36 per cent had porotic hyperostosis and/or orbital lesions (Table 5.1). Among the subadults, a general pattern of increased prevalence with advancing age is shown to around 15 years of age (Figure 5.7). In the adolescents and adults a similar pattern of increased prevalence is evident, however, this is not as steep as in the subadults. It should be noted that the sample sizes are highly variable among the subadults which is probably reflected in the prevalence rates. Chi-square tests showed no statistically significant differences between the age groups.

Table 5.2 shows the severity of orbital lesions in subadults and adults. There is a general decrease with age in the proportion of Slight lesions in the subadults. However, after 15 years of age the proportion of Slight lesions increases to remain constant at around 50 per cent in the adults (Figure 5.8). The proportion of Severe lesions is highest in the 1–5 year olds. Severe lesions are absent in Young adults and observed in only one individual each in the Mid and Old adults.

In the subadults, a high proportion of orbital lesions were active at time of death (Table 5.2 and Figure 5.9). However, the proportion of active lesions decreases throughout childhood. The proportion of remodelled lesions increases steadily from the 1–5 year age group until 20 years. All adults had remodelled lesions at time of death.

The proportion of vault and orbital lesions between the sexes is comparable overall and in the Young and Old adults (Table 5.3 and Figure 5.10). However, a higher proportion of females were affected in the Mid adults. Chi-square analysis showed that this difference was not statistically significant.

5.4.2 Tonga

The Tongan samples are considerably smaller than the Taumako sample. This is particularly so in Mound 2 where many of the adults had no cranial material present. The prevalence of lesions in all age groups is presented in Table 5.4. Overall, 11 per cent of the individuals at Mound 1 and 31 per cent at Mound 2 had cribra orbitalia. At Mound 1, two young children between one and five years of age had Slight orbital lesions and one had Moderate lesions. All were active at the time of death. The one affected adult had Moderate remodelled lesions. At Mound 2, the two children in the 1–5 year age group had Moderate active lesions and the other two affected subadults also had active lesions. The one individual in the 16–20 year age group had Moderate lesions, while the seven year old had Slight lesions. All lesions in the adults at Mound 2 were remodelled at death. Three of these adults had Moderate lesions and one had Slight lesions. The single affected adult at Mound 1 was female and of the four affected adults at Mound 2, three were female and one male. Figure 5.11 illustrates that no subadults under one year or over five years old were affected at Mound 1, while all ages, except infants were affected at Mound 2.

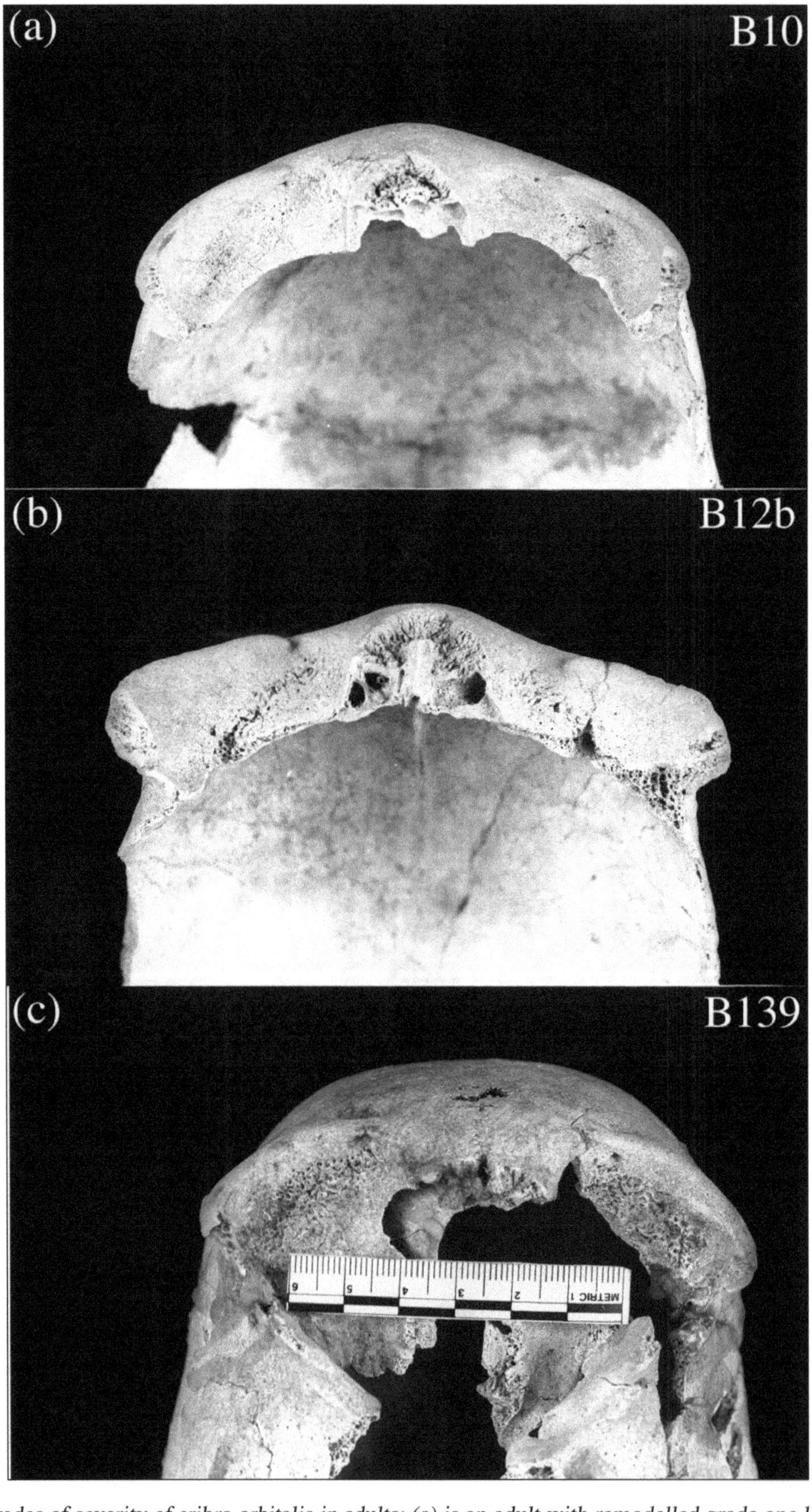

Figure 5.4. *Grades of severity of cribra orbitalia in adults; (a) is an adult with remodelled grade one lesions; (b) is an adult with remodelled grade two lesions; (c) is an adult with remodelled grade three lesions. Photos by author.*

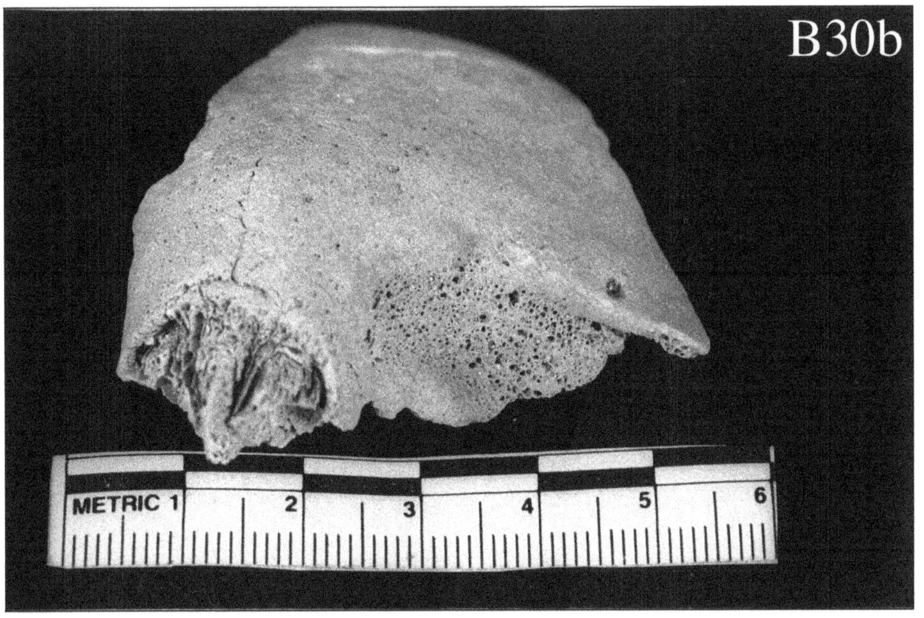

Figure 5.5. *Active grade three lesions of the left orbit of a subadult. Photo by author.*

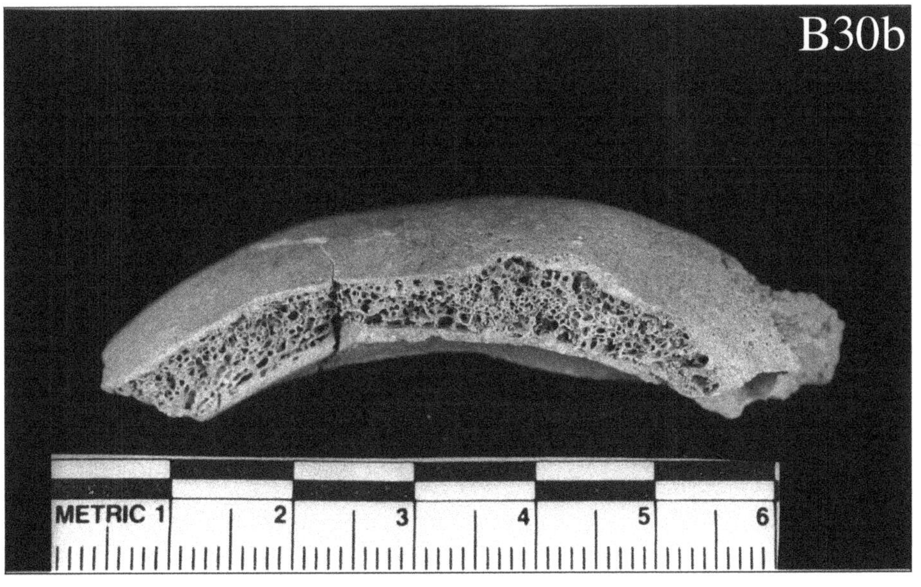

Figure 5.6. *Diploic expansion of the cranial vault in a subadult. Photo by author.*

Table 5.1. *Prevalence of cribra orbitalia and/ or porotic hyperostosis: Taumako*

Age (yrs)	N	Affected	Per cent
Subadult			
Foetal	2	0	0
0–0.9	19	5	26
1–5	27	12	44
6–15	9	6	67
16–20	10	3	30
Subtotal	67	26	39
Adult			
Young	49	14	29
Mid	32	11	34
Old	30	13	43
Subtotal	111	38	34
Total	**178**	**64**	**36**

N = Number of observations; Affected = Number of individuals with cribra orbitalia and/or porotic hyperostosis

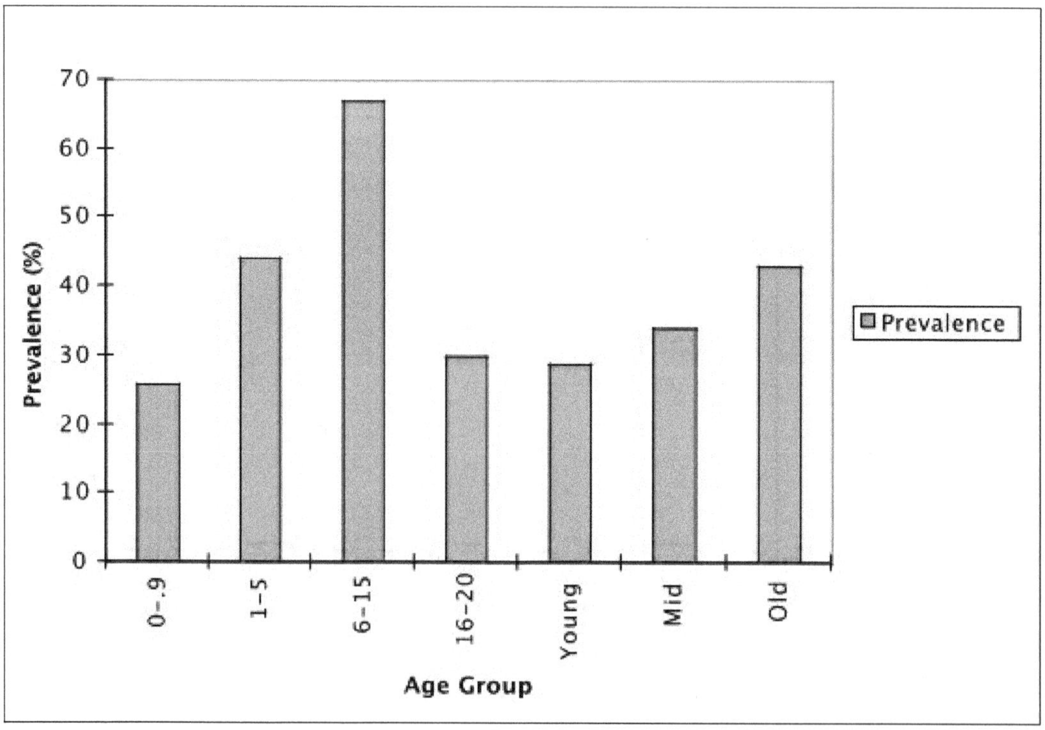

Figure 5.7. *General prevalence of porotic hyperostosis and orbital lesions: Taumako.*

Table 5.2. *Prevalence of cribra orbitalia and/ or porotic hyperostosis: Taumako*

Age (yrs)	CO*	1	Per cent	2	Per cent	3	Per cent	A	Per cent	R	Per cent
Subadult											
Foetal	0										
0–0.9	2	1	50	0	0	1	50	2	100	0	0
1–5	8	2	25	1	13	5	63	7	88	1	13
6–15	6	1	17	3	50	2	33	5	83	1	17
16–20	3	1	33	1	33	1	33	2	67	1	13
Subtotal	19	5	26	5	26	9	47	16	84	3	16
Adult											
Young	13	7	54	6	46	0	0	0	0	13	100
Mid	11	6	55	4	36	1	9	0	0	11	100
Old	11	6	55	4	36	1	9	0	0	11	100
Subtotal	35	19	54	14	40	2	6	0	0	35	100
Total	**54**	**24**	**44**	**19**	**35**	**11**	**20**	**16**	**30**	**38**	**70**

* Only individuals with cribra orbitalia are presented in this table
CO = cribra orbitalia; 1 = slight lesions; 2 = moderate lesions; 3 = severe lesions; A = active lesions; R = remodelled lesions

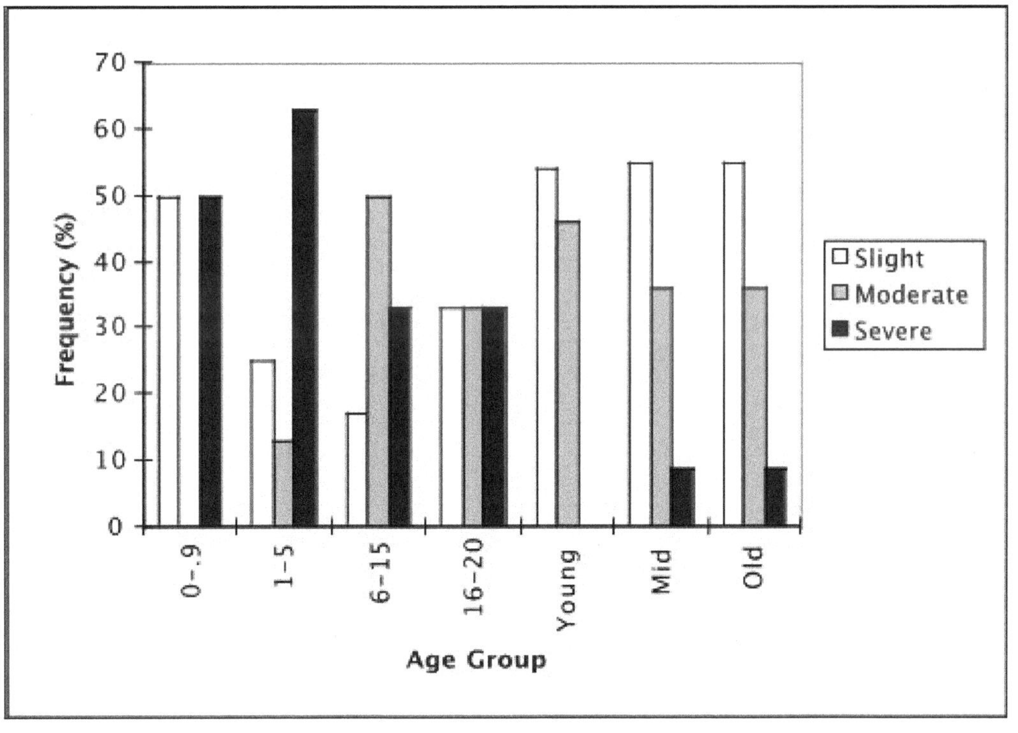

Figure 5.8. *Overall severity of orbital lesions: Taumako.*

Table 5.3. *Prevalence of cribra orbitalia and/ or porotic hyperostosis: Taumako.*

Age	Male	Affected	Per cent	Female	Affected	Per cent
Young	27	8	30	34	9	26
Mid	14	3	21	16	8	50
Old	11	5	45	22	8	36
Total	**52**	**16**	**31**	**72**	**25**	**35**

Affected = number of individuals with lesions

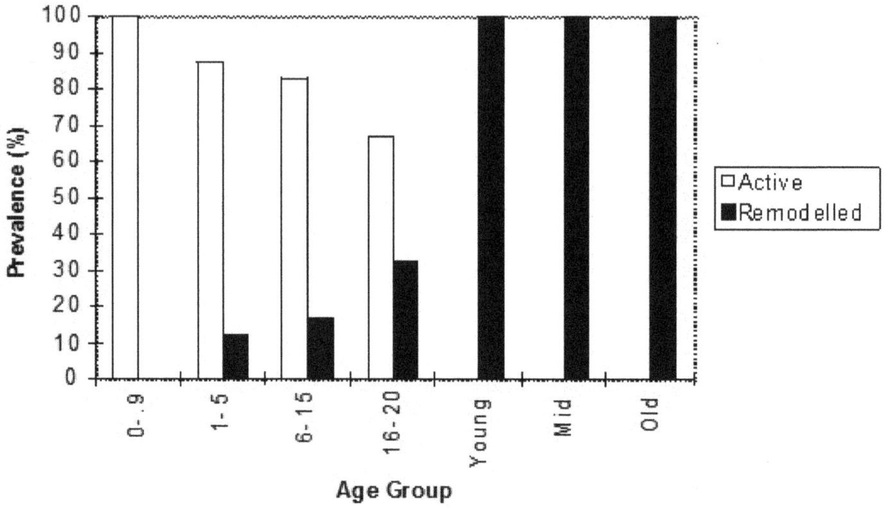

Figure 5.9. *Status of orbital lesions: Taumako.*

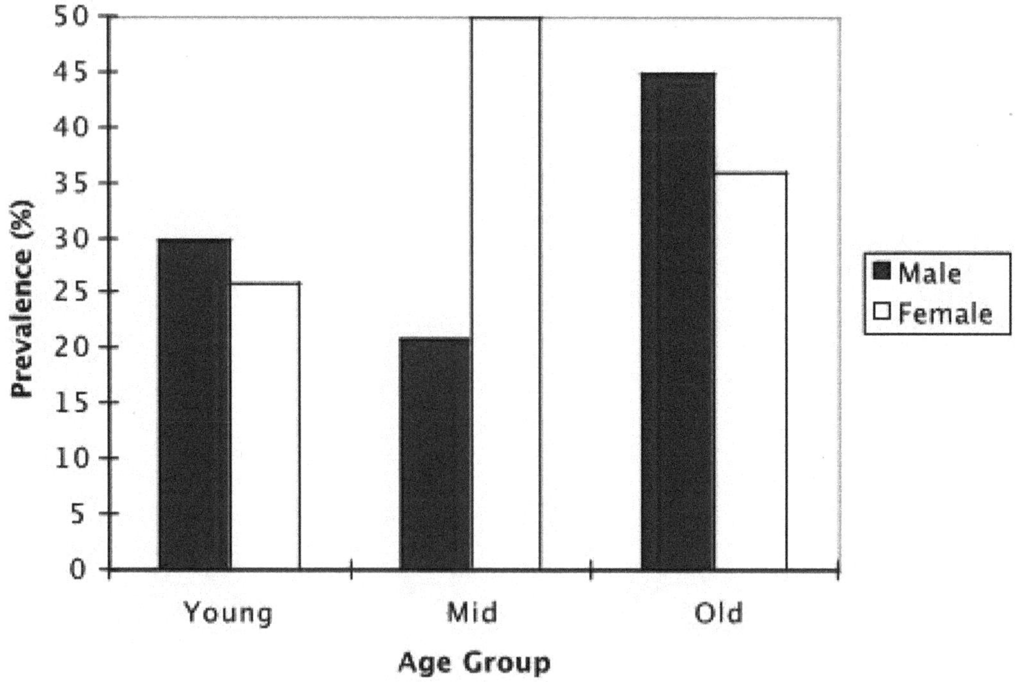

Figure 5.10. *Comparison of proportion of cranial and vault lesions between the sexes: Taumako.*

Table 5.4. *General prevalence of porotic hyperostosis and orbital lesions: Mounds 1 and 2, Tonga.*

Age (yrs)	N	Affected	Per cent
Mound 1			
0–0.9	7	0	-
1–5	10	3	30
6–15	5	0	-
16–20	1	0	-
Subtotal	23	3	13
Adult			
Young	9	1	11
Mid	4	0	-
Old	0	0	-
Subtotal	13	1	8
Total	**36**	**4**	**11**
Mound 2			
0–0.9	2	0	0
1–5	3	2	66
6–15	2	1	50
16–20	2	1	50
Subtotal	9	4	44
Adult			
Young	8	1	13
Mid	4	1	25
Old	5	2	40
Subtotal	17	4	24
Total	**26**	**8**	**31**

N = Number of observations; Affected = Number of individuals with cribra orbitalia and/or porotic hyperostosis

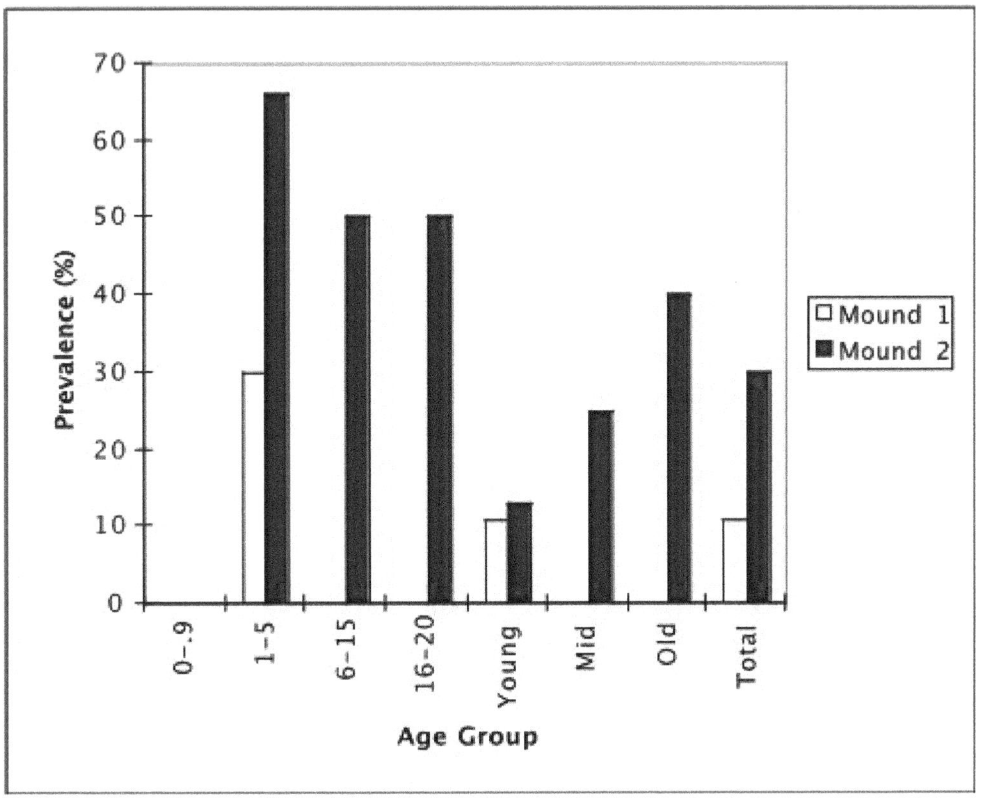

Figure 5.11. *Prevalence of orbital lesions at Mound 1 and 2: Tonga.*

5.5 Results Summary

5.5.1 Taumako

Overall, 36 per cent of the sample had lesions indicative of anaemia. The subadults and adults were affected in similar proportions. Most subadult lesions were active while all adult lesions were remodelled. More severe lesions were observed in the younger subadults than in older children and adults. Finally, Chi-square statistical analysis showed no significant differences between the sexes which indicates that the underlying anaemic condition affected all individuals of the population to the same degree.

5.5.2 Tonga

Mound 1

Overall, 11 per cent of the sample had lesions. More subadults than adults were affected. Only three 1–5 year old subadults were affected while a single young adult had lesions.

Mound 2

Overall, 31 per cent of the sample had lesions. All age groups except infants were affected, although in most cases in only a single individual. Nearly twice as many subadults were affected with lesions than adults.

5.5.3 Comparison of cribra orbitalia and porotic hyperostosis between Taumako and Tonga

There is a significant difference (chi-square *p*-value = 0.0243) between the higher overall prevalence at Taumako and that at Mound 1. The prevalence of cribra orbitalia at Taumako and Mound 2 is similar (Figure 5.12). When the Tongan data were combined, higher overall subadult and adult prevalences were found at Taumako (Table 5.5). Chi-square statistical analysis showed that none of these differences were significant.

At Taumako all age groups were affected, while at Mound 2 all ages except infants were affected. This pattern is considerably different from Mound 1, where only 1–5 year olds and one young adult was affected (Figure 5.13).

No Severe lesions were recorded at either of the Tongan mounds, while quite high frequencies were recorded in the subadults from Taumako. The small samples of the Tongan mounds preclude any worthy comparison of lesion status at death. However, all Tongan subadults had active lesions, and all adults had remodelled lesions. This pattern is different to Taumako where some remodelling of lesions occurred in the 1–5 year age group.

5.6 Conclusions

Part of the third objective of this thesis has been achieved by the analysis carried out in this chapter. Based on the evidence presented above it can be concluded that Taumako had more evidence of childhood anaemia than Tonga. This different pattern of ages affected and greater severity at Taumako may indicate different causal factors in the development of iron-deficiency in this population. It would also seem that the anaemic episodes of the Taumako children were more severe and prolonged than at Tonga. The possible differences in causal factors of iron-deficiency in these two environments and the implications of these results with respect to the third aim of thesis will be considered in chapter seven.

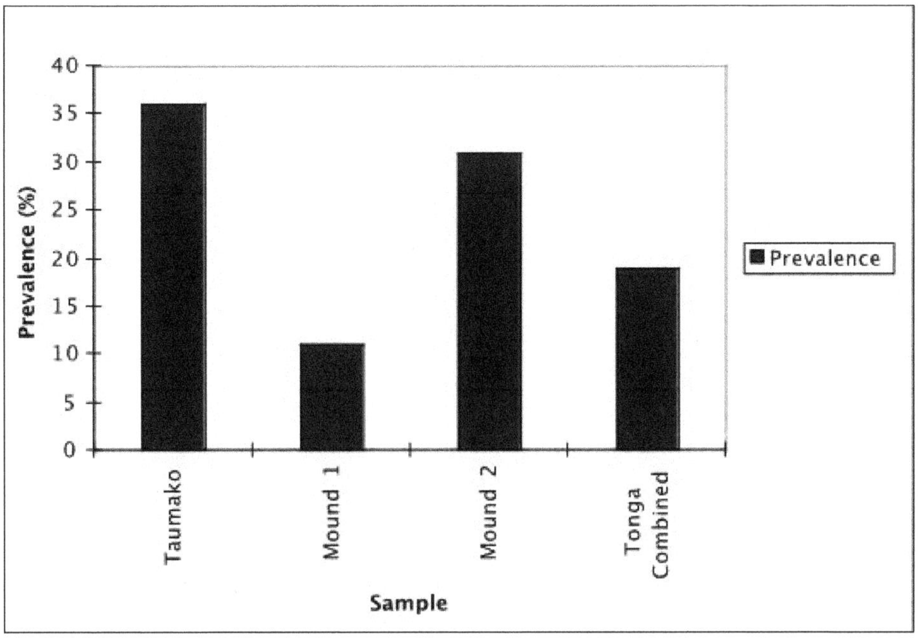

Figure 5.12. *Comparison of the general prevalence of cribra orbitalia between Taumako and Tonga.*

Table 5.5. *Prevalence of cribra orbitalia at Taumako and Tonga.*

Group	Taumako		Tonga combined	
	A/n	Per cent	A/n	Per cent
Overall	64/178	36	12/62	19
Subadults	26/67	39	7/32	22
Adults	38/111	34	5/30	17

A = number of individuals affected; *n* = number of individuals

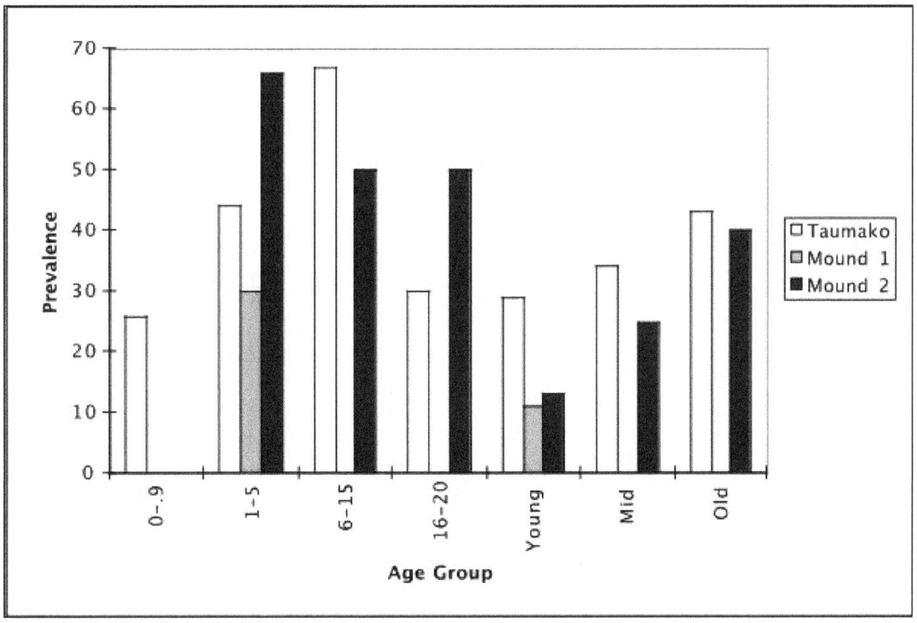

Figure 5.13. *Comparison of age groups affected with cribra orbitalia between Taumako and Tonga.*

Chapter 6

Non-Malarial Infectious Disease

The second part of the third research aim of this thesis is to assess whether the presence of malaria at Taumako exacerbated the impact of infectious disease, reflected in the prevalence of skeletal lesions. In order to address this aim the third objective is to recognise, record, and analyse all evidence of skeletal pathology, other than anaemia.

As outlined in chapter two, bone tissue is subject to continual renewal and resorption. A pathological state is considered present when the balance between renewal and resorption is disrupted and one process overtakes the other (Resnick and Niwayama, 1981). The morphology of subadult bone is primarily the same as adult bone. However, in children, the periosteum is thicker, more vascular and not as firmly attached to the bone as it is in adults. This difference in periosteal anatomy means that insults which might promote production of subperiosteal new bone in subadults may be slighter than those which produce similar changes in adults (Jaffe, 1972). For example, in premature infants incidental trauma during normal handling, such as diaper changing, can stimulate the periosteum to lay down new bone (Carpenter, 1999). Recognising a pathological process in subadult bone, particularly in young children, is more problematic than in adults (Buikstra and Cook, 1980). This is primarily because the bones of a young child are growing more rapidly than the bones of an adult. Therefore an understanding of the normal morphological variation of subadult bones is essential when attempting to identify a pathological process (Silverman, 1985).

The methods for recording and diagnosing pathology in prehistoric human skeletal remains are numerous and varied. As Waldron (1994) states, even with the benefits of a patient history and sophisticated diagnostic technology, modern clinicians still make incorrect diagnoses of bone pathology. Palaeopathologists do not have the benefit of a patient history and must rely upon the bones themselves to tell the story. A main cause of confusion in diagnosing bone pathology is the limited way in which bone can respond to a pathological insult. When a condition or disease is in place, bone tissue can respond either by the production of new bone, the resorption of bone or a combination of the two processes (Ortner, 2003). Therefore, when recording pathological lesions in dry bone, the two processes must be clearly distinguished before they can aid in diagnosis.

6.1 Methods: Skeletal Pathology

6.1.1 Methods of determining the proportion of individuals with pathology

It was decided to estimate the prevalence of bone lesions by using both the *individual* and *skeletal elements* as denominators. This is primarily because the criteria set for the individual analysis of skeletal pathology yielded very small samples, particularly in the Tongan samples. The analysis of pathology by skeletal element is considered less subjective than the individual analysis and more quantitative. The individual analysis gives more of an indication of the pattern of bone involvement within the *individual*, while the analysis by skeletal element provides information of the pattern of bone involvement within a *population*. It is believed that the use of the individual as a denominator can give a more sensitive estimate of the epidemiological impact of a disease on the population, rather than relying solely on bone elements. For these reasons, both methods were employed in this study.

An 'individual' was defined by the presence of certain bone elements that are most frequently affected by systemic disease. Therefore, only when multiple bones from the same 'individual' are available is a diagnosis suggested. As described below, the bones that were selected for the minimum criteria of defining an 'individual' were adapted from published accounts of palaeopathological studies of the diseases included in the differential diagnosis (Steinbock, 1976; Powell, 1988; Resnick, 1995b; Trembly, 1996; Stodder, 1997; Aufderheide and Rodriguez-Martin, 1998).

Based on the criteria of these studies the bone elements selected for a minimum criteria of an 'individual' for this study are: Femur, Tibia, Clavicle, Ulna, and either the Hands or Feet. The femur and the tibia are included in the criteria based on the work of Powell (1988) who followed the conclusions of Cook (1976). In a study of a large sample of prehistoric Amerindians Cook (1976) found a high prevalence of treponematosis. She found that the tibiae were predominantly affected and to a lesser extent all other lower limb bones. Similarly, Stodder (1997) based her diagnosis of yaws in a skeletal sample from Micronesia on the presence of at least one tibia affected by subperiosteal new bone formation, and at least two types of diagnostic lesions for treponemal disease in two other skeletal elements. The presence of affected tibiae have been used by other researchers as a minimum criteria for the diagnosis of treponemal disease in dry bone (Hackett, 1976; Rothschild and Rothschild, 1996). In treponemal disease, lesions of the clavicle, ulna, humerus and hands and/or feet have also been suggested as commonly involved bones, after the tibia, fibula and femur (Steinbock, 1976; Ortner, 2003). Furthermore, bilateral periostitis of the distal third of the tibia and fibula are found in up to 78 per cent of leprosy cases that involved bone (Aufderheide and Rodriguez-Martin, 1998).

The overwhelming presence of neurotrophic lesions of the hands and feet in osseous leprosy (Möller-Christensen, 1967) and frequent involvement of the hands and feet in treponemal disease and tuberculosis of infants and children (Ortner, 2003) necessitated their inclusion in the criteria. In tuberculosis, infection of the diaphysis is rare in adults. However, diaphyseal involvement is more common in children and will affect firstly the tibia, then the ulna, radius, humerus, femur and fibula in order of decreasing importance (Aufderheide and Rodriguez-Martin, 1998).

As mentioned above, children's bones are more readily affected by pathological change than adult bones (Buikstra and Cook, 1980). Therefore, for subadults, the selection criteria for inclusion in the individual analysis was loosened compared to adults. This is because the subadult remains in both samples were fragmentary and in the Taumako sample no complete subadult postcranial skeletons were present. Due to the relative lack of subadult material, it was decided that if a subadult burial included the tibia and any other two postcranial bones they could be included as an individual.

In both subadults and adults, if either the left or right bone was present then the individual was included. This was considered appropriate because the skeletal lesions of the diseases included in the differential diagnosis of this thesis generally affect the bones bilaterally. The exception is tuberculosis, in which bone involvement is usually unilateral. However, limb joint involvement in tuberculosis is relatively infrequent when compared with the destructive lesions of the vertebrae (Aufderheide and Rodriguez-Martin, 1998). Therefore, this was considered to represent a minimal bias. It is recognised that the prevalence of tuberculosis of the appendicular skeleton may be under-stated by using this method of individual selection. Lesions of the vertebrae are considered in a following section.

Those burials defined as individuals were then assessed for the presence of cranial material, either complete (the presence of all cranial elements whether articulated or fragmentary), partial (the presence of some cranial elements, may be missing the crucial element of the frontal), or absent (no cranial material present). In the cranium the presence of the frontal and a partial or complete naso-maxillary region was used as ideal criteria for inclusion in the individual group. However, a burial was retained in the individual group regardless of cranial presence. This is because the diseases considered in the differential diagnosis usually affect the cranial bones only after a considerable time (Steinbock, 1976). A possible exception to this is leprosy. However, it was decided that the appendicular postcranial bones are a more sensitive indicator of disease presence during life.

If the bones representing a burial did not meet the criterion of an individual, they were included in the analysis of lesions by skeletal element. From the Taumako sample, four individuals who qualified for inclusion on the basis of skeletal representation were excluded from the individual analysis. In the adult sample, three had purely resorptive lesions of the hands and/or feet. Similarly, one subadult (Burial 5; 13.5 yrs old) had purely resorptive lesions in the left hip joint. Because the lesions of these individuals may represent a disease of a different aetiology to the larger sample, they were not included further in this analysis. In addition, fifteen individuals had proliferative lesions of the appendicular limb bones with concurrent resorptive lesions of the hands and/or feet. These were included in the individual analysis, however they are also considered as a separate group below.

6.1.2 Methods for recording skeletal lesions

It is known that certain diseases may affect some bones more than others. For example, treponemal disease is known for its predilection for the tibiae (Heathcote et al., 1998). In order to remove the possibility of a bias in initial assumptions of the disease present, the evidence of pathological bone resorption or production was recorded in all available skeletal material following the cautions of Buikstra and Cook (1980). Therefore, the *pattern* of skeletal involvement could be utilised to aid in diagnosis of the pathological condition responsible for producing the lesions (Steinbock, 1976; Ortner, 2003).

The skeletal lesions were identified by macroscopic examination of the bones. The presence of subperiosteal new bone, representing an osteoblastic response to infection or other mechanism, was identified as a hypervascular bone surface indicated by diffuse pitting (Ortner and Ericksen, 1997; Ortner et al., 1999). In most cases a layer of new woven bone could be clearly distinguished from the underlying cortex. A distortion from the normal bone shape was also noted as pathological, even in the absence of hypervascularity (Ortner, 1992). Lesions that develop as the result of osteoclastic action were identified by lytic foci in the cortical bone and/or trabeculae. Active (or unremodelled) and remodelled lesions were coded as such based on the criteria of Mensforth et al. (1978). Generally, active lesions display a fibrous, vascular, porous, and irregular layer of new bone which has a scab-like appearance over the normal smooth cortex. Remodelled lesions exhibit resorption and remodelling of the new bone into mature or lamellar bone. The remodelled bone is usually smooth in appearance and more organised than new woven bone (Mensforth et al., 1978).

Firstly, each burial was assessed to record which bones were present. All bones and bone fragments were examined for pathological changes of bone tissue as described above, including the large paper bags full of fragments in the Taumako sample. Each type of bone change was recorded, and the bone side for paired bones was noted. Initially, I recorded whether the proximal, diaphyseal, distal or all of the appendicular long bones were affected. This detailed method of recording lesions produced an enormous amount of data. The reason for originally recording the presence of lesions on each section of bone was based on the idea of following the pathogenesis of a disease within an individual bone and extrapolating this to the individual. For example, if the distal tibiae were affected by remodelled lesions, while the proximal portion had active lesions, it could be argued that the infection originated in the distal portion of the bone. However, it was decided that analysis of these data would not add enough additional information to aid in differential diagnosis for the current study and a simpler reported technique was eventually adopted here.

Each bone was given a numerical code (30 = present; 40 = absent) and the type of pathological change was also given a numerical code. The numerical code indicates what type of pathological change was present, whether it was primarily the result of osteoblastic activity, osteoclastic activity, or a mixed response (Tables 6.1–6.3 and Figures 6.1–6.7). Radiographs were used to aid in recognition of the type of lesion present.

The osteolytic lesions in Table 6.2 represent the gumma that are pathognomonic of treponemal disease. Gumma are granulomatous lesions specific to tertiary treponemal disease (Ortner, 2003). These graded changes are only ever observed in conjunction with the osteoblastic changes and form in the new bone surrounding the original cortex. They are distinct from the osteolytic changes listed in Table 6.3.

6.1.3 Pattern of individual skeletal involvement

The focus of the analysis of individual pathology is one of differential diagnosis. Therefore, the methods of analysis of the lesions were designed so that recognition of the typical pattern of bone involvement in a certain disease was possible. As outlined above, the tibia is the most commonly affected bone in all the diseases included in the differential diagnosis therefore several questions were asked;

1. How many individuals were affected in multiple bones including the tibia? The involvement of multiple bones indicates a systemic disease.
2. How many individuals were affected in the tibia only? This is expected to reflect those with an earlier manifestation of the same disease as the larger group or it may indicate a non-specific infection.
3. How may individuals are affected in any single bone that is not the tibia? This selection will differentiate those individuals with bone lesions that may be of a different aetiology from the group affected in multiple bones.

6.1.4 Status and severity of individual postcranial lesions

The status and severity of the pathology among the individuals was then assessed. The 'status' of the pathology is defined as whether the lesions were active or remodelled at the time of death. An active lesion is assumed to represent a skeletal response to an existing disease process at the time of death. Remodelled lesions indicate resorption and remodelling of the new bone into mature or lamellar bone and would suggest that the disease had ceased before death (Table 6.4).

The individual's status of infection was assessed as a single unit. So, if several bones were affected and all lesions were active at death, the individual was coded as Status 1. Status 2 was defined as an individual with active lesions, but with remodelling present at the time of death. This may indicate an immune response to the disease, as in treponematosis or tuberculosis, which ordinarily have a latency period. Remodelled lesions indicate a cessation of the disease.

The severity of pathology was then assessed. Severity is based on the assumption that the amount of new bone production may be a reflection of the severity and duration of a disease. The greater the volume of bone produced, the greater length of time the condition has been present. Severity grades 5–8 are defined as the presence of gummatous lesions at death (Table 6.5).

6.1.5 Individual cranial pathology

The presence of lesions in the cranium were recorded. The coding criteria of Hackett (1976) were used for recording classic *caries sicca* lesions of the cranial vault that are pathognomonic of treponemal disease (Figure 6.8). As described in chapter two, *caries sicca* lesions are characterised by mixed resorptive and reparative changes of the cranial bones and are pathognomonic of tertiary treponemal disease. Below are the codes utilised for recording bone changes other than *caries sicca* and in subadults (Table 6.6 and Figures 6.9–6.12). Non-*caries sicca* proliferative changes were also recorded on the crania.

6.1.6 Methods for analysis of skeletal lesions with the skeletal element as the denominator

A further analysis of these samples by using the skeletal element as the denominator was carried out. It should be noted that in this section the interpretation of the prevalence of skeletal lesions relates to the number of elements affected and is in no way meant to be a reflection of patterns of pathology within individuals. For this analysis, all bones present were included. This includes partial elements. Left and right elements were combined to increase the sample sizes. The bones were then divided into age groups. The bone elements of all individuals in the previous analyses were also included in this section.

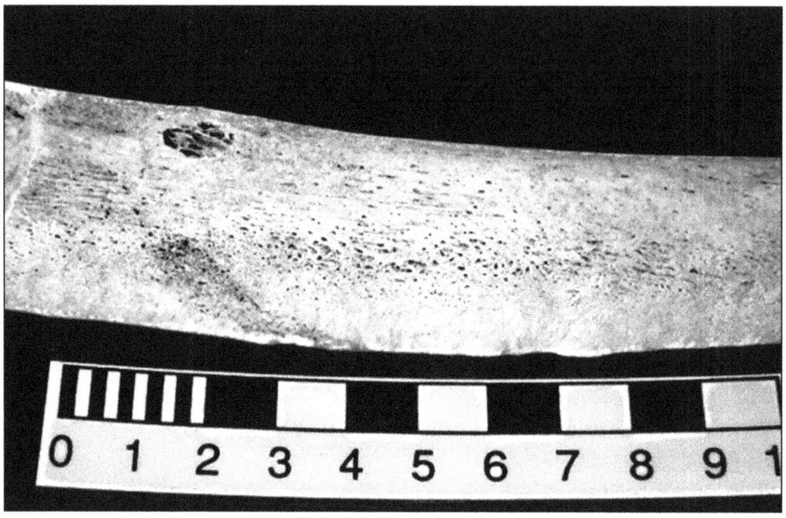

Figure 6.1. *Remodelled grade OB1 of the tibia. This illustrates the abnormal porosity of the cortical bone indicative of hypervascularity and new bone production. The smooth edges of these pores indicate some remodelling has occurred. Photo by author.*

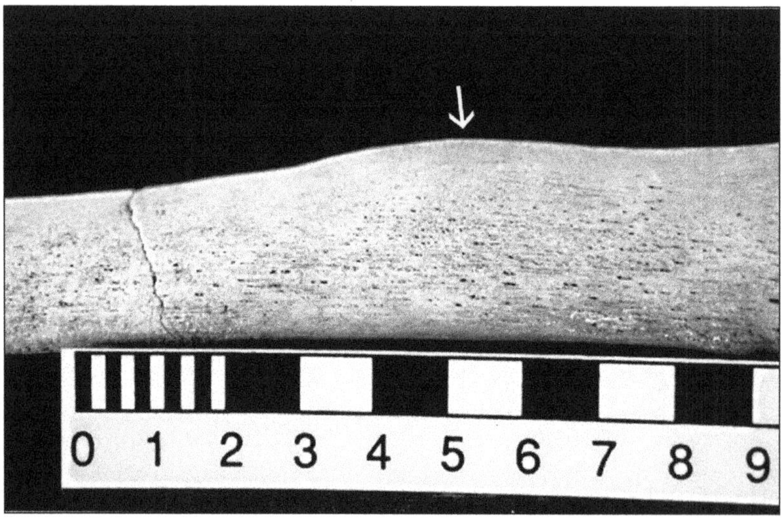

Figure 6.2. *Remodelled grade OB2 of the tibia. The enlarged cortical margin (arrows) indicates production of subperiosteal new bone. The smooth edges of the pores indicates some remodelling has occurred. Photo by author.*

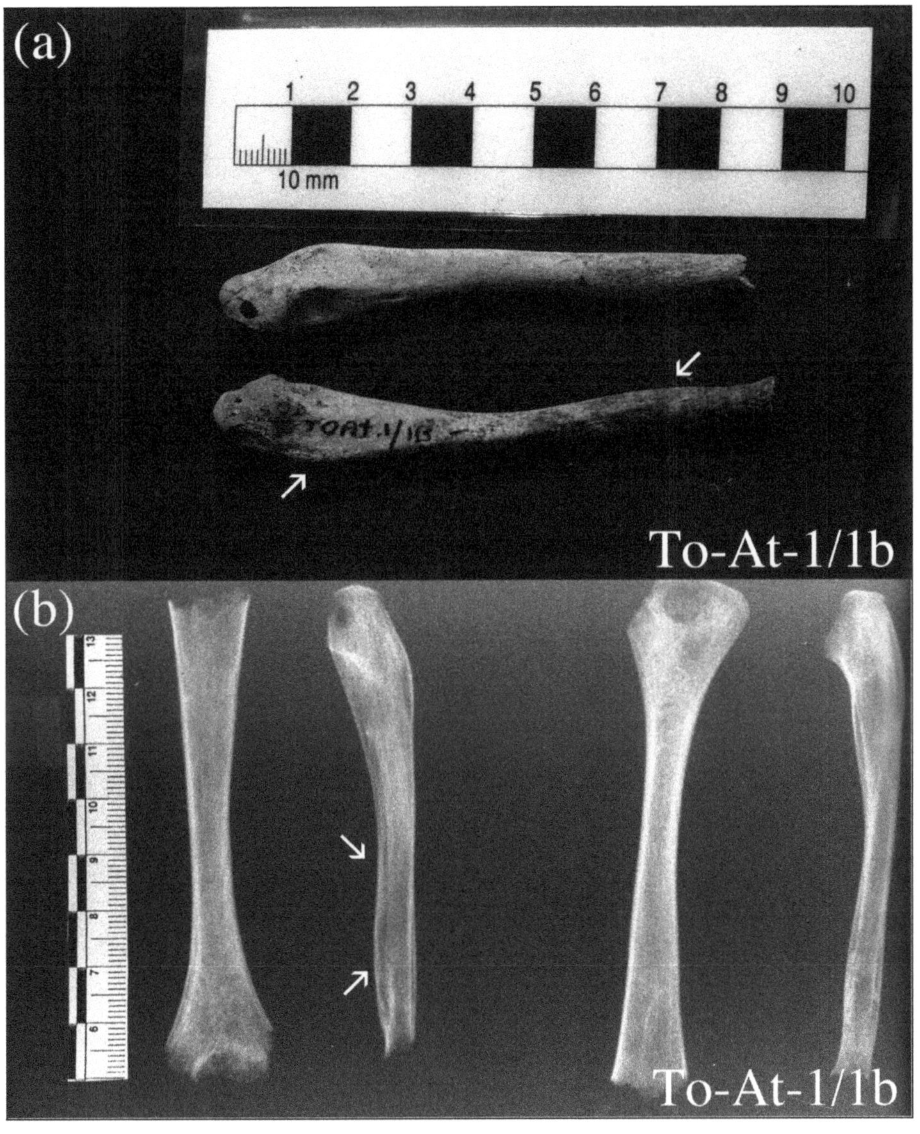

Figure 6.3. *(a) Active grade OB1 of ulnae of a subadult. The subperiosteal new bone has formed on the distal and proximal ends of the bone (arrows). (b) A radiograph of the ulnae and humerus of the same individual. The subperiosteal new bone is seen as a shell encasing the original cortical margin. Images adapted from Buckley (2000b).*

Table 6.1. *Definition of the grades used in osteoblastic pathological lesions.*

Osteoblastic (OB) reactions of bone tissue:

OB1
Defined as the formation of new bone at the subperiosteal level or periostitis. Periostitis can be characterised by an uneven distribution of stratified new bone on the cortical surface of the bone. The new bone is irregular in size and thickness, and its appearance is marked by uneven small or large pores due to hypervascularity of the bone tissue (Ortner, 2003). This grade of bone tissue reaction is differentiated from the more severe grades by the lack of involvement of the underlying cortical bone.
OB2
A diffuse apposition of subperiosteal new bone with incipient signs of cortical involvement characterised by an 'expansion' of the endosteal and/or subperiosteal margins of the bone.
OB3
A gross and diffuse apposition of new bone on the subperiosteal and endosteal margins of the bone, encroaching on the medullary space of the diaphysis. The endosteal and/or subperiosteal, margins of the bone will largely be intact, but some resorption is evident.
OB4
Complete closure of the medullary cavity due to cortical and endosteal expansion of loose cancellous bone tissue (Hackett, 1976). By this stage of the infection, the endosteal and/or subperiosteal margins have been completely resorbed.
Anterior Diaphyseal Bowing
This bony change is a typical osseous manifestation of treponemal infections. It is characterised by gross cortical expansions and periosteal apposition of the anterior aspect of particularly the tibiae. The increase of bone on the middle third of the anterior aspect lends the bone a 'bowed appearance', while actual bowing does usually not occur (Hackett, 1936b; Jones, 1972; Ortner, 2003).
The anterior bowing of long bones will be recorded as present or absent, and will be differentiated from the lateral diaphyseal bowing that may occur as a result of rickets (Ortner, 2003).

Table 6.2. *Definition of the grades used in osteolytic pathological lesions.*

Osteolytic changes in inflammatory conditions (modified from Hackett 1976):

OL1
Superficial cavitation of the new bone on the cortical surface. The opening of the cavity is smaller in relation to the bell-shaped cavity. Best recorded radiographically.
OL2
The opening of the cavity becomes wider, and the cortical margin may begin to be compromised, but *only* in conjunction with osteoblastic activity of the endosteal margin i.e.: OB3. The edges of the cavity are sharp.
OL3
The cavities coalesce to form large openings with the edges beginning to remodel, thus indicating remodelling of the lesion.
OL4
The cavities are remodelling from the inferior surface, so-called 'filling in'. The edges of the cavity are thick and round. The outer margin of the lesion is greater in diameter than the cavity.

Where there is a combination of OB and OL grades, the median grade for each site of the bone was recorded.

Table 6.3. *Resorptive lesions not associated with osteoblastic lesions.*

Primarily Resorptive
A lesion of primarily osteolytic origin. No associated osteoblastic changes.
Resorptive with Osteoblastic
A lesion of primarily osteolytic origin with associated osteoblastic changes.

Table 6.4. *Codes for lesion status.*

Status 1 = Solely active lesions at death
Status 2 = Active with some remodelling at death
Status 3 = Solely remodelled at death

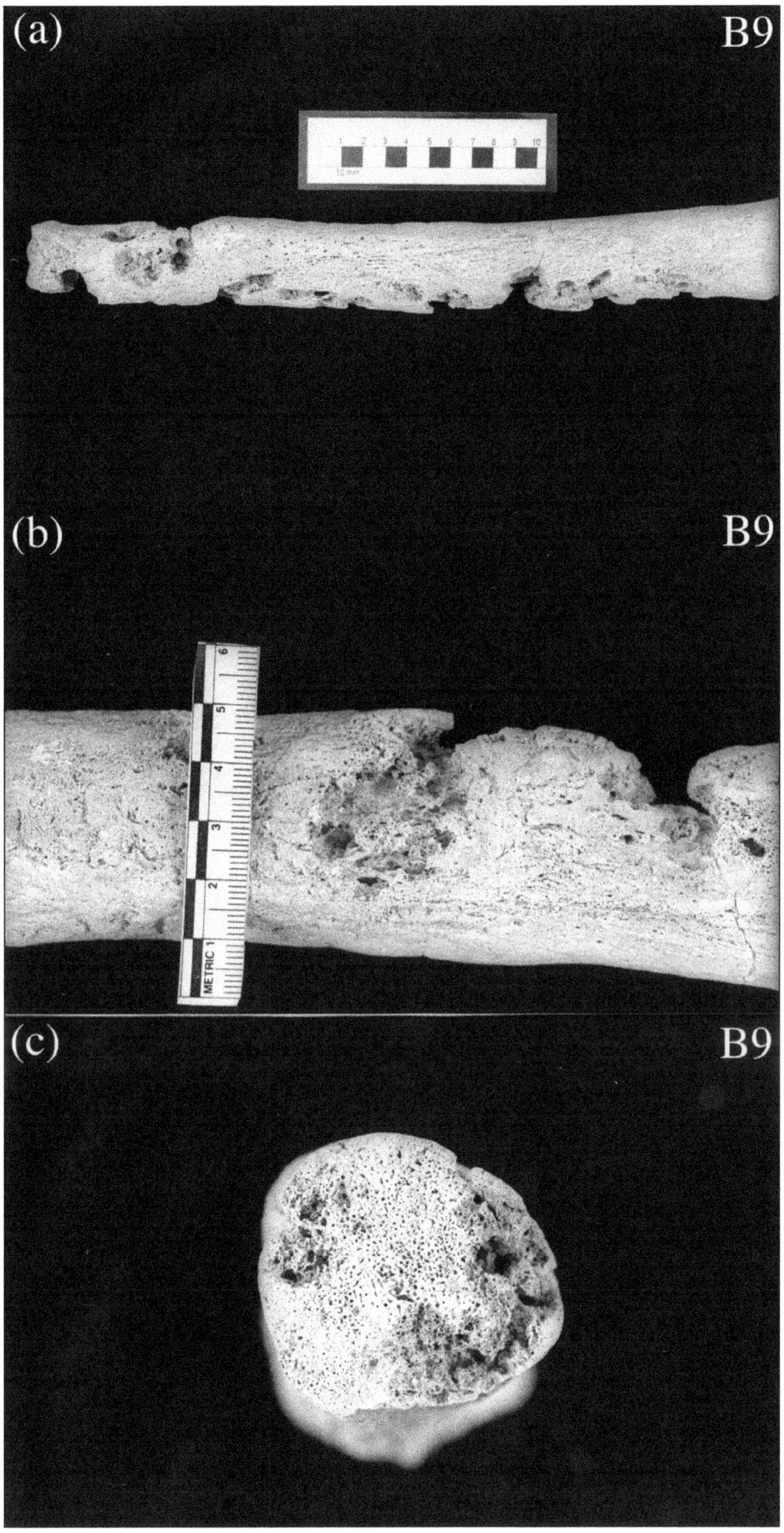

Figure 6.4. *(a) Grade OB4 of the tibia with concurrent gummatous lesions. (b) Detail of gummatous lesions from the same individual as (a). The largest lesion in the center of the image is Grade OL2. (c) Indicates the complete closure of the medullary cavity of grade OB4 changes. Also note the lytic foci indicative of gumma around the outer margins of the bone. B9 from Taumako. Photos by author.*

Table 6.5. *Codes for lesion severity.*

1–4 = Solely new bone production. These grades are synonomous with OB grades 1–4.
5–8 = Gummatous involvement (particular to treponemal disease). These grades are synonomous with the osteolytic grades.

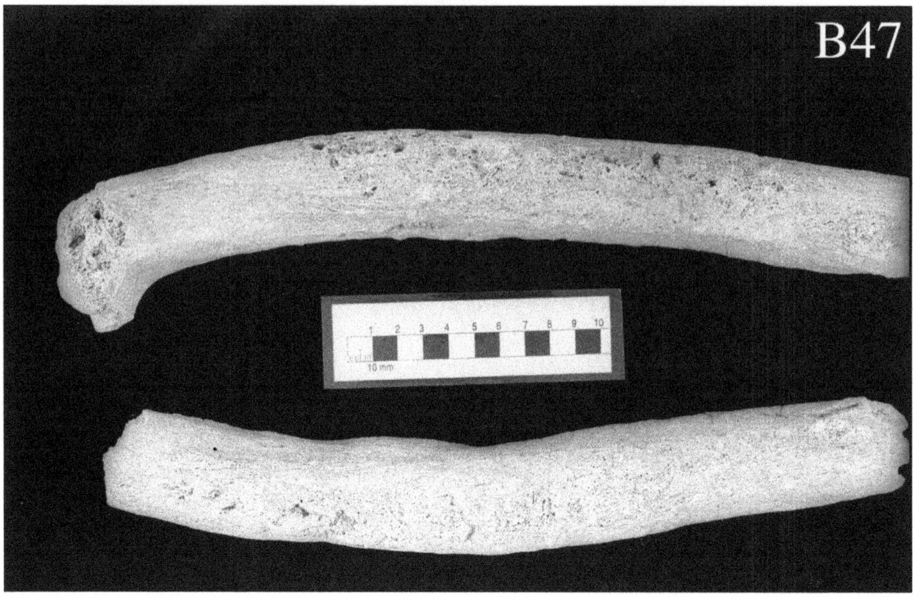

Figure 6.5. *Anterior bowing of the tibial diaphyses. These tibiae also illustrate grade OB3 and grades OL1 and OL2. B47 from Taumako. Photos by author.*

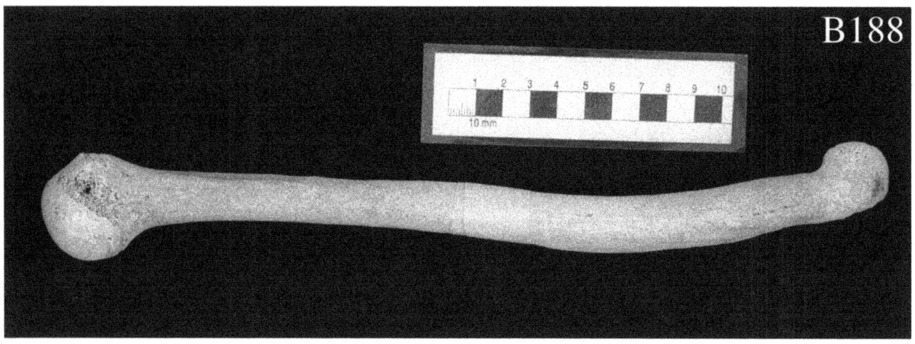

Figure 6.6. *Bowing of the distal humerus. B188 from Taumako. Photo by author.*

Table 6.6. *Codes for recording cranial lesions other than* caries sicca.

ENDOCRANIAL	
FRONTAL = 26	REMODELLED = 29
PARIETAL = 27	BI-LATERAL = 9
OCCIPITAL = 28	
ECTOCRANIAL	
AS WITH POSTCRANIAL CODES FOR OSTEOBLASTIC AND OSTEOCLASTIC CHANGES	
NASAL	
RESORPTIVE = 12	INFERIOR = 14-ACTIVE
PROLIFERATIVE = 13	SUPERIOR = 15-ACTIVE
	REMODELLED = 16
PALATAL	
PROLIFERATIVE = 17	PERFORATION INITIAL = 20
SUPERIOR = 18	PERFORATION COMPLETE = 21
INFERIOR = 19	
ZYGOMAE	
PROLIFERATIVE = 23	
BI-LATERAL = 9	
REMODELLED = 25	

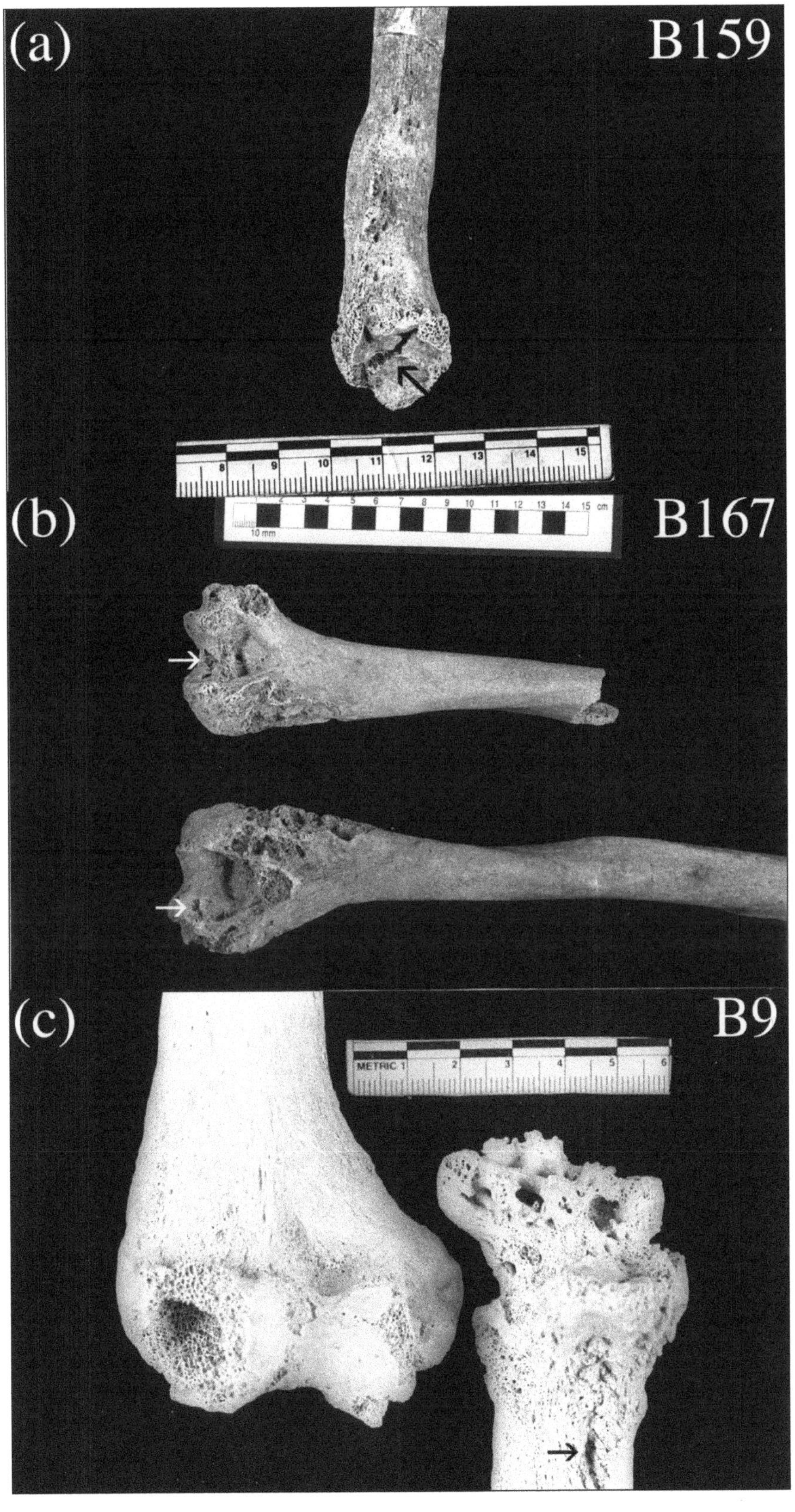

Figure 6.7. *(a) A primarily resorptive lesion of the radiocarpal joint (arrow). (b) Lytic lesion of the elbow joint with associated osteoblastic lesions and gumma (arrow). Note the resorptive lesions associated with the common flexor attachment and extensor attachments. (c) A more advanced form of lesion as shown in (b). the olecranon of the ulna is almost compltely resorbed. The smooth edges of these lesions indicate a chronic condition. Note the remodelled, spiculated new bone associated with this lesion (arrow). Burials from Taumako. Photos by author.*

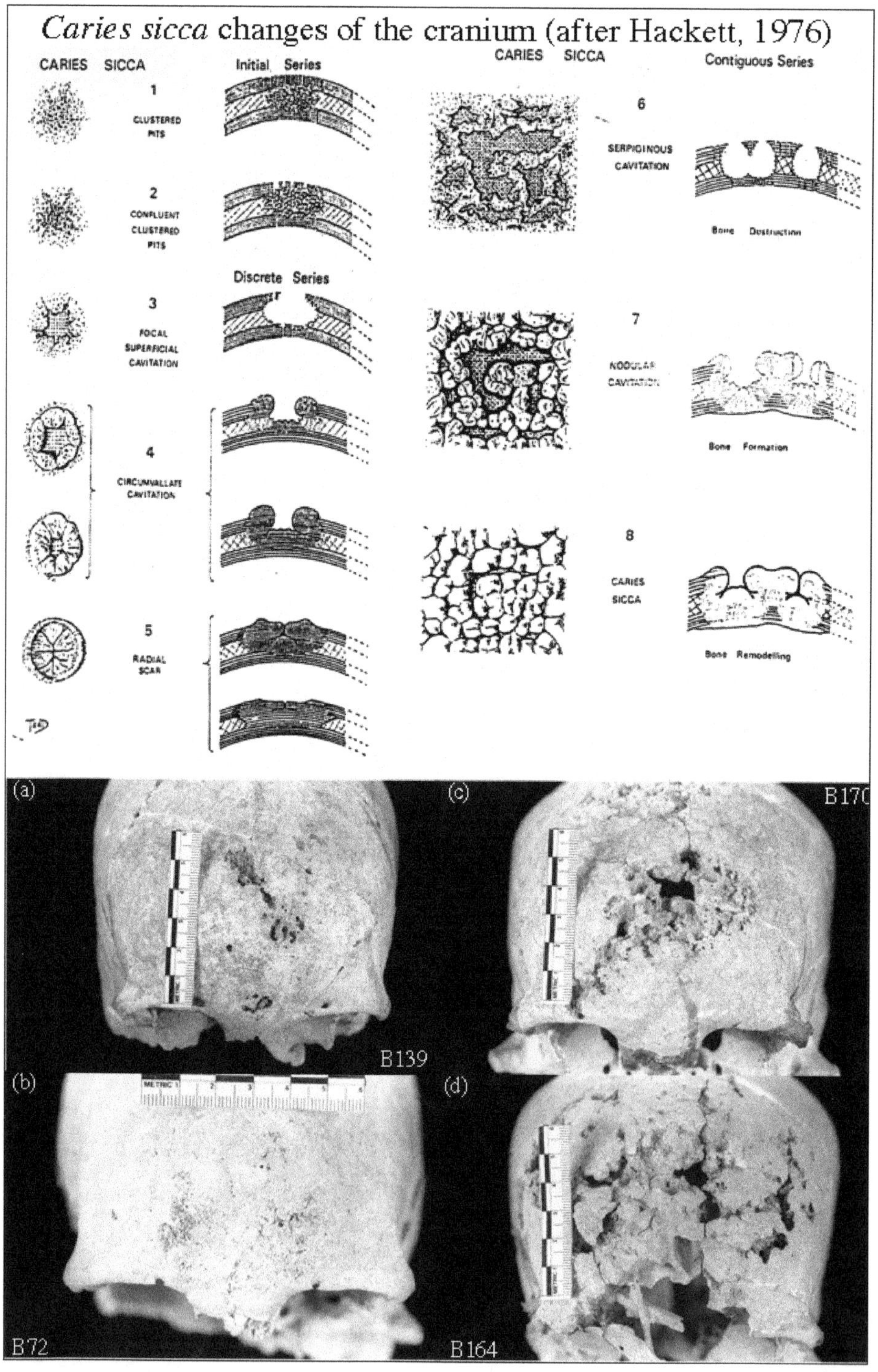

Figure 6.8. *The caries sicca changes of treponemal disease. Initial series: (a) caries sicca grades one through three of the frontal of an 18-year-old female. (b) caries sicca grades four through five of the frontal. Middle-aged female. Contiguous series: (c) caries sicca grades six through seven of the frontal. Old male. (d) caries sicca grades seven through eight of the frontal. Middle-aged male. The initial series of caries sicca grades are characterised by bone destruction and eventual remodelling of the lesions. The contiguous series are lesions which develop over the sites of initial bone destruction with more florid bone production during remodelling of the lesions. Burials from Taumako. Photos by author.*

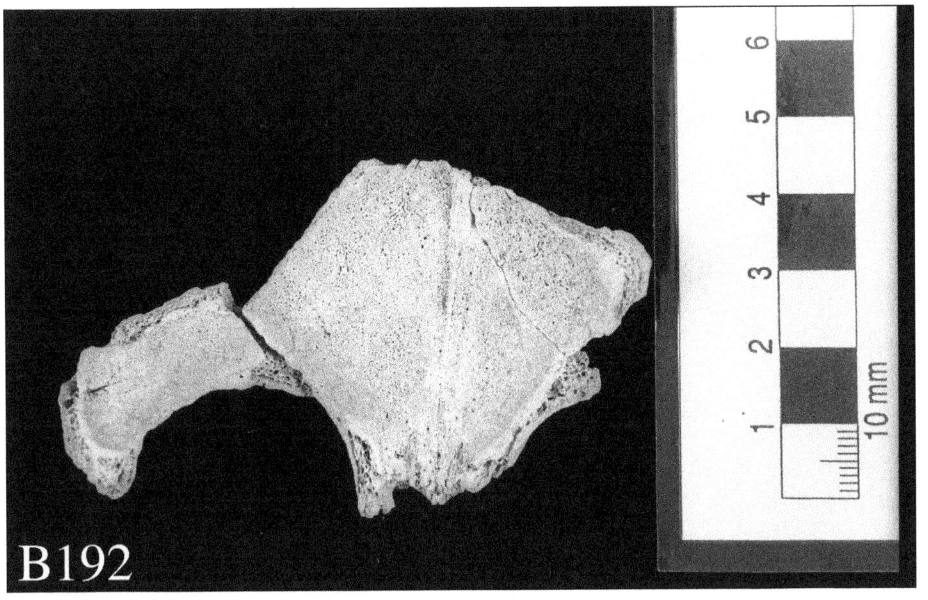

Figure 6.9. *Active endocranial new bone of a subadult. The porosity of the endocranial bone indicates hypervascularity and new bone production. B192 from Taumako. Photo by author.*

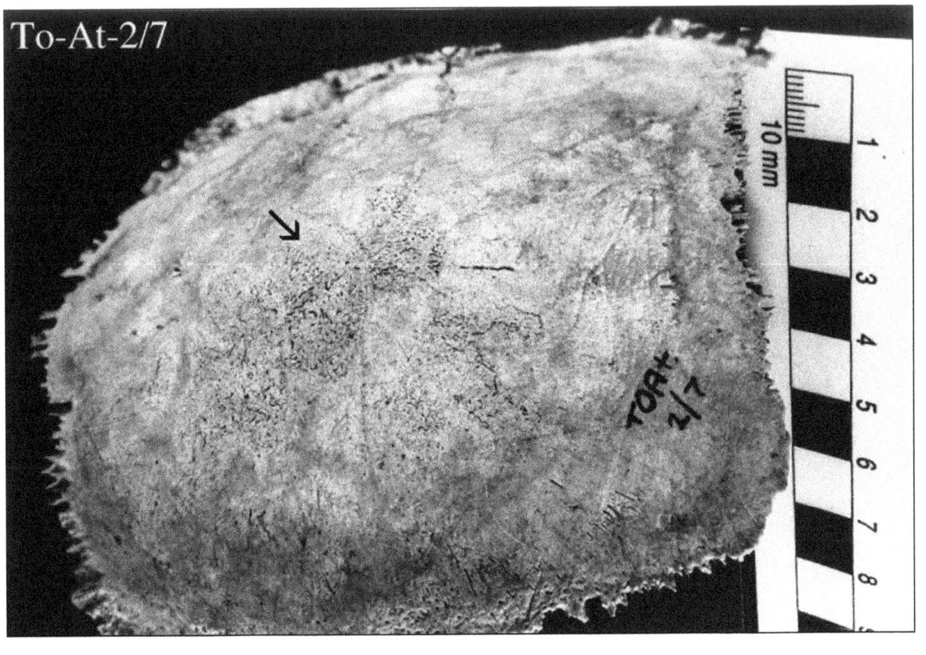

Figure 6.10. *Remodelled endocranial new bone of a subadult (arrow). Remodelled lesions are characterised by smoother surfaces of the new bone and a more vascular appearance than seen in Figure 6.9. These patches of new bone are distinct from the normal appearence of endocranial bone of a young child (arrowhead). Photo by author.*

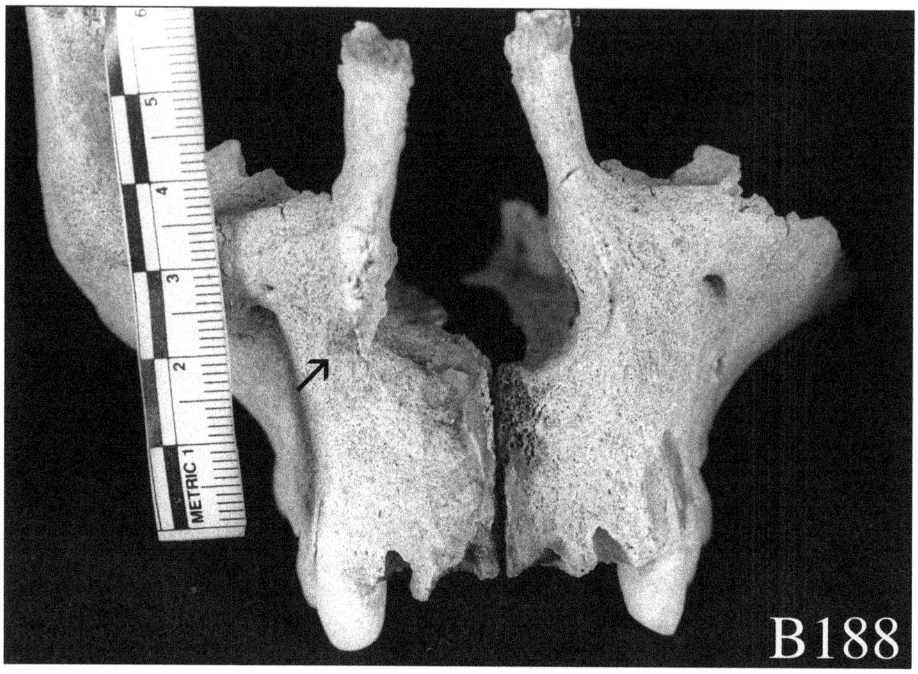

Figure 6.11. *Active proliferative lesions of the nasal region (arrows) anterior view. B188 from Taumako. Photo by author.*

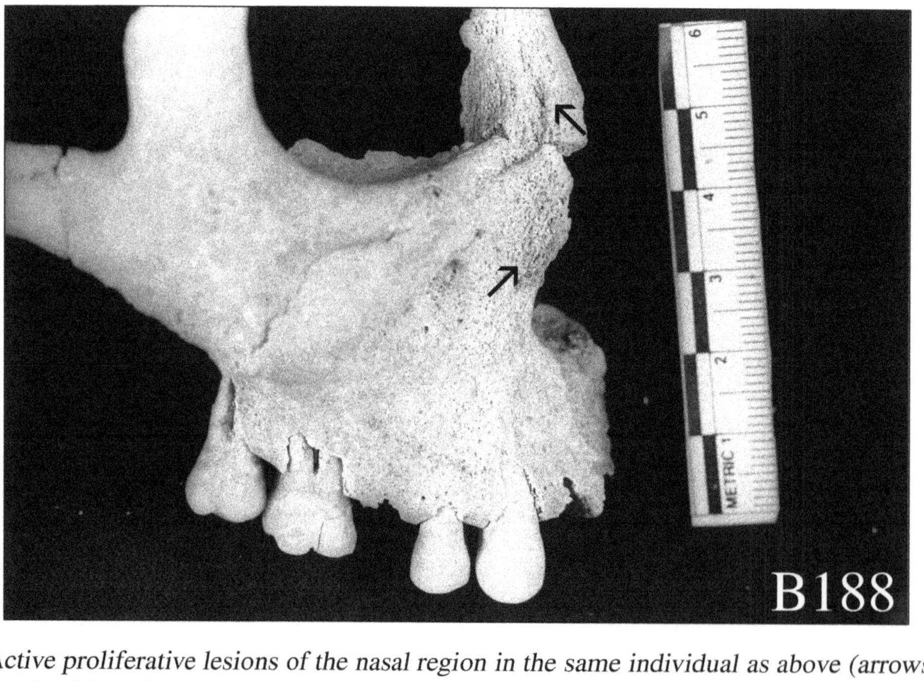

Figure 6.12. *Active proliferative lesions of the nasal region in the same individual as above (arrows) lateral view. B188 from Taumako. Photo by author.*

The status of the lesions were then assessed. The method of assessing the status of lesions was altered slightly from the individual analysis. The lesions were classed into three main groups; active, remodelled or gummatous. This classification of lesions was performed for ease of presentation, and also to increase the sample sizes in each category. The number of bones with evidence of anterior diaphyseal bowing are also presented here. An analysis of any sex differences in skeletal pathology was not performed in this section as this information is provided in the individual analysis.

The cranial lesions were considered separately. The cranial lesions in the adults are presented by the *caries sicca* grade, as described previously. Any lesions that did not fit within the *caries sicca* criteria of Hackett (1976) are presented as a separate group. Subadult cranial lesions are presented as either occurring on the endocranial, ectocranial or facial aspect of the crania. The adult age groups were defined in the same manner as the postcranial pathology.

6.1.7 Methods for consideration of individuals with resorptive lesions of the appendicular joints and lesions of the vertebral column

The majority of the lesions observed in this study were osteoblastic in character. Therefore a detailed quantitative analysis of these lesions is presented. However, a group of individuals with primarily osteolytic lesions of the appendicular, including those excluded from the individual analysis of proliferative lesions in the Taumako sample. The axial skeleton from both samples were also separately analysed. Lesions which were primarily osteolytic were less frequent, therefore they are treated separately.

Firstly, a group of individuals with erosive lesion of the vertebrae are considered. Secondly, those individuals with lytic lesions of the appendicular joints are assessed. These lesions were not associated with the diffuse osteoblastic bone changes previously dealt with. The reason for dealing with the erosive lesions separate from the proliferative lesions is because they may represent a different aetiologic agent to that responsible for the osteoblastic lesions. It may also be that these lesions are part of the milieu of the same disease causing the osteoblastic changes.

This section is more descriptive that quantitative. This is primarily because the vertebral material was largely fragmented and identification of specific elements was difficult in most cases. Some of the individuals with erosive lesions of the appendicular skeleton were not included in the individual analysis. Therefore, the number of skeletal elements available for these individuals vary considerably and negates any exhaustive quantitative analysis. The differential diagnosis of the vertebral and appendicular osteolytic lesions includes; normal variation in vertebral anatomy, vascular abnormality, tumours, osteomyelitis, rheumatoid arthritis, one of the seronegative arthritides, brucellosis, fungal disease, echinococciasis, leprosy, and tuberculosis.

6.1.8 Methods for analysis of vertebral lesions

Individuals with reasonably complete vertebral columns were selected for analysis of the frequency of lytic lesions in the vertebral body and endplates. A reasonably complete vertebral column was defined as: four or more cervical vertebrae; six or more thoracic vertebrae; and two or more lumbar vertebrae.

In order to provide a rigorous description of the lesions Buikstra (1976) argues that three factors must be considered. These are: the age at death of the individual, lesion morphology, location of the lesion on the vertebra and the association of extravertebral osteolytic lesions (Buikstra, 1976). Where possible these recommendations were followed in the analysis.

The location of the lesion on the vertebra was recorded as either on the anterior body or in the superior or inferior endplate of the vertebral body. All lesions recorded were osteolytic in nature and the morphology of the lesions is described in the results section below. The association of extravertebral lesions in the appendicular skeleton and the association of lesions of the ribs were also assessed. The ribs in both samples were extremely fragmented and difficult to identify to anatomical position. Therefore the presence of one or more rib fragment with individuals with vertebral material was utilised in the analysis.

It should be noted that the term 'lesion' is used in relation to the vertebral abnormalities only for want of a better word. It is possible these changes are not pathological but are part of the continuum of normal vascular anatomy of the vertebrae. A differential diagnosis of axial and appendicular osteolytic lesions will be addressed in chapter seven.

6.2 Results: Taumako Skeletal Pathology

6.2.1 Individual analysis

Prevalence of individual pathological lesions

The total sample of burials from Taumako comprised 227 interments that contained human remains. Of the 227 burials, 71 adults and 30 subadults were selected as individuals according to the criteria outlined above.

It is recognised that the sample sizes in the following results are small. This is partly an artefact of the selection criteria for individual analysis, and also the preservation of the sample.

6.2.2 Cranial pathology

Of the individuals, 69 adults and 24 subadults had crania present for analysis. Of this subset, three adults and three subadults were affected in the cranium only, but the rest of the affected individuals with cranial lesions also had lesions in the postcranial skeleton.

The results show there is a dramatic decrease in prevalence of cranial involvement with age in subadults and again after 20 years of age (Table 6.7). Of the subadults, 43 per cent were affected while only 17 per cent of adults had cranial lesions (Figure 6.13). All five infants with cranial material were affected. This difference is statistically significant when compared with all age groups of the adults (chi-square p-value = 0.0474). More young adult individuals than old had lesions of the crania. The highest proportion of affected females was in the 16–20 year age group while no males were affected under 20 years of age (Table 6.7 and Figure 6.14). Chi-square statistical analysis showed there were no significant differences in the proportions of females or males affected between the age groups.

6.2.3 *Type of cranial pathology*

Ten subadults had cranial lesions, six of these were under two years of age. The predominant type of bony change in these infants was new bone formation of the endocranial vault. Other cranial and facial bones were affected with slight new bone deposition or unusual porosity (Table 6.8). Only subadults over 16.5 years of age had lesions that were in accordance with *caries sicca*-type changes.

The Young adult individuals with cranial lesions all had changes of *caries sicca* type. These lesions range from Grade 1 to 5 of the codes listed above (Table 6.9). All affected Mid-aged adults had a similar pattern of lesions to the young individuals (Table 6.10). The old male with cranial lesions had the most advanced *caries sicca* changes of the adult group (Table 6.11).

6.2.4 *Postcranial pathology*

Of the 101 individuals, 53 per cent were affected with lesions of the postcranial appendicular skeleton eight of these had no cranial material. Of the affected individuals, 67 per cent were affected in the postcranial skeleton only. The highest prevalence of postcranial lesions was in the 1–5 year age group (75 per cent), but 60 per cent of infants also had lesions of the postcranial material. Overall, the subadults and adults are similarly affected (Table 6.12). In the adults, individuals who died younger had more lesions (Figure 6.15). In the young and old adults, a higher proportion of females were affected compared to males, with the young females having the highest prevalence of all adults at 69 per cent (Figure 6.16). Chi-square statistical tests showed there were no significant differences between age groups or sex in the prevalence of postcranial lesions.

6.2.5 *Pattern of skeletal involvement*

The majority (69 per cent) of individuals were affected in two or more skeletal elements including the tibia (Table 6.13). More adults than subadults were affected in multiple bones. All males and females between 16 and 20 years of age were affected in multiple bones. Overall, 11 per cent of individuals were affected in the tibia only. This constitutes six individuals, three of whom were under 20 years of age at death (Table 6.14). Nearly twice as many subadults as adults were affected in the tibia only (Table 6.15). In total, 20 per cent of individuals were affected in a skeletal element other than the tibia. Three subadults were affected in more than one bone other than the tibia. Figure 6.17 illustrates the differing patterns of skeletal involvement in these individuals. The difference between the overall frequency of individuals with lesions of a bone other than the tibia and multiple bones was significant (Fisher's exact p-value 0.0030) and the difference between the frequency of individuals with only the tibia affected and those with multiple bones affected was highly significant (FET p-value ≤ 0.0001).

6.2.6 *Status and severity of lesions*

Lesion status

Seventy-five per cent of subadults had lesions which were solely active at death, with some evidence of remodelling in the 1–5 and 16–20 year age groups (Table 6.16). In the adults, the status of lesions was varied throughout the different grades. However, the predominant type of lesion for all age groups was active with some remodelling (Table 6.17).

Lesion severity

The predominant (69 per cent) type of lesion in subadults under 16 years of age was grade 1–2 new bone formation. Gummatous lesions were found in the 16–20 year olds (Table 6.18). Similarly, in the adults 66 per cent of the lesions were grade 1–2, however more adults had lesions of greater severity than the subadults. The highest frequency of gummatous lesions was found in the young adults (Table 6.19).

6.3 Skeletal Element Analysis: Taumako

In this section, individuals aged 15 years and older were included in the young adult group. This was decided on the basis of results from the individual analysis where the pattern of pathology in the 16–20 years olds was similar to the young adults. Therefore, by merging these older subadults with the young adults, the sample size of this age group was increased. Subadults under 15 years of age were merged into a single group, again to increase sample sizes. When cranial pathology was examined, the subadult ages were divided into 0–3.9 years and 4–15 years old. This division was carried out in order to test whether infants and young children were more frequently affected in the crania than older children. All skeletal lesions discussed in the following section were primarily osteoblastic in nature. All types of skeletal elements were affected, in varying degrees.

Table 6.7. *The prevalence of cranial lesions in subadult and adult individuals: Taumako.*

Age (yrs)	Male		Female		Total	
	A/n	**Per cent**	**A/n**	**Per cent**	**A/n**	**Per cent**
Subadults						
0–0.9					5/5	100*
1–5					1/6	17
6–15					1/6	17
16–20	0/2	0	3/5	60	3/7	43
Subtotal	0	0	3	60	10/24	43
Adult						
Young	5/21	24	3/13	23	8/34	24
Mid	0/4	0	3/13	23	3/17	18
Old	1/11	9	0/7	0	1/18	6
Subtotal	6/36	17	6/33	18	12/69	17
Total	**6/38**	**16**	**9/38**	**24**	**22/93**	**24**

A = Number of individuals affected; *n* = number of observations; per cent = percentage of individuals affected; * Statistically significant compared with the Young adults (chi-square *p*-value = 0.0423), Mid adults (chi-square *p*-value = 0.0410), and Old adults (Fisher's exact (FET) *p*-value = 0.0105).

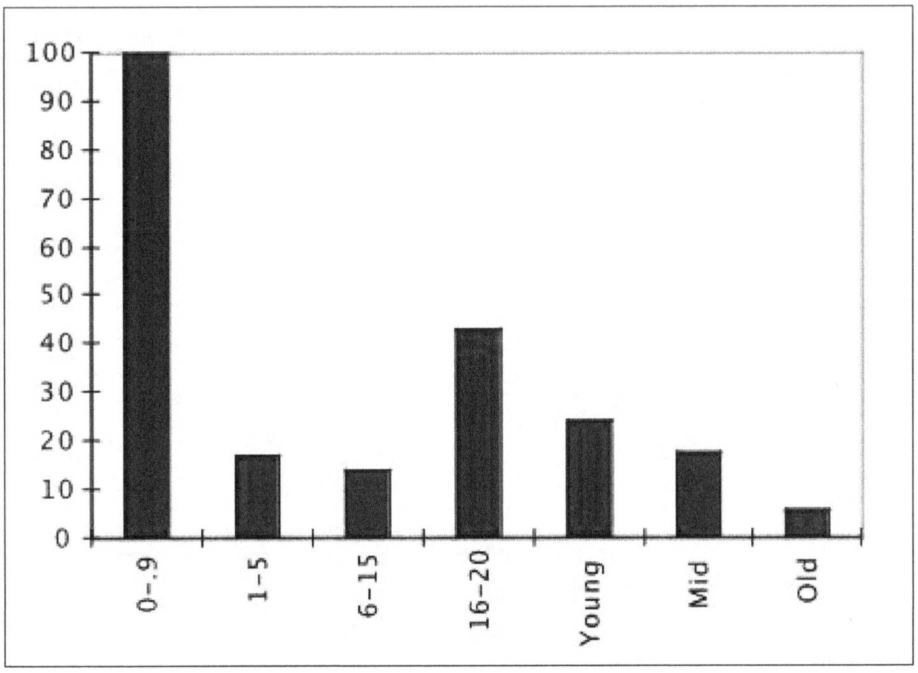

Figure 6.13. *Prevalence of cranial lesions in subadult and adult individuals: Taumako.*

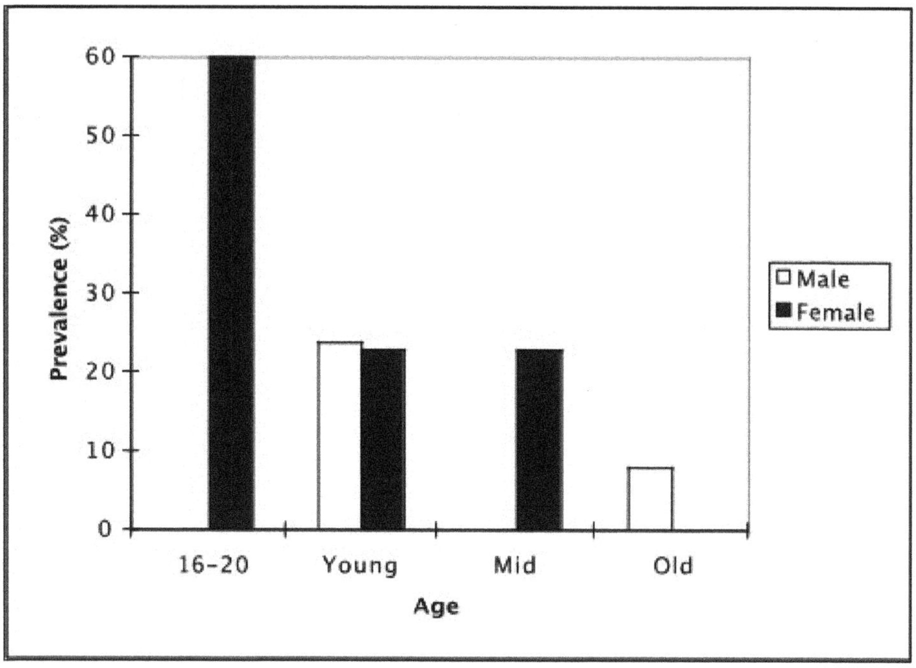

Figure 6.14. *Prevalence of cranial lesions in male and female individuals: Taumako.*

Table 6.8. *The type of cranial pathology in subadult individuals: Taumako.*

Burial	Age	PC	Type of cranial pathology
53.2	0	0	Remodelled endocranial new bone of parietals
125	0	3	Active endocranial new bone of frontal
61	0.25	4	01 of mandible
136	0.75	0	Remodelled endocranial new bone of parietals
124	0.75	3	Active endocranial new bone of the frontal, partietals, and occipital. Some patches of remodelled new bone of parietals. Also ectocranial new bone of zygomae, orbits, temporals and sphenoid
92	2	3	Endocranial new bone of frontal, and remodelled endocranial new bone of parietals and occipital. Ectocranial new bone of orbits, temporals, sphenoid and mandible
165	12	1	Exostosis of the superior palate, porosity of superior palate and resorption of the nasal aperture
137	16.5	0	Remodelled ectocranial new bone of frontal and parietals. Also active new bone of the inferior nasal aperture
139.1	18	7	*Caries sicca* changes grade 1–3 of frontal and grade 2 of parietals. Also, remodelled new bone of inferior nasal bones, and zygomae and active new bone of the superior palate
188	18.5	7	*Caries sicca* changes grade 1 and remodelled new bone of both nasal bones. Also an initial perforation of the palate from the inferior aspect with concurrent active new bone

PC = number of postcranial bones with concurrent proliferative lesions.

Table 6.9. *Type of cranial pathology in Young adult individuals: Taumako.*

Burial	Sex	PC	Type of Pathology
9	M	3	*Caries sicca* grade 3–4: frontal
36.1	M	3	Nasal resorption
47	M	3	*Caries sicca* grade 5: frontal
85	M	0	*Caries sicca* grade 1: frontal
187	M	0	Proliferative of frontal and zygoma
115	F	3	Proliferative of zygoma
121	F	0	*Caries sicca* grade 2–3: frontal
141.1	F	3	Nasal proliferative, zygoma

PC = number of postcranial bones with concurrent proliferative lesions.

Table 6.10. *The type of cranial pathology in Mid adult individuals: Taumako.*

Burial	Sex	PC	Type of Pathology
16.1	F	3	Nasal resorptive, remodelled maxillary sinusitis and palatal perforation
53.1	F	3	*Caries sicca* grade 3–5: frontal, zygoma, nasal proliferative with high degree of alveolar erosion
109	F	3	*Caries sicca* grade 1–2: parietal

PC = number of postcranial bones with concurrent proliferative lesions.

Table 6.11. *The type of cranial pathology in Old adult individuals: Taumako.*

Burial	Sex	PC	Type of Pathology
170	M	3	*Caries sicca* grade 6–8: Frontal, parietal and resorption of nasal

PC = number of postcranial bones with concurrent proliferative lesions.

Table 6.12. *Prevalence of postcranial pathology in subadult and adult individuals: Taumako.*

Age (yrs)	Male		Female		Total	
	A/n	**Per cent**	**A/n**	**Per cent**	**A/n**	**Per cent**
Subadults						
0–0.9					3/5	60
1–5					6/8	75
6–15					4/10	40
16–20	1/2	50	2/5	40	3/7	43
Subtotal	1	50	2	40	16/30	53
Adults						
Young	12/21	57	9/13	69	21/34	62
Mid	2/4	50	7/13	54	9/17	53
Old	4/13	31	4/7	57	8/20	40
Subtotal	18/38	47	20/33	61	38/71	54
Total	**19/40**	**48**	**22/38**	**58**	**54/101**	**53**

A = number of individuals affected;*n* = number of observations; per cent = percentage of individuals affected.

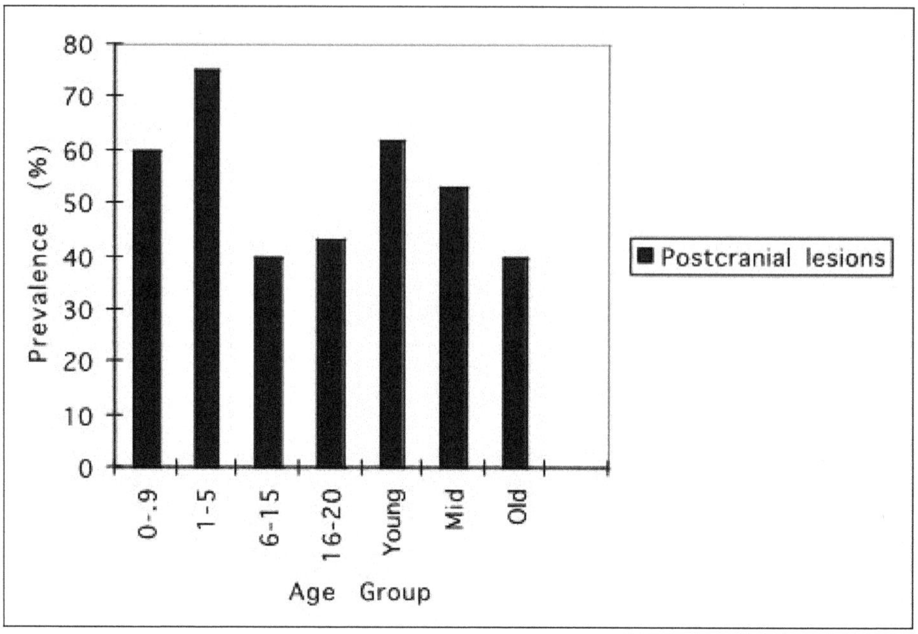

Figure 6.15. *Prevalence of postcranial lesions in subadult and adult individuals: Taumako.*

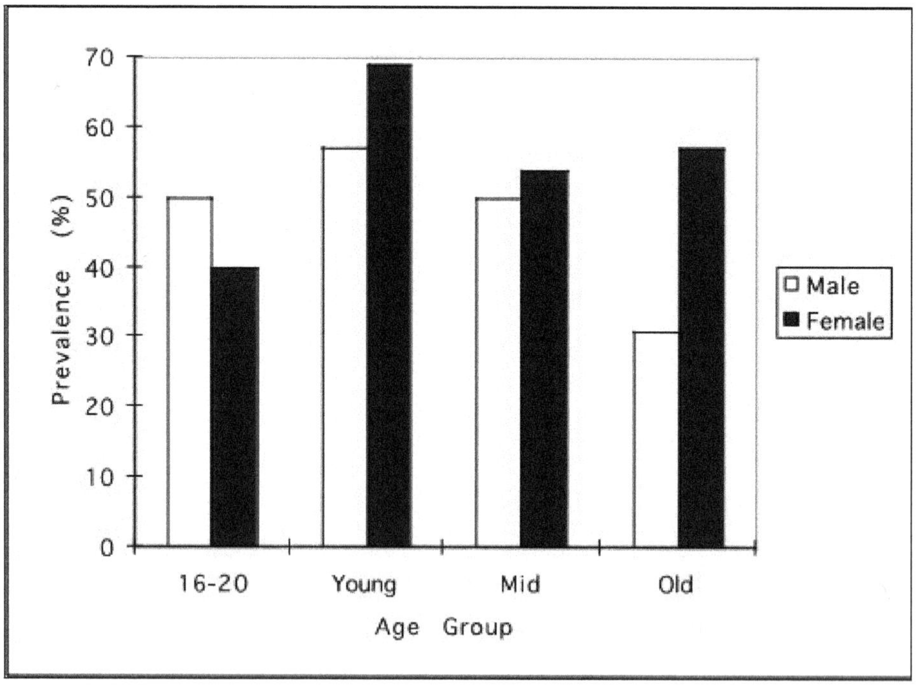

Figure 6.16. *Prevalence of lesions in males and female individuals: Taumako.*

Table 6.13. *Prevalence of individuals with two or more skeletal elements affected which includes the tibia: Taumako*

Age (yrs)	Male		Female		Total	
	A/n	**Per cent**	**A/n**	**Per cent**	**A/n**	**Per cent**
Subadults						
0–0.9					1/3	30
1–5					3/6	50
6–15					2/4	50
16–20	1/1	100	2/2	100	3/3	100
Subtotal	1/1	100	2/2	100	9/16	56
Adult						
Young	9/12	75	8/17	47	17/21	81
Mid	1/2	50	4/7	57	5/9	56
Old	4/4	100	2/4	50	6/8	75
Subtotal	14/18	78	14/20	70	28/38	74
Total	**15/19**	**79**	**16/22**	**73**	**37/54**	**69**

A = number of individuals affected;*n* = number of observations; per cent = percentage of individuals affected.

Table 6.14. *Prevalence of individuals with only the tibia affected: Taumako.*

Age (yrs)	Male		Female		Total	
	A/n	**Per cent**	**A/n**	**Per cent**	**A/n**	**Per cent**
Subadults						
0–0.9					0/3	0
1–5					2/6	33
6–15					1/4	25
16–20					0/3	0
Subtotal	0/1	0	0/2	0	3/16	19
Adults						
Young	0/12	0	0/9	0	0/21	0
Mid	0/2	0	2/7	29	2/9	22
Old	0/4	0	1/4	25	1/8	13
Subtotal	0/18	0	3/20	15	3/38	8
Total	**0/19**	**0**	**3/22**	**14**	**6/54**	**11**

A = number of individuals affected;*n* = number of observations; per cent = percentage of individuals affected.

Table 6.15. *Prevalence of a skeletal element other than tibia affected: Taumako.*

Age (yrs)	Male		Female		Total	
	A/n	Per cent	A/n	Per cent	A/n	Per cent
Subadults						
0–.9					2/3	67
1–5					1/6	17
6–15					1/4	25
16–20					0/3	0
Subtotal	0/1	0	0/2	0	4/16	31
Adults						
Young	3/12	25	1/9	11	4/21	19
Mid	1/2	50	1/7	14	2/9	22
Old	0/4	0	1/4	25	1/8	13
Subtotal	4/18	22	2320	10	7/38	18
Total	**4/19**	**21**	**2/22**	**9**	**11/54**	**20**

A = number of individuals affected; *n* = number of observations; per cent = percentage of individuals affected.

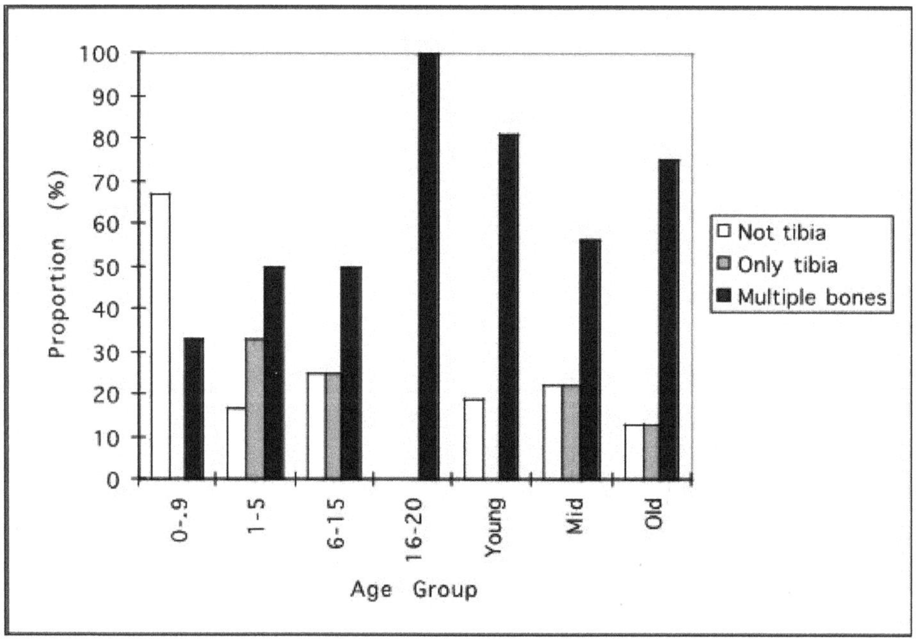

Figure 6.17. *Proportion of individuals with different numbers of skeletal elements affected: Taumako.*

Table 6.16. *Prevalence of a skeletal element other than tibia affected: Taumako.*

Age (yrs)	N	Grade 1	Per cent	Grade 2	Per cent	Grade 3	Per cent
0–0.9	3	3	100	0	-	0	-
1–5	6	5	83	0	-	1	17
6–15	4	4	100	0	-	0	-
16–20	3	0	-	0	-	3	100
Total	**16**	**12**	**75**	**0**	**0**	**4**	**25**

N = number of affected individuals.

Table 6.17. *Status of lesions in adult individuals: Taumako.*

Age (yrs)	N	Grade 1	Per cent	Grade 2	Per cent	Grade 3	Per cent
Young	21	3	14	14	67	2	19
Mid	9	1	11	6	67	3	22
Old	8	3	38	4	50	2	12.5
Total	**38**	**7**	**18**	**24**	**47**	**7**	**18**

N = number of affected individuals.

Table 6.18. *Severity of postcranial appendicular lesions in subadult individuals: Taumako.*

Age (yrs)	N	Grade 1–2	Per cent	Grade 3–4	Per cent	Grade 5–6	Per cent	Grade 7–8	Per cent
0–0.9	3	3	100	-	-	-	-	-	-
1–5	6	5	83	1	17	-	-	-	-
6–15	4	3	75	1	25	-	-	-	-
16–20	3	-	-	0	-	1	33	2	66
Total	**16**	**11**	**69**	**2**	**12.5**	**1**	**6**	**2**	**12.5**

Grade 1–2 = severity grades equivalent to OB1 and OB2; Grade 3–4 = severity grades equivalent to OB3 and OB4; Grade 5–6 = severity grades equivalent to gummatous lesions OL1 and OL2; Grade 7–8 = severity grades equivalent to gummatous lesions OL3 and OL4.

Table 6.19. *Severity of postcranial appendicular lesions in subadult individuals: Taumako.*

Age	N	Grade 1–2	Per cent	Grade 3–4	Per cent	Grade 5–6	Per cent	Grade 7–8	Per cent
Young	21	13	62	3	14	2	10	3	14
Mid	9	6	67	2	22	1	11	-	-
Old	8	6	75	2	38	-	-	-	-
Total	**38**	**25**	**66**	**7**	**18**	**3**	**8**	**3**	**8**

Grade 1–2 = severity grades equivalent to OB1 and OB2; Grade 3–4 = severity grades equivalent to OB3 and OB4; Grade 5–6 = severity grades equivalent to gummatous lesions OL1 and OL2; Grade 7–8 = severity grades equivalent to gummatous lesions OL3 and OL4.

6.3.1 *Cranial pathology*

Overall, 26 per cent (n = 47/179) of crania from the Taumako sample had lesions of various types. Forty-three per cent of subadult crania had lesions; of these 51 per cent were under 4 years at death (Table 6.20). Of those under four years of age at death, eleven were infants (<1yr) and eleven were between one and three years old. The infants had predominantly endocranial lesions with some concurrent external porosity and one newborn had concurrent proliferative new bone of the nasal region. The older children, between one and three years of age had a similar pattern of lesions, with endocranial predominating. The difference between young children and children over four years of age was significant (FET p-value = 0.0003). Of the lesions found, 95 per cent percent of the children under 4 years of age had endocranial lesions of the vault, while the two older children had ectocranial and facial lesions (Table 6.21). Of the younger children with endocranial lesions, four had concurrent proliferative new bone of the nasal region.

Of the adult crania, 19 per cent were affected (Table 6.22). There were no significant differences in the number of crania affected between the age groups. Of the young adults 58 per cent had lesions of *caries sicca* Grade 1–4, while 60 per cent of mid adults had more severe *caries sicca* lesions (Table 6.23). Fisher's exact statistical analysis showed there were no significant differences between the age groups of adults in relation to severity of lesions. The relative likelihood of more subadults than adults having lesions of the crania was tested and this difference was statistically significant (chi-square p-value = 0.026).

6.3.2 *Postcranial pathology*

Table 6.24 provides the data of the proportion of each skeletal element affected in all age groups. The hands, feet, and scapula were least affected of all skeletal elements in adults. A similar pattern was observed in the subadults, except 16 per cent of subadult hands were affected. The young adults had a higher proportion of bones affected than the older adults, although this was not marked. The upper limb bones were affected in similar proportions throughout, while the lower limb bones had a higher percentage affected than the upper limbs. The three bones of the upper limb (humerus, radius and ulna) and the three bones of lower limb (femur, tibia and fibula) were merged in order to test if the higher percentage of lower limb involvement was significant. The results were highly significant (chi-square p-value ≤ 0.0001).

The tibia was the most affected bone in all age groups, particularly the subadults (Figure 6.18 and Table 6.26). Chi-square statistical tests showed that the involvement of the tibia was significantly greater against all other bones except the fibula (Table 6.24). This difference is most extreme when compared to the upper limb bones. The subadults had a slightly greater proportion of most bones involved or similar levels to adults. The frequency of involvement between subadults and all adults is shown in Table 6.26.

The status of the lesions for each skeletal element was then assessed. Nearly all lesions (99 per cent) in subadult bones were active at time of death, while in all adult age groups 56 per cent of lesions were remodelled at death (Table 6.27). Gummatous lesions were found only in adult bones, with more in the young adults than the older age groups (Figure 6.19). Only the hands and feet had no gummatous involvement. The humerus, tibia and fibula were affected with anterior diaphyseal bowing in some cases.

6.3.3 *Lytic vertebral and extravertebral lesions: Taumako*

Description of observed vertebral lesions

Osteolytic lesions of the vertebrae were observed in the Taumako sample. In the vertebral bodies the severity of the bony resorption ranged from an enlarged nutrient foramina of the anterior or lateral vertebral body (Figure 6.20), to resorption of a significant portion of the cortical bone and trabeculae of the body (Figure 6.21). The morphology of the lesions varied between individuals and within a single element. Some were sharp edged with no apparent remodelling of the cavity created in the trabeculae, while others were smooth edged with well remodelled walls within the trabeculae.

Some individuals had lesions of the endplates (Figure 6.22). These lesions were found on the posterior, anterior and lateral margins of the disc space. An erosive or infectious origin of lesion was differentiated from a Schmorl's node on the basis of lesions morphology. Schmorl's nodes (of varied non-infectious aetiology) were identified as having a well remodelled margin and lesion floor (Rogers and Waldron, 1995) whereas an erosive lesion was defined as a cavity with little or no remodelling of the walls and floor. Lesions predominated in the lower thoracic and lumbar, however, some were also observed in the cervical vertebrae.

Vertebral lesions

Thirty seven individuals had enough vertebral elements present to be included in this analysis. This analysis is separate from the previous 'individual' analysis. Some of these will have been included in the individual analysis, but the the vertebral lesions were not assessed. Of the 37 with vertebrae, 33 (89 per cent) had osteolytic lesions of one or more vertebrae. The descriptive information on these individuals is provided in Table 6.28. The predominant type of lesions were enlarged nutrient foramina of the anterior thoracic and lumbar vertebral bodies. Only one individual had involvement of the thoracic endplate, while 11 had lesions of the superior or inferior lumbar endplate. Ten (30 per cent) of these individuals had osteolytic lesions at extravertebral sites, such as the elbow, wrist, and or hands and feet. Five of the individuals with vertebral lesions had changes to the ribs. The bony changes to the ribs consisted of new bone in some cases or lytic lesions (Figure 6.23).

Table 6.20. *Prevalence of lesions in subadult and adult crania: Taumako.*

Age (yrs)	N	Affected	Per cent
0–3.9	43	22	51
4–14	13	2	14
Subtotal	56	24	43
Adult	123	23	19
Total	**179**	**47**	**26**

N= number of subadult crania; Affected= number of subadult crania with lesions

Table 6.21. *Type of lesions in subadult crania: Taumako.*

Age (yrs)	Affected	Endocranial n/Per cent	Ectocranial n/Per cent	Facial n/Per cent	Resorptive n/Per cent
0–3.9	22	21/95	0/0	1/5	0/0
4–14	2	0/0	1/50	0/0	1/50
Total	**24**	**21/88**	**1/4**	**1/4**	**1/4**

Affected= number of subadult crania with lesions; Facial= number with lesions of the facial bones including the nasal, sphenoid and mandible; Resorptive= number with resorptive lesions of the facial region.

Table 6.22. *Prevalence of lesions in adult crania: Taumako.*

Age	N	Affected	Per cent
Young	59	12	19
Mid	34	5	15
Old	30	7	23
Total	**123**	**23**	**19**

N = number of adult crania; Affected = number of adult crania with lesions.

Table 6.23. *Status of lesions in adult crania: Taumako.*

Age	Affected	Caries sicca 1–4 n/Per cent	Caries sicca 5–8 n/Per cent	Other n/Per cent
Young	12	7/58	2/17	3/25
Mid	5	1/20	3/60	1/20
Old	7	2/29	4/57	1/14
Total	**23**	**10/43**	**9/39**	**5/22**

Affected = number of adult crania with lesions; Other = number of adult crania with lesions of the nasal, zygoma, and palatal bones which are described in the descriptive tables of the individual analysis.

Table 6.24. *General prevalence of skeletal elements affected (upper limbs): Taumako.*

Bone	Age	A/n	Per cent	Chi-square Tibia p-value
Scapula	Subadult	2/47	4	
	Young	4/84	5	
	Mid	2/46	4	
	Old	2/50	4	
	Total	10/227	4	<0.0001
Clavicle	Subadult	12/60	20	
	Young	13/70	19	
	Mid	6/39	15	
	Old	7/45	16	
	Total	38/214	18	0.0004
Humerus	Subadult	6/48	20	
	Young	17/98	17	
	Mid	8/58	14	
	Old	6/57	11	
	Total	37/261	14	<0.0001
Ulna	Subadult	8/41	19	
	Young	20/83	20	
	Mid	6/53	11	
	Old	3/49	6	
	Total	37/226	16	<0.0001
Radius	Subadult	6/45	13	
	Young	17/85	20	
	Mid	7/51	14	
	Old	4/48	8	
		34/229	15	<0.0001
Subtotal Upper Limb		**136/716**	**19**	
Hands	Subadult	3/19	16	
	Young	1/89	1	
	Mid	0/43	0	
	Old	0/40	0	
	Total	4/191	2	<0.0001

Table 6.25. *General prevalence of skeletal elements affected (Lower Limbs): Taumako.*

Bone	Age	A/n	Per cent	Chi-square Tibia p-value
Pelvis	Subadult	2/41	5	
	Young	3/104	3	
	Mid	2/49	4	
	Old	0/56	0	
	Total	7/250	3	<0.0001
Femur	Subadult	3/54	6	
	Young	35/120	29	
	Mid	12/59	17	
	Old	13/56	23	
	Total	63/289	22	0.0033
Tibia	Subadult	15/25	60	
	Young	34/96	36	
	Mid	19/50	38	
	Old	16/51	31	
	Total	84/222	38	
Fibula	Subadult	5/22	23	
	Young	22/84	26	
	Mid	8/38	21	
	Old	14/44	32	
	Total	49/188	26	ns: 0.0686
Subtotal Lower Limb		**196/699**	**28**	
Feet	Subadult	0/13	0	
	Young	2/77	3	
	Mid	2/40	5	
	Old	2/45	4	
	Total	6/175	3	<0.0001
Total skeletal elements (This table and Table 6.24)		**369/2472**	**15**	

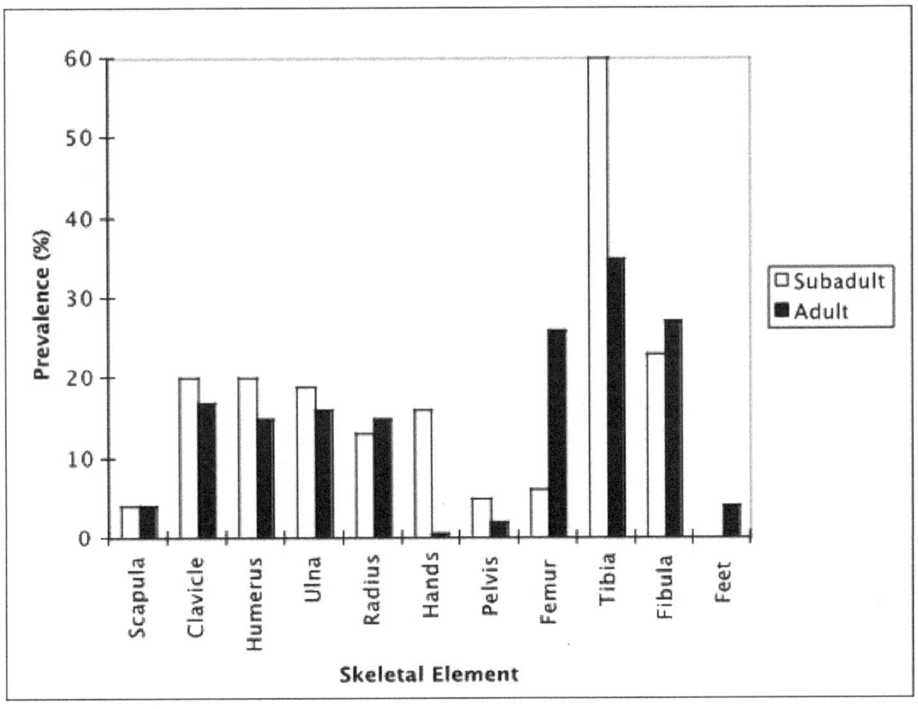

Figure 6.18. *Prevalence of lesions in postcranial skeletal elements of subadults and adults: Taumako.*

Table 6.26. *Comparison of prevalence of skeletal elements affected between subadults and adults: Taumako.*

Bone	Age	A/n	Per cent
Scapula	Subadult	2/47	4
	Adult	8/180	4
Clavicle	Subadult	12/60	20
	Adult	26/154	17
Humerus	Subadult	6/48	20
	Adult	31/213	15
Ulna	Subadult	8/41	19
	Adult	29/185	16
Radius	Subadult	6/45	13
	Adult	28/184	15
Hands	Subadult	3/19	16
	Adult	1/172	0.5
Pelvis	Subadult	2/41	5
	Adult	5/209	2
Femur	Subadult	3/54	6
	Adult	60/235	26
Tibia	Subadult	15/25	60
	Adult	69/197	35
Fibula	Subadult	5/22	23
	Adult	44/166	27
Feet	Subadult	0/13	0
	Adult	6/162	4
Total		**369/2472**	**15**

A = number of skeletal elements with lesions; *n* = number of skeletal elements observed; ns = no significant difference between subadult and adult skeletal involvement

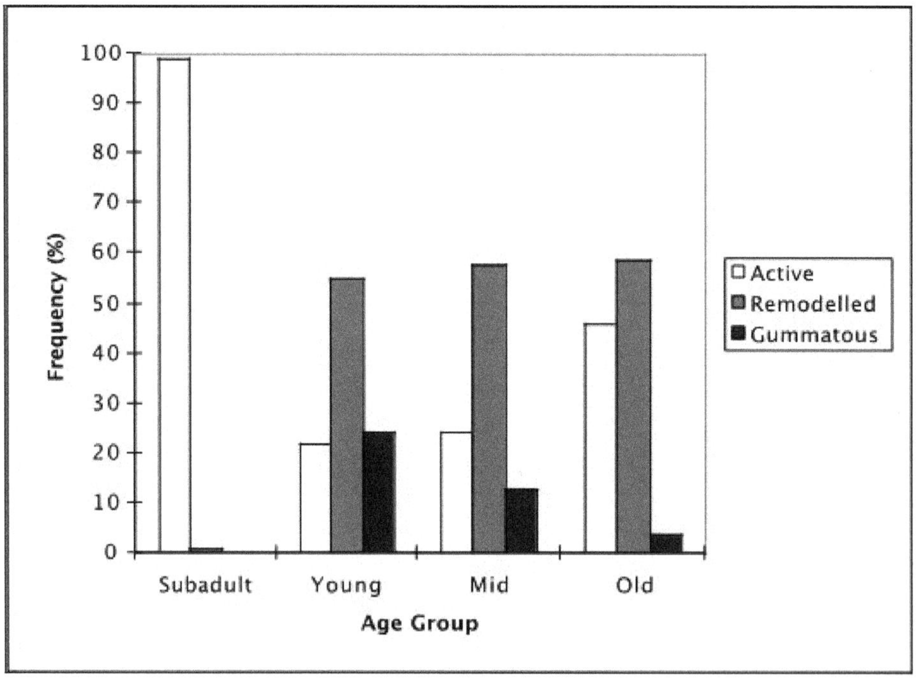

Figure 6.19. *The frequency of the status of lesions between the age groups.*

Table 6.27. *Status of lesions in subadult and adult skeletal elements: Taumako.*

Bone	Age	N	Active n/Per cent	Remodelled n/Per cent	Gummatous n/Per cent	Anterior Bowing n/Per cent
Scapula	Subadult	2	2/100	-	-	
	Young	4	1/25	1/25	2/50	
	Mid	2	1/50	-	1/50	
	Old	2	1/50	1/50	-	
Clavicle	Subadult	12	12/100	-	-	
	Young	13	3/23	6/46	4/31	
	Mid	6	2/33	4/67	-	
	Old	7	-	7/100	-	
Humerus	Subadult	6	6/100	-	-	
	Young	17	3/18	10/59	4/24	8/47
	Mid	8	3/60	5/40	-	
	Old	6	2/33	2/33	2/33	3/50
Ulna	Subadult	8	7/88	1/12	-	
	Young	20	1/5	10/50	9/45	
	Mid	6	1/17	3/50	2/33	
	Old	3	2/67	1/33	-	
Radius	Subadult	6	6/100	-	-	
	Young	17	6/35	9/53	2/12	
	Mid	7	2/29	4/57	1/14	
	Old	4	3/75	1/25	-	
Hands	Subadult	3	3/100	-	-	
	Young	1	-	1/100	-	
	Mid	-	-	-	-	
	Old	-	-	-	-	
Pelvis	Subadult	2	2/100	-	-	
	Young	3	2/67	-	1/33	
	Mid	2	2/100	-	-	
	Old	-	-	-	-	
Femur	Subadult	3	3/100	-	-	
	Young	35	8/23	24/69	3/9	
	Mid	12	2/17	10/83	-	
	Old	13	2/15	11/85	-	
Tibia	Subadult	15	15/100	-	-	4/26
	Young	34	7/21	16/47	11/32	7/21
	Mid	19	2/11	13/68	4/21	
	Old	16	8/50	8/50	-	
Fibula	Subadult	5	5/100	-	-	
	Young	22	4/18	14/64	4/18	2/9
	Mid	8	2/25	5/63	1/13	
	Old	14	7/50	6/43	1/7	
Feet	Subadult	-	-	-	-	
	Young	2	1/50	1/50	-	
	Mid	2	1/50	1/50	-	
	Old	2	-	2/100	-	
Total Subadults		67	66/99	1/1	-	-
Adults						
Young		170	37/22	93/55	40/24	19/11
Mid		80	19/24	46/58	10/13	-
Old		69	32/46	41/59	3/4	3/4
Total Adults		319	88/28	180/56	53/17	22/7

A = number of skeletal elements with lesions; *n* = number of skeletal elements observed; ns = no significant difference between subadult and adult skeletal involvement

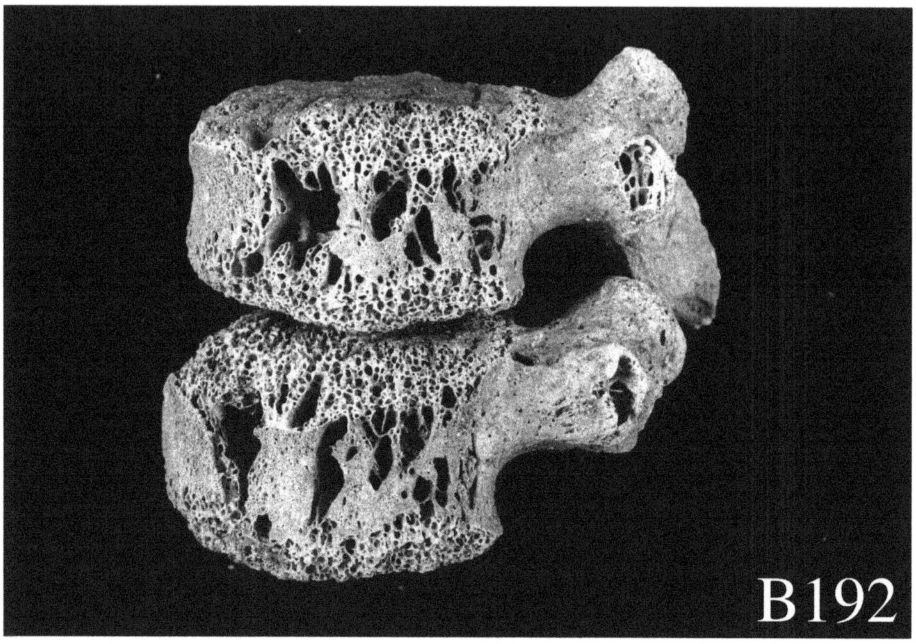

Figure 6.20. *Enlarged nutrient foramina of two lumbar vertebral bodies of a subadult. The smooth remodelled edges of the lesions may indicate a chronic condition. B192 from Taumako. Photo by author.*

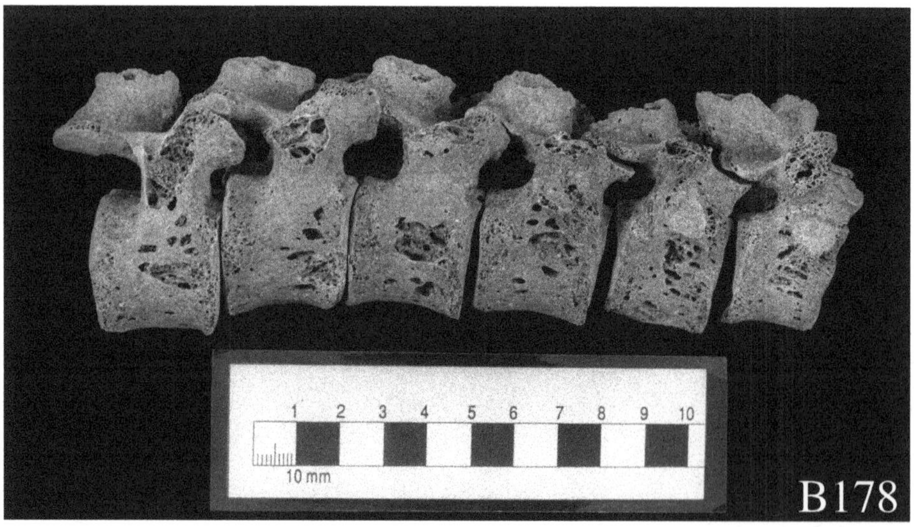

Figure 6.21. *Vertically aligned lytic foci of the vertebral bodies of T10-L3 of an adult. Some lesions have sharp edges which may indicate an active condition or rapid bone destruction. B178 from Taumako. Photo by author.*

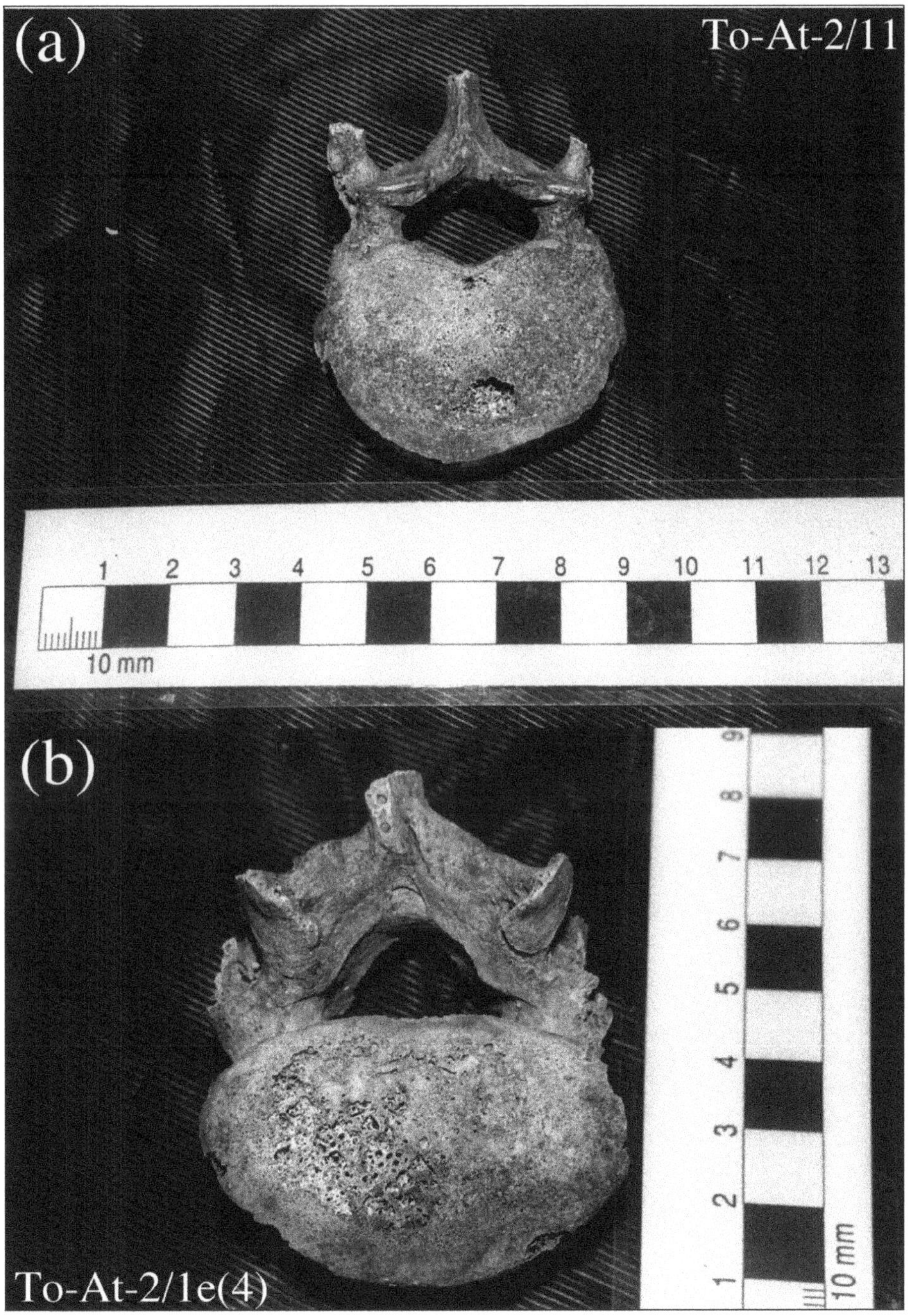

Figure 6.22. *(a) A discrete lesion of the superior aspect of L3 in an adult. (b) More diffuse lytic lesion of the inferior aspect of L4 in an adult. Photos by author.*

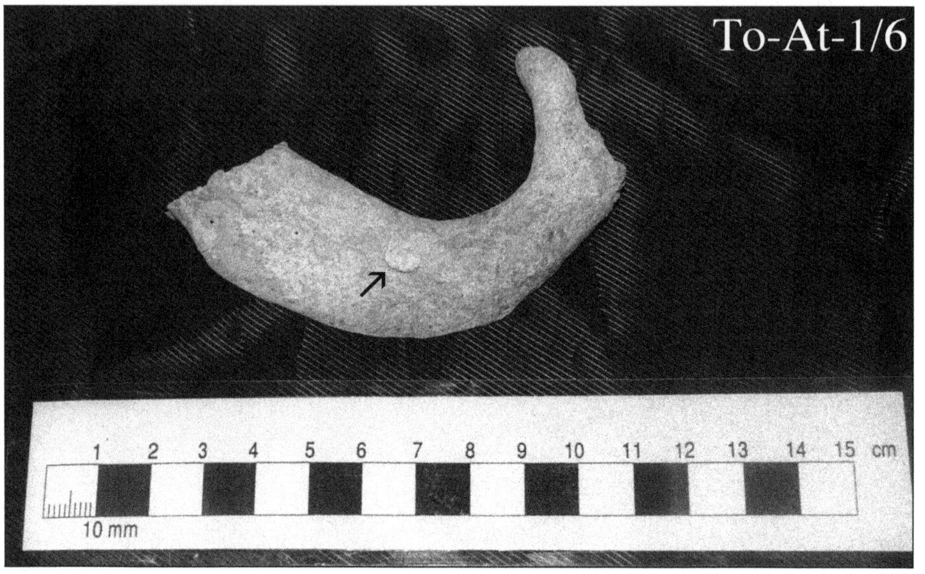

Figure 6.23. *A small plaque of new bone on the superior aspect of the first rib (arrow). Photo by author.*

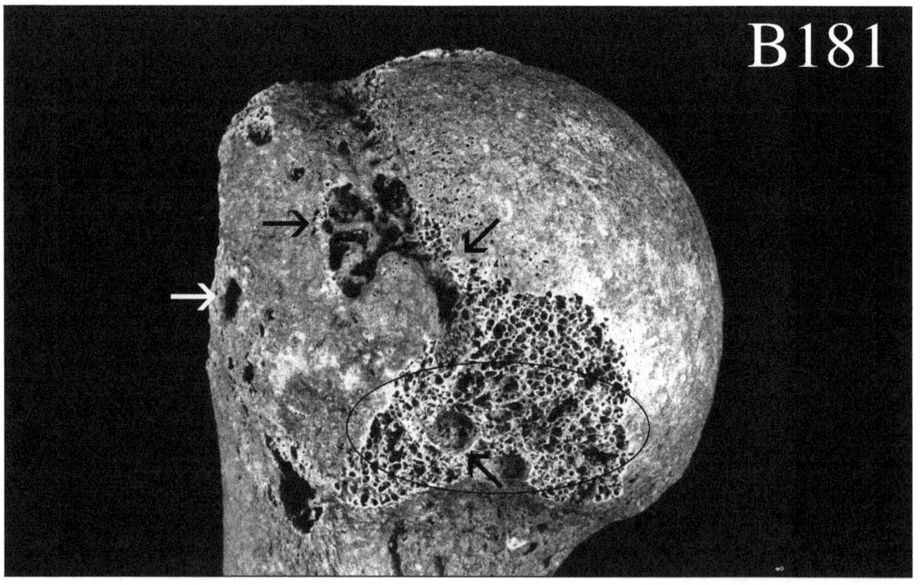

Figure 6.24. *Lytic lesions marginal to the joint on the superior humerus (arrows). The region within the ellipse where the trabeculae are exposed is probably due to postmortem damage. But note the small cyst-type lesions within this area of postmortem damage (arrow). B181 from Taumako. Photo by author.*

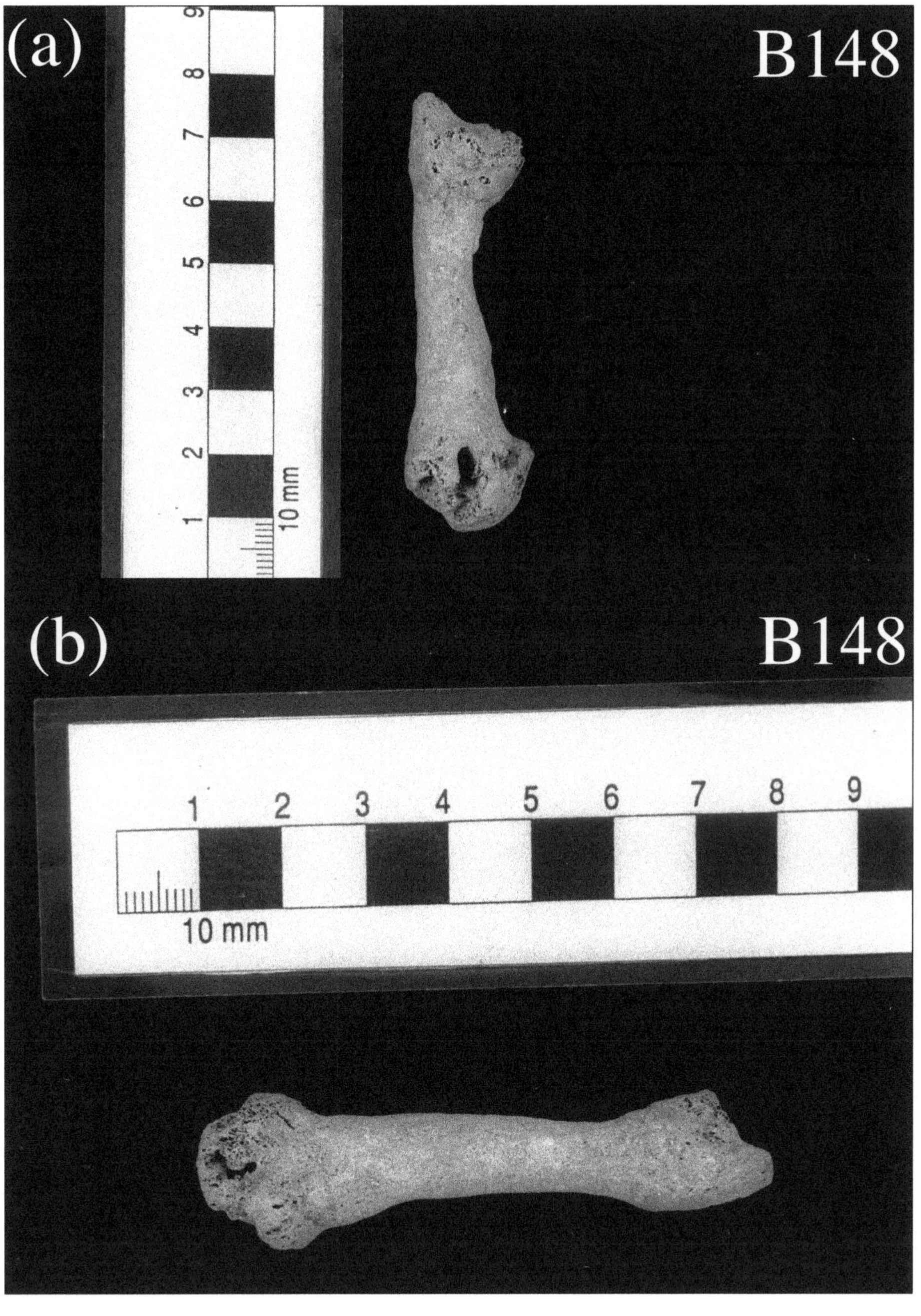

Figure 6.25. *Lytic lesions of the articular space of the distal third metacarpal. (a) Lateral view (b) superior view. B148 from Taumako. Photos by author.*

Table 6.28. *Individuals with osteolytic lesions of the vertebrae: Taumako.*

Burial	Age	Sex	Extrav	Tdisc	Tbody	Ldisc	Lbody	Ribs	CO	Other
102	10		no		x		x		yes	
192	13.5		no		x		x		yes	sacroiliac
89	16	F	no		x		x		yes	
137	16.5		no		x				yes	
139	18	F	yes		x			x	yes	OB
188	18.5	F	yes		x		x	x	no	OB
9	Young	M	no				x	x	no	OB
11	Young	M	no			x	x		no	OB
42	Young	M	no	x	x				no	
73	Young	M	no		x		x		no	
85	Young	M	no		x				yes	
108	Young	M	yes		x		x		no	
126	Young	M	no				x		no	OB
145	Young	M	no				x		yes	
148	Young	M	yes		x	x	x	x	yes	sternum
149	Young	M	no					x	no	OB
185	Young	M	no				x		no	
91	Young	F	no				x		no	OB
150	Young	F	yes			x			no	OB
178	Young	F	no		x	x	x		yes	
184	Mid	M	no				x		no	
23	Mid	F	no		x				no	OB
88	Mid	F	yes		x		x		no	OB
109	Mid	F	yes		x	x	x		no	OB
14	Old	M	no			x	x		yes	OB
22	Old	M	no			x	x		yes	
94	Old	M	yes		x		x		yes	OB
104	Old	M	no		x	x	x		no	
172	Old	M	yes		x	x	x		no	OB
196	Old	M	no		x		x		no	
90	Old	F	yes		x				no	OB
110	Old	F	no			x	x		no	
160	Old	F	no			x	x		yes	

Extrav = osteolytic lesion of appendicular joints; Tdisc = lesion of thoracic endplate; Tbody = lesion of thoracic body; Ldisc = lesion of lumbar disc; Lbody = lesions of lumbar body; Ribs = osteolytic or osteoblastic lesion of ribs; CO = individuals with cribra orbitalia; Other = lists whether individuals have osteoblastic lesions of the appendicualr skeleton or crania and names the site of other osteolytic lesions of the axial skeleton

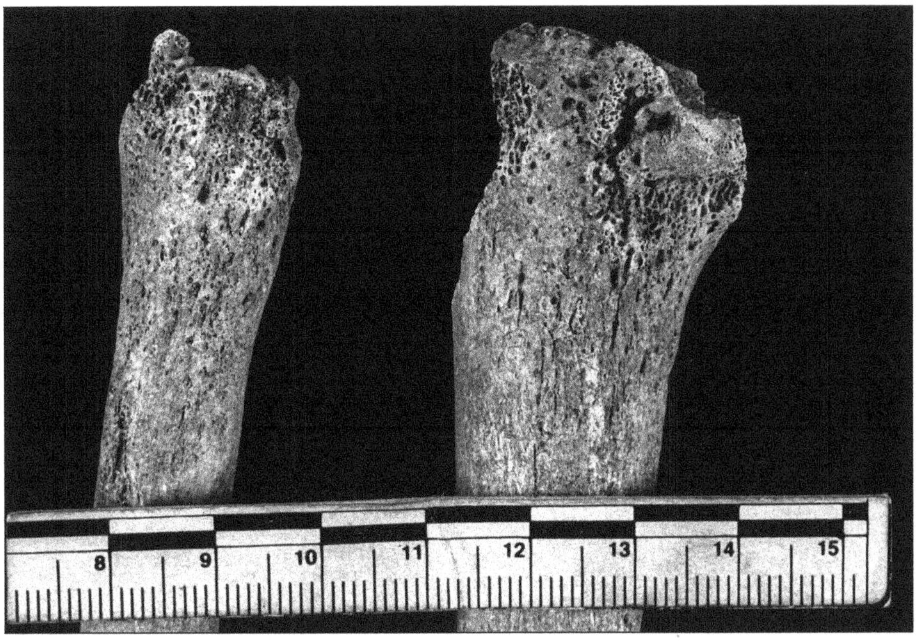

Figure 6.26. *Lytic lesions of the left and right distal ulna. The radio-ulnar on the left has completely resorbed with some dimunition of the cortical bone. The styloid process of the ulna on the right has resorbed completely with some marginal erosion surrounding the joint. Photo by author.*

The aspect of the ribs on which the lesions were situated varied.

The age structure of the affected individuals is given in Table 6.29. Of the 31 individuals over 15 years of age, over half of those affected were Young adults. Thirty of the adults had sex estimations and eighteen of those (55 per cent) affected were males.

All individuals with vertebral lesions had postcranial material present. The association of either osteoblastic postcranial lesions or cribra orbitalia with the vertebral lesions was then assessed. Twenty-eight (85 per cent) of the individuals with vertebral lesions had either osteoblastic changes or cribra orbitalia. Fifteen (45 per cent) of the individuals had osteoblastic changes of one or more limb bone and 39 per cent had cribra orbitalia. The pattern observed was that most individuals with vertebral lesions had either osteoblastic changes or cribra orbitalia concurrent with lytic lesions of the vertebrae, but few had these changes together. Of those individuals with no evidence of cribra orbitalia, five had no cranial material present and another three had broken facial regions which precluded observation of the orbits. There seemed to be no particular pattern of age or sex in relation to these other changes (Table 6.27).

Individuals with extravertebral osteolytic lesions

Because of the varied preservation of these individuals, which would affect the representation of lesions, this section is predominantly descriptive. Twenty-nine individuals had lytic lesions of the appendicular skeleton (Table 6.29). This comprises 30 per cent (28/93) of the total sample of adults with a reasonably complete postcranial skeleton. It is recognised this frequency is highly subjective and is again separate from the previous 'individual analysis'. Only one individual was under 15 years at death. The lesions observed varied from extensive lysis of multiple appendicular joints in an individual or a lytic lesion of a single element. Lesion types consisted of marginal erosions around the articular space (Figure 6.24)) and purely resorptive lesions within an articular space (Figure 6.25). In some cases both marginal and joint space erosion occurred concurrently (Figure 6.26). In order of frequency the wrists, elbows and hands and feet were the most commonly affected joints. However, other lower limb joints were affected in some individuals. There was little difference in involvement between the sexes, with 15 females and 13 males affected. Seven (25 per cent) of the individuals were Mid aged at death and 29 per cent of the affected adults were Old at death.

The pattern of lesions was then assessed to ascertain whether joints affected were bilateral or unilateral. The pattern of involvement was unable to be assessed in eleven (38 per cent) of the affected individuals because of missing elements. However, 31 per cent (n = 10/29) of affected individuals had unilateral lesions, while the remaining eight (24 per cent) individuals had bilateral involvement of one or more joints. Figures 6.27 to 6.29 illustrate the different patterns of involvement. The frequencies are not given in these figures because some individuals were affected in multiple joints.

Table 6.29. *Age structure of individuals with osteolytic lesions of the vertebrae: Taumako.*

Age	N	Per cent	Male	Per cent	Female	Per cent
<15	2	6	-	-	-	-
Young	18	55	11	65	6	35
Mid	4	12	1	25	3	75
Old	9	27	6	67	3	33
Total	33		18		12	36

N = number of individuals affected with osteolytic lesions of the vertebrae; includes those under 20 years with sex estimate

Table 6.30. *Subadults and young adults with extravertebral osteolytic lesions: Taumako.*

Burial	Age	Sex	OB	Axial	Lytic	Hands	Feet	Pattern
5	13.5	-	no	no	hip			U
139	18	F	yes	yes	elbow and wrist		foot, with OB	B
188	18.5	F	yes	yes	elbow and wrist	OB of Mc		B
13	Young	F	no	no		carpal		U
54	Young	F	yes	no			Metatarsals	N
117	Young	F	no	no	ribs		proximal phalanx	N
141	Young	F	yes	no	elbow, wrist and ribs	phalanges, OB Mc	Metatarsals	B
150	Young	F	yes	no		Mc marginal		U
36	Young	M	yes	no			marginal Mt1, Phalanges	N
108	Young	M	no	yes			Mt 3 and 4, marginal	U
148	Young	M	no	yes	Ribs& sternum	Mc3 joint		U
167	Young	M	yes	no	elbow, wrist, and knee		right talus	B
181	Young	M	no	no	shoulder, elbow, wrist	semilunar, Mc, distal phalanges	Mt 1 distal	B
194	Young	M	yes	no	elbow and wrist			U

F = female; M = male; OB = concurrent osteoblastic changes; Axial = concurrent lytic lesions of vertebrae; Osteolytic = site of extravertebral lytic lesion; Mc = metacarpal; Mt = metatarsal; U = unilateral involvement; B = bilateral involvement; N = the antimere was not present or could not be identified. Therefore the pattern of involvement could not be evaluated.

Table 6.31. *Mid and Old adults with extravertebral osteolytic lesions: Taumako.*

Burial	Age	Sex	OB	Axial	Lytic	Hands	Feet	Pattern
88	Mid	F	yes	yes	elbow			N
109	Mid	F	yes	no			Mt marginal	N
159	Mid	F	no	no	wrist			U
171	Mid	F	no	no			Mt marginal	N
193	Mid	F	yes	no	humerus			N
147	Mid	M	yes	no	wrist	Mc3 marginal and joint, scaphoid	Mt5 new bone	U
164	Mid	M	yes	no			Mt and phalanges, marginal and joint	N
39	Old	F	no	no	widespread app.			B
90	Old	F	yes	yes	shoulder, wrist, and elbow	carpals	Phalange tufts resorbed	B
146	Old	F	yes	no	wrist	Mc2	OB of Mt 3	N
170	Old	M	yes	yes		Ankylosis of left wrist and carpals		N
172	Old	M	yes	yes		Mc, distal phalanx	Right Mt 5 joint and marginal	U
176	Old	M	no	no	wrist	Ankylosis of left carpals		B
182	Old	M	no	no	wrist		Mt5 right marginal	U
94	Old	M	yes	no		carpals		N

F = female; M = male; OB = concurrent osteoblastic changes; Axial = concurrent lytic lesions of vertebrae; Osteolytic = site of extravertebral lytic lesion; Mc = metacarpal; Mt = metatarsal; U = unilateral involvement; B = bilateral involvement; N = the antimere was not present or could not be identified. Therefore the pattern of involvement could not be evaluated.

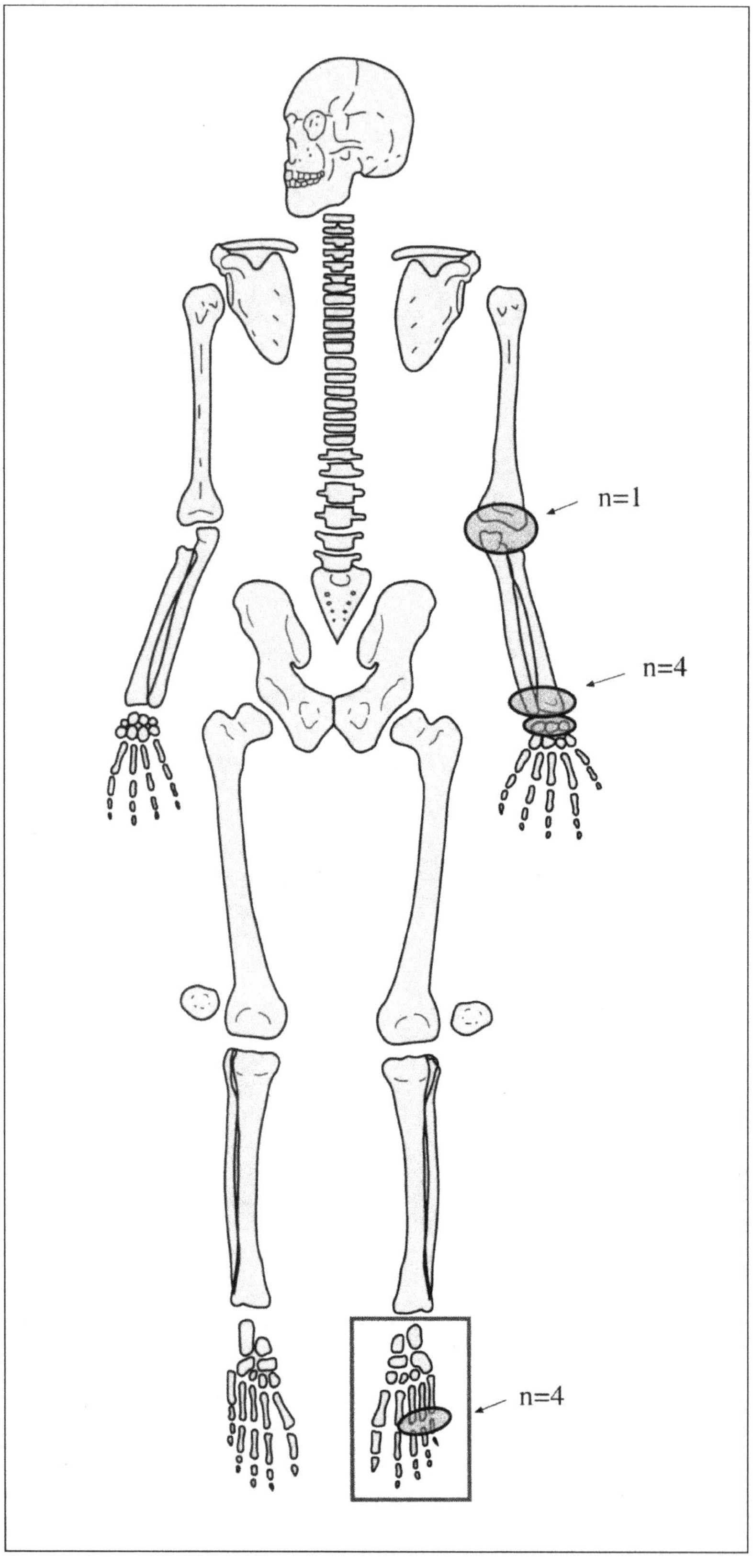

Figure 6.27. *Eleven individuals had lesions of the appendicular joints. The pattern of involvement was unable to be assessed. The numbers given in this illustration are the numbers of joints affected. The proportion of joints affected is not presented.*

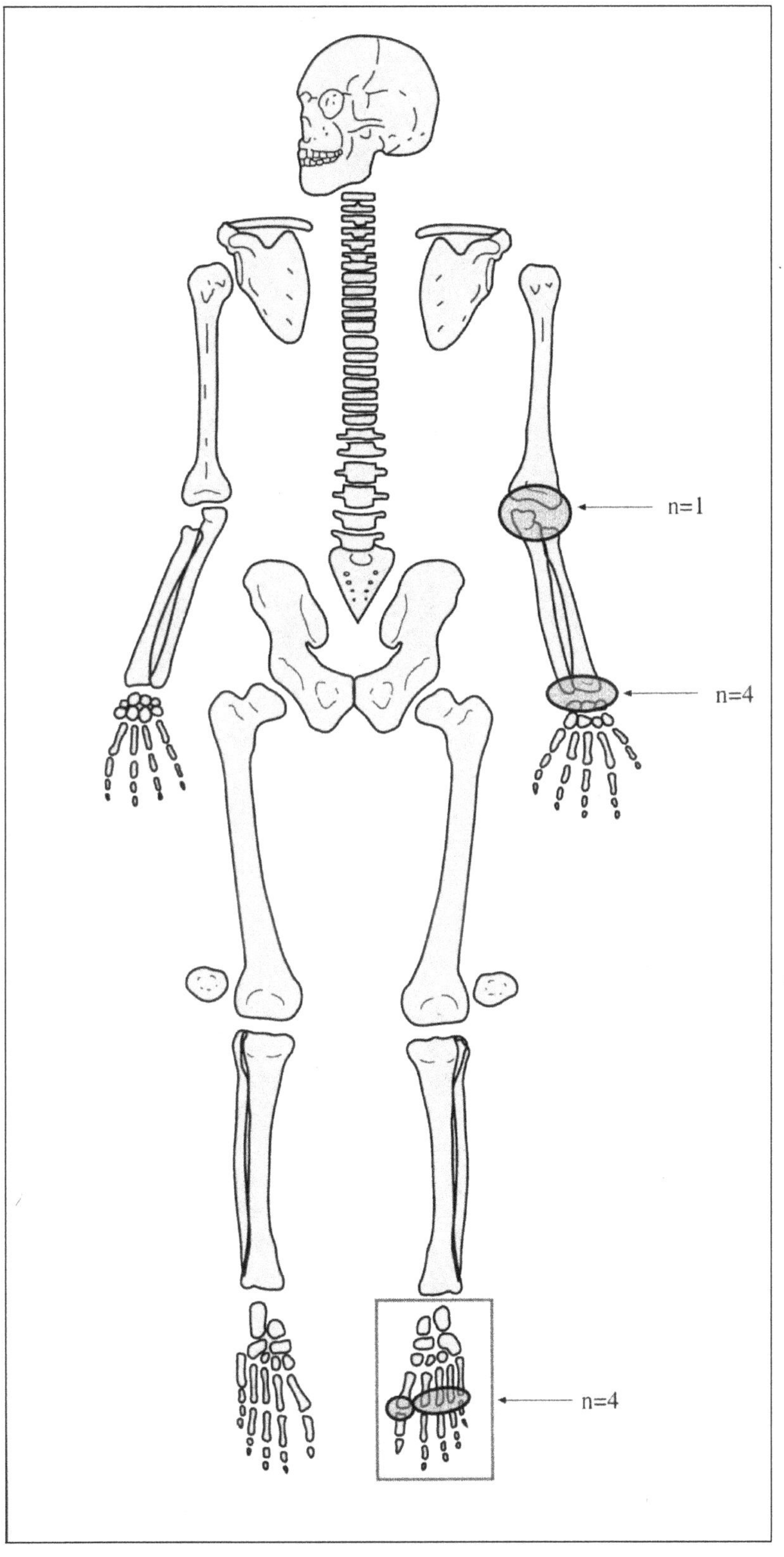

Figure 6.28. *Ten individuals had unilateral lesions of the appendicular joints. The numbers given in this illustration are the number of joints affected. The proportion of joints affected is not presented.*

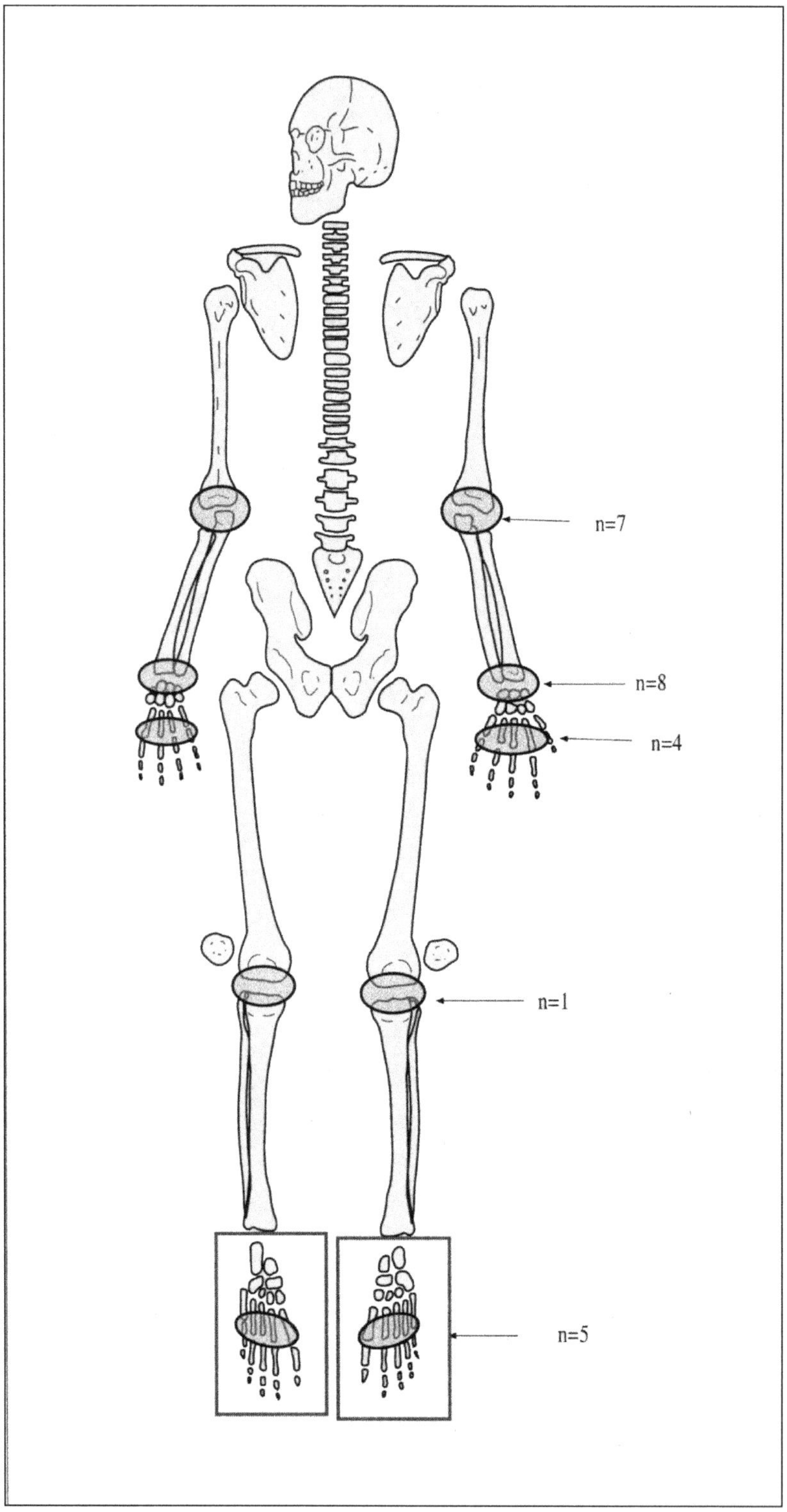

Figure 6.29. *Eight individuals had bilateral lesions of the appendicular joints. The numbers given in this illustration are the number of joints affected. The proportion of joints affected is not presented.*

6.4 Results: Tonga Skeletal Pathology

6.4.1 Sampling issues in the Tongan sample

The individual pathology of the Tongan sample was analysed following the same method as the Taumako sample. However, very few of the burials fitted the criteria for inclusion in the individual analysis (Tables 6.32 and 6.33). The reasons for this are undoubtedly due to the sampling techniques of the excavation discussed in chapter two. This has particularly affected the representation of skeletal elements from Mound 2. Because of these small samples it was decided that further analysis of the Tongan skeletal pathology would concentrate on the analysis of lesions in skeletal elements.

6.4.2 Individual analysis

A description of the type of skeletal lesions observed in the individuals is given in Table 6.34 through to Table 6.37. The type of pathology in the adults from both the Tongan mounds was largely osteoblastic. One young adult female from Mound 1 had a gummatous lesion of the tibia. A similar pattern of new bone formation was observed in the subadult individuals from both mounds.

Further analysis of the skeletal pathology from the Tongan sample will be carried out in the following section. The results from this analysis will provide quantitative data to compare the prevalence of lesions between Taumako and Tonga which the individual analysis could not provide.

6.4.3 Skeletal element analysis

As with the previous section, the samples from the Tongan mounds are considerably smaller than those from Taumako. Therefore, these results are viewed with caution. Also because of the small samples, adult skeletal elements have been merged into one group and all subadults between 4 and 14 years of age have been merged into one group.

6.4.4 Cranial pathology

Overall 33 per cent of all crania from Mound 1 had some form of pathology (Table 6.38). Crania of subadults under four years of age at death had the most lesions. Sixty-three per cent of crania from individuals less than four years old had lesions of the facial bones (Table 6.39). These lesions included abnormal porosity as well as new bone formation. The one affected subadult over four years old with cranial pathology was an 11 year old child with marked porosity of the temporalis muscle attachment on the frontal bone. This child also had new bone formation of the alveolar ridge of the maxilla. Of the adult crania from Mound 1 with lesions, two were young females with Grades 1–5 *caries sicca* lesions of the frontal and parietals and the third individual was an adult female with grade 5 *caries sicca* of the frontal bone. Fisher's exact statistical analysis showed there were no significant differences between the degree of cranial involvement in subadults compared to the adults.

At Mound 2, 9 per cent (n = 2/23) of crania had lesions. Only two out of four (50 per cent) of the subadult crania had lesions and no adult crania (n = 0/19) had lesions at this site. The cranial lesions of the two affected subadults (B1a and B7) have been described in the previous section.

6.4.5 Postcranial pathology

Most types of bones had lesions to varying degrees at both Tongan mounds. Overall, 11 per cent of skeletal elements were affected at Mound 1. At Mound 1, all bones except the pelvis were affected (Table 6.40). In all bones, except the tibia, a higher proportion of subadult bones were affected when compared to adults (Figure 6.30). Because the involvement of the skeletal elements was so variable and the samples small, no statistical analysis was performed on the data to test the greater involvement of subadult bones.

There is little difference in the proportion of bones involved between the upper limb (humerus, ulna, and radius) and the lower limb (femur, tibia and fibula). When bones from each limb were combined, similar prevalences of involvement were found between the upper and lower limbs (14 per cent and 15 per cent respectively). A greater proportion of tibiae were involved than any other bone.

At Mound 2, no lesions were found in the scapulae, humerus, or pelvis, with a patchy involvement between the subadults and adults (Table 6.41 and Figure 6.31). Overall, 9 per cent of all skeletal elements were affected. Where bones were involved in both subadults and adults, the subadult skeletal elements were more consistently affected than adult bones. The pattern of involvement was higher in the lower limbs of the adults. When the limb bones were combined, a higher prevalence of involvement was found in the lower limbs than the upper limbs (18 per cent and 7 per cent respectively). This difference was significant (Chi-square p-value = 0.024).

The status of the lesions at both mounds were then assessed (Tables 6.42 and 6.43). All lesions of subadult skeletal elements from both mounds were active at death. In the adult bones, lesions were evenly distributed between active and remodelled. Gummatous lesions were found only at Mound 1. No evidence of anterior diaphyseal bowing of the tibiae or humerus was observed at either of the Tongan mounds.

6.4.6 Resorptive lesions

Vertebral lesions

Osteolytic lesions of the vertebrae were observed in both the Tongan mounds. At Mound 1, six individuals had enough vertebral material to assess. Of those four (67 per cent) individuals had some form of lysis of the vertebrae (Table 6.44). Two of the individuals had lesions of the thoracic endplates, while Burial 9 had ankylosis of the right side of the thoracic

Table 6.32. *Number of burials included in individual analysis from Mound 1, Tonga.*

Age	Individuals	Burials	Affected
Subadult	13	25	4
Adult	5	20	3
Total	18	45	7

Table 6.33. *Number of burials included in individual analysis from Mound 2, Tonga.*

Age	Individuals	Burials	Affected
Subadult	6	15	2
Adult	3	39	1
Total	8	54	3

Table 6.34. *Description of affected subadult individuals: Mound 1, Tonga.*

Burial	Age	Postcranial Pathology	Cranial Pathology
2	0.25	OB1 of humerus	No cranium
5	0.25	OB1 of scapula, in supraspinatous fossa	Diffuse OB1 of mandible
3	1	OB1 of humerii and ulnae, and R radius. OB1 of tibiae and fibulae	Remodelled endocranial new bone of frontal, and mixed active and remodelled OB1 of parietals.
35	1	OB1 of R ulna and tibiae.	Active OB1 of nasal bones. Remodelled OB1 of endocranial frontal, parietal and temporal bones.
R = right; L = left			

Table 6.35. *Description of affected subadult individuals: Mound 1, Tonga.*

Burial	Age	Sex	Postcranial Pathology	Cranial Pathology
19	Young	Female	OB1 of R humerus diaphysis	No pathology
23	Young	Female	OB2 and OL1 of L tibia; Remodelled OB2 of R tibia	No pathology
29.1	Mid	Male	Remodelled OB2 of L tibia; Remodelled OB1 of L femur with a cystic lesion of distal medial entheses	No pathology
R = right; L = left				

Table 6.36. *Description of affected subadult individuals: Mound 2, Tonga.*

Burial	Age	Postcranial Pathology	Cranial Pathology
7	1	Upper and lower limb bones have OB1 and enlarged cortical margins with associated porosity. Radiography showed marked diminution of all cortices associated with porosity. All nutrient foramina of limbs and digits were enlarged	Active OB1 of endocranial parietals and occipital. OB1 of orbital aspect of the zygomae
17	1	OB1 of all limb bones present. Local areas of node-like deposits	No cranium
R = right; L = left			

Table 6.37. *Description of affected adult individual: Mound 2, Tonga.*

Burial	Age	Sex	Postcranial Pathology	Cranial Pathology
34	Adult	Male	Remodelled OB1 of bilateral tibia	No Cranium

R = right; L = left

Table 6.38. *Prevalence of lesions in subadult and adult crania: Mound 1, Tonga.*

Age (yrs)	N	Affected	Per cent
0–3.9	16	8	50
4–14	5	1	20
Subtotal	21	9	43
Adult	15	3	20
Total	36	12	33

N = number of crania; Affected = number of crania with lesions

Table 6.39. *Type of lesions in subadult crania: Mound 1, Tonga.*

Age (yrs)	Affected	Endocranial n/Per cent	Ectocranial n/Per cent	Facial n/Per cent	Resorptive n/Per cent
0–3.9	8	3/38	0/0	5/63	0/0
4–14	1	0/0	1/100	0/0	0/0
Total	9	3/33	1/11	5/56	0/0

Affected = number of crania with lesions; Facial = number of crania with lesions of the facial bones including; nasal, zygoma, maxilla, sphenoid and mandible

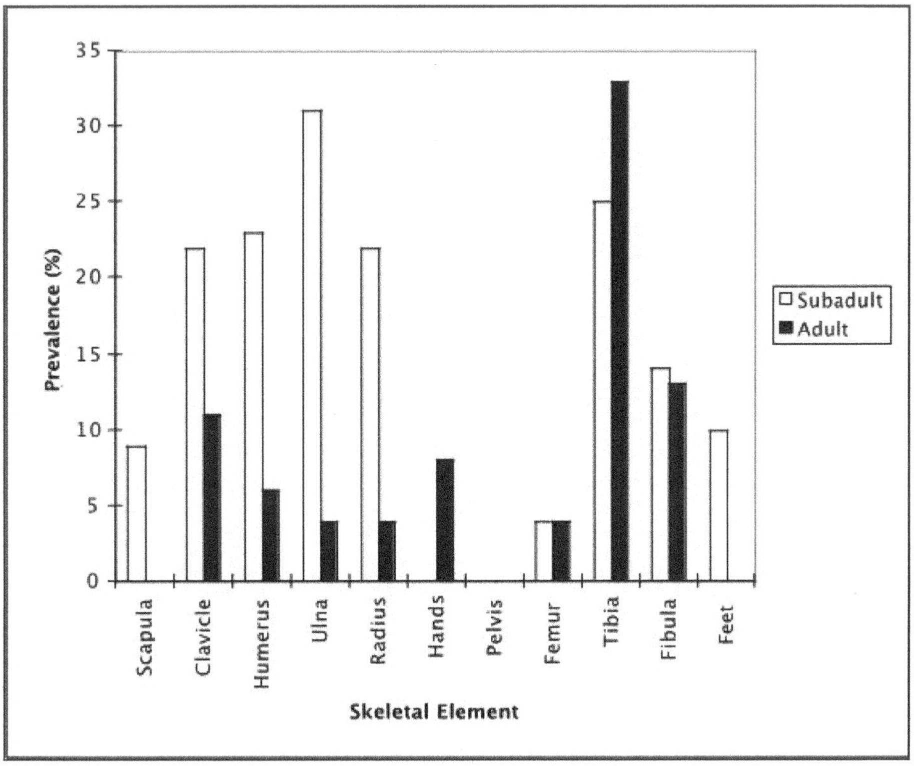

Figure 6.30. *Prevalence of lesions subadult and adult skeletal elements: Mound 1, Tonga.*

Table 6.40. *Prevalence of lesions in subadult and adult crania: Mound 1, Tonga.*

Bone	Age	A/n	Per cent
Scapula	Subadult	2/23	9
	Adult	0/25	0
	Total	2/52	4
Clavicle	Subadult	6/27	22
	Adult	3/27	11
	Total	9/54	17
Humerus	Subadult	7/30	23
	Adult	2/33	6
	Total	9/63	14
Ulna	Subadult	9/29	31
	Adult	1/27	4
	Total	10/56	18
Radius	Subadult	6/27	22
	Adult	1/28	4
	Total	5/57	11
Subtotal Upper limb (Humerus, ulna, radius)		24/176	14
Hands	Subadult	0/11	0
	Adult	1/13	8
	Total	1/24	4
Pelvis	Subadult	0/24	0
	Adult	0/17	0
	Total	0/41	0
Femur	Subadult	1/26	4
	Adult	1/23	4
	Total	2/49	4
Tibia	Subadult	5/20	25
	Adult	6/18	33
	Total	11/38	29
Fibula	Subadult	2/14	14
	Adult	2/16	13
	Total	4/30	13
Subtotal lower Limb (Femur, tibia, fibula)		17/117	15
Feet	Subadult	1/10	10
	Adult	0/6	0
	Total	0/16	0
Total		**53/480**	**11**

A = number of skeletal elements with lesions;*n* = number of skeletal elements

Table 6.41. *Prevalence of lesions in subadult and adult crania: Mound 1, Tonga.*

Bone	**Age**	**A/n**	**Per cent**
Scapula	Subadult	0/9	0
	Adult	0/31	0
	Total	0/40	0
Clavicle	Subadult	2/6	33
	Adult	0/30	0
	Total	2/36	6
Humerus	Subadult	0/10	0
	Adult	0/32	0
	Total	0/42	0
Ulna	Subadult	4/12	33
	Adult	2/31	6
	Total	6/43	14
Radius	Subadult	3/7	43
	Adult	0/30	0
	Total	3/37	8
Subtotal Upper Limb (Humerus, ulna, radius)		9/122	7
Hands	Subadult	1/3	33
	Adult	0/14	0
	Total	1/17	6
Pelvis	Subadult	0/7	0
	Adult	0/22	0
	Total	0/29	0
Femur	Subadult	0/7	0
	Adult	2/36	6
	Total	2/43	5
Tibia	Subadult	3/7	43
	Adult	9/41	22
	Total	12/48	25
Fibula	Subadult	3/6	50
	Adult	6/29	21
	Total	9/35	26
Subtotal Lower Limb (Femur, tibia, fibula)		23/126	18
Feet	Subadult	0/2	0
	Adult	1/20	5
	Total	1/22	5
Total		**36/392**	**9**

A = number of skeletal elements with lesions; *n* = number of skeletal elements

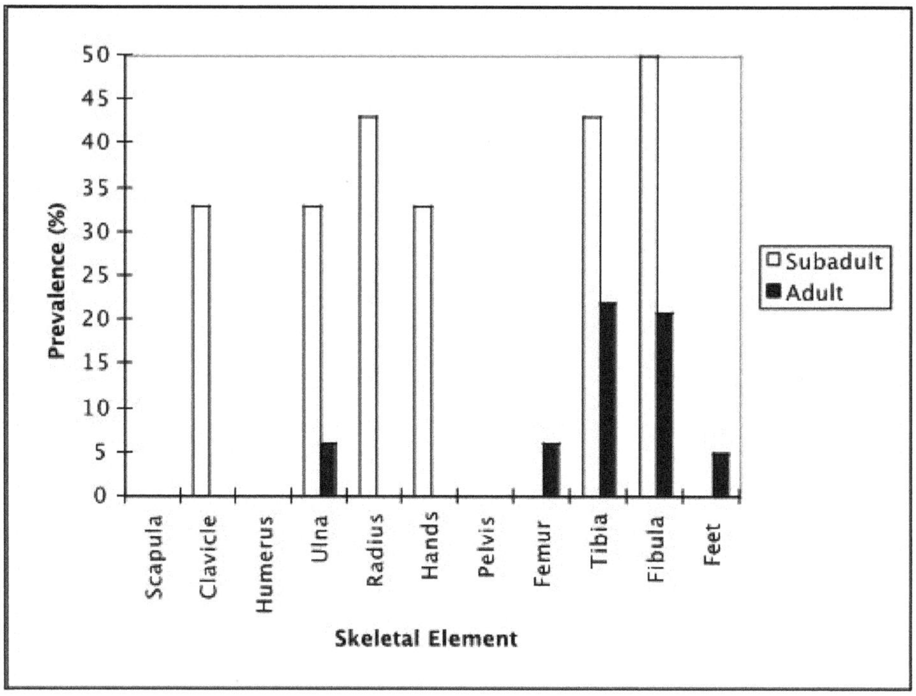

Figure 6.31. *Prevalence of lesions subadult and adult skeletal elements: Mound 2, Tonga.*

Table 6.42. *Status of lesions of subadult and adult skeletal elements: Mound 1, Tonga*

Bone	Age	Affected	Active n/Per cent	Remodelled n/Per cent	Gummatous n/Per cent
Scapula	Subadult	2	2/100		
	Adult	0			
Clavicle	Subadult	6	6/100		
	Adult	3	3/100		
Humerus	Subadult	7	7/100		
	Adult	2	2/100		
Ulna	Subadult	9	9/100		
	Adult	1			1/100
Radius	Subadult	6	6/100		
	Adult	1		1/100	
Hands	Subadult	0			
	Adult	1		1/100	
Pelvis	Subadult	0			
	Adult	0			
Femur	Subadult	1	1/100		
	Adult	1		1/100	
Tibia	Subadult	5	5/100		
	Adult	6		3/50	3/50
Fibula	Subadult	2	2/100		
	Adult	2		2/100	
Feet	Subadult	0			
	Adult	1	1/100		

N = number of skeletal elements with lesions; Active = number of skeletal elements with active lesions; Remodelled = number of skeletal elements with remodelled lesions; Gummatous = number of skeletal elements with gumma

Table 6.43. *Status of lesions of subadult and adult skeletal elements: Mound 1, Tonga*

Bone	Age	Affected	Active n/Per cent	Remodelled n/Per cent	Gummatous n/Per cent
Scapula	Subadult	0			
	Adult	0			
Clavicle	Subadult	2	2/100		
	Adult	0			
Humerus	Subadult	0			
	Adult	0			
Ulna	Subadult	4	4/100		
	Adult	2	2/100		
Radius	Subadult	3	3/100		
	Adult	0			
Hands	Subadult	1	1/100		
	Adult	0			
Pelvis	Subadult	0			
	Adult	0			
Femur	Subadult	0			
	Adult	2	2/100		
Tibia	Subadult	3	3/100		
	Adult	9	4/44	5/56	
Fibula	Subadult	3	3/100		
	Adult	6	4/67	2/33	
Feet	Subadult	0			
	Adult	1	1/100		

N = number of skeletal elements with lesions; Active = number of skeletal elements with active lesions; Remodelled = number of skeletal elements with remodelled lesions; Gummatous = number of skeletal elements with gumma vertebral bodies and associated enlarged nutrient foramina. Burial 14 had new bone formation of the anterior thoracic bodies.

At Mound 2, 46 per cent ($n = 6/13$) of those individuals with vertebrae had lytic lesions. Three had disc lesions of the lumbar vertebrae with associated enlarged foramina (Table 6.44). The remaining six individuals had enlarged nutrient foramina of the thoracic vertebrae. None of the individuals from either mound had concurrent osteoblastic lesions of the postcranial skeleton, while two individuals from Mound 2 had cribra orbitalia as well as vertebral lesions.

Extravertebral osteolytic lesions

Primarily resorptive lesions were observed in a few individuals from both the Tongan mounds. Table 6.45 presents a description in the lesions of these individuals.

6.5 Results Summary

6.5.1 Taumako: Proliferative skeletal lesions

Individual analysis

Over half of the Taumako sample had proliferative lesions of the post-cranial skeleton. While the prevalence of cranial lesions was lower than postcranial lesions at 26 per cent for the total sample, all infants had cranial lesions. Postcranial lesions were highest in the 1–5 year old (75 per cent) subadults and lowest in older children and Old adults. In the adults, cranial and postcranial involvement was highest in the Young adults. Most (94 per cent) individuals with cranial lesions also had lesions of the postcranial skeleton. There were no significant differences between the adult age groups in postcranial or cranial involvement. No significant sex differences were found in cranial or postcranial involvement, although Young females had the highest prevalence of postcranial lesions.

Most individuals (69 per cent) were affected in multiple bones, but this was higher in adults than subadults. More subadults than adults were affected in only the tiba and a bone other than the tibia. Lesions were predominantly active in subadults and mixed active and remodelled in adults. The severity of lesions increased with age. The degree of new bone production did not progress past grades 1–2 in most subadults. This stage of bone involvement was also high in adults, although 21 per cent had more severe grades 3-4 bone changes. These more severe lesions were highest in Old adults.

Gummatous involvement of postcranial bones was not observed in any subadults under 16 years of age, but observed in all adults age groups except Old adults. *Caries sicca*-type cranial changes were observed in all adult age groups but were not found in any subadults under 16 years of age. Subadult cranial involvement predominantly affected the endocranial aspect although some new bone production was observed on the facial bones of these infants and children.

Skeletal element analysis

The skeletal element analysis showed different results to the individual analysis. Overall, 26 per cent of crania were affected while only 15 per cent of all postcranial skeletal elements had lesions. Most crania of subadults under four years of age had lesions of the endocranial aspect and more Old adult crania were affected than the younger adult age groups. This larger sample of adult crania showed that a similar proportion of lesions advanced to *caries sicca* grades 6–8 than in the individual analysis. As with the individual analysis, the crania of older adults had these more advanced *caries sicca* lesions. A statistically significant difference was found between the involvement of the crania in children younger than four years of age compared to older children and a statistically greater number of subadults were affected in the crania compared to adults.

Postcranial skeletal element analysis showed that the lower limb bones were significantly more affected than upper limb bones. The tibia was the most frequently affected bone in subadults and adults. More subadult tibiae were affected than adults, while a greater proportion of adult femora were affected compared to subadults. The status of lesions was similar in the skeletal element analysis to the results of the individual analysis where nearly all subadult postcranial lesions were active at death, while over half of the adult lesions had some evidence of remodelling. Again, no evidence of gummatous lesions was observed in subadults under 16 years of age and the prevalence was highest in Young adults.

Resorptive lesions of the vertebrae and appendicular skeleton

The analysis of lesions of the vertebral column found a high prevalence (89 per cent) of lytic lesions in the sample. One third of the individuals with lytic vertebral lesions had concurrent resorptive lesions of the appendicular skeleton. A high proportion (85 per cent) of those with vertebral lesions had either concurrent proliferative changes of the limbs or cribra orbitalia. Overall, more males than females had lytic changes of the vertebrae, and two thirds were either Young adults or between 10 and 18 years of age at death.

A more qualitative analysis of individuals with resorptive lesions of appendicular joints was also carried out. Nearly half of these individuals had lytic lesions of the wrist joints. The material was too fragmented to assess the pattern of involvement with any certainty, but an almost equal number had unilateral and bilateral involvement of the affected joints.

Table 6.44. *Individuals with osteolytic lesions of the vertebrae: Mound 1 and 2, Tonga.*

Burial	Age	Sex	Tbody	Tdisc	Lbody	Ldisc	Ribs	Other
Mound 1								
14	13.5		x					
7	Young	Male		x			x	Sternum
9	Young	Female	x					Cervical
Sternum								
6	Mid	Male	x	x			x	Cervical
Mound 2								
1e(4)	15.5		x		x			
11	Young	Female	x		x			
4	Mid	Male				x		costostern
13.2	Mid	Male	x					Cervical
costostern								
1.3	Mid	Female	x		x	x		humerus
18	Old	Male	x		x	x		Cervical

Tdisc = lesion of thoracic disc; Tbody = lesion of thoracic body; Ldisc = lesion of lumbar disc; Lbody = lesions of lumbar body; Ribs = osteolytic or osteoblastic lesion of ribs; Other = names the site of other osteolytic lesions of the axial skeleton

Table 6.45. *Individuals with extravertebral osteolytic lesions of the postcranial skeleton: Mound 1 and 2, Tonga.*

Burial	Age	Sex	Lesion	Pattern
Mound 1				
Burial	Age	Sex	Lesion	Pattern
21	Young	Female	Lytic lesion of right wrist, OB of Mc and humerus, cyst of ribs	N
29a	Mid	Male	Lytic lesion of distal medial entheses of left femur, OB of left tibia	N
Mound 2				
21	Adult	Female	Lytic lesions of metatarsals	N
22b	Adult	Female	Left metatarsal with lytic lesions, OB of left fibula	N
13c	Adult	Female	Cyst of distal femur	N

Lesion = site of extravertebral lytic lesion; N = the antimere was not present or could not be identified. Therefore the pattern of involvement could not be evaluated.

6.5.2 *Tonga*

Individual analysis

Individual analysis of the Tongan sample yielded samples too small for any quantitative analysis. However, the skeletal analysis presented a picture of diffuse proliferative lesions of subadult postcranial and cranial bones in both mounds. The predominant type of cranial lesion was of the endocranial aspect, with porosity and new bone production of the facial bones in some individuals. Adult lesions were not severe, with only one gummatous lesion in a Mid aged male from Mound 1. None of the adult individuals with crania had any cranial lesions.

Skeletal element analysis

Overall, 33 per cent of crania from Mound 1 were affected. Of the subadult crania from Mound 1, 43 per cent had lesions, half of which were under four years of age at death, while 20 per cent of adult crania had lesions. Two thirds of the crania from subadults under four years had lesions of the facial bones. The three adult crania from Mound 1 all had *caries sicca* changes. No adult crania from Mound 2 were affected.

Overall, 11 per cent of postcranial elements were affected at Mound 1. Similar proportions of upper and lower limb bones were affected, although the tibia had the highest prevalence of lesions in subadult and adults. At Mound 2, 9 per cent of skeletal elements were affected. A significantly higher number of lower limb bones were involved at this mound. Subadult bones were more frequently affected than adults bones. All subadult lesions at both mounds were active at death, with some remodelling in lesions of adult skeletal elements. Gummatous lesions were observed only at Mound 1.

Resorptive lesions

Resorptive lesions of the vertebrae were observed at both the Tongan mounds. None of the affected individuals had concurrent proliferative changes of the appendicular skeleton, while two individuals from Mound 2 had cribra orbitalia. Extravertebral lesions were observed in a number of adults from both mounds. These lesions were mostly of the lower limb.

6.5.3 *Comparison of skeletal pathology between Taumako and Tonga*

An analysis of lesions in skeletal elements provided comparative data with which to assess the different patterns of pathological lesions in cranial and postcranial skeletal material from each sample. Because of the relative paucity of data from the Tongan samples the comparisons made below are by necessity very general.

Cranial pathology

Overall, Mound 1 had a similar proportion of crania affected (33 per cent) to Taumako (26 per cent) while Mound 2 (9 per cent) was considerably lower. Of the adults, Taumako and Mound 1 had similar proportions of crania affected while no ($n = 0/19$) adult crania from Mound 2 had lesions. With the Tongan data combined, prevalences were similar to Taumako, except in adults (Table 6.46). A significantly higher number of adult crania were affected at Taumako compared to combined Tongan mounds (chi-square p-value < 0.00001).

A greater proportion of subadults had lesions of the crania compared to adults in all samples. A difference in the type of cranial lesions was found between Taumako and Tonga. The young subadults from Tonga had more facial lesions than endocranial lesions, while the young children from Taumako had predominantly endocranial lesions. The *caries sicca* lesions of adults from Taumako were more severe than those at Mound 1, although the lack of older adults at Mound 1 may explain this difference in severity.

Postcranial pathology

Overall, the prevalence of skeletal elements affected was higher at Taumako (Table 6.47). This difference was statistically significant between Taumako and Mound 2. No statistical difference was found between Taumako and Mound 1 or between the two Tongan mounds. However, a significant difference was found between Taumako and the combined Tongan data (Table 6.48).

The total number of limb bones were then pooled. These data are considered to represent the more important skeletal elements which aid in differential diagnosis of the disease causing the lesions (Table 6.49). Figure 6.32 demonstrates that a greater proportion of skeletal elements were affected at Taumako than either of the Tongan mounds. The differences between Taumako, both the Tongan mounds, and the combined Tongan data were statistically significant (Table 6.50).

The pooled data from subadult limb bones were then compared to test whether the subadults at Taumako were more at risk of developing skeletal lesions than subadults from Tonga (Table 6.51 and Figure 6.33). These data show that less subadult bones from Taumako were affected than at Tonga. The prevalence of lesions in the combined Tongan data was significantly higher compared to Taumako (Table 6.52).

Table 6.52 and Figure 6.33 show that a higher proportion of adult limb bones from Taumako were affected than at both the Tongan mounds and with the Tongan data combined. These differences were significant and are presented in Table 6.54. No significant differences were observed between the Tongan mounds.

Table 6.46. *Prevalence of lesions in crania from Taumako and Tonga.*

Group	Taumako		Tonga Combined	
	A/n	Per cent	A/n	Per cent
Overall	47/179	26	14/59	24
Subadults	24/56	43	11/25	44
Adults	23/123	19	3/34	9

A = number of affected elements;*n* = number of elements observed

Table 6.47. *Total prevalence of all postcranial skeletal elements from Taumako and Tonga.*

Sample	A/n	Per cent
Taumako	369/2472	15
Mound 1	53/480	11
Mound 2	36/392	9
Tonga Combined	89/872	10

A = number of skeletal elements with lesions;*n* = number of observations.

Table 6.48. *Chi-square p-values of statistical differences of prevalence of lesions in postcranial skeletal elements between Taumako and Tonga.*

	Mound 1	Mound 2	Tonga Combined
Taumako	ns	0.0074	0.0022
Mound 1		ns	

Table 6.49. *Prevalence of lesions in major limb bones from Taumako and Tonga.*

Sample	A/n	Per cent
Taumako	307/1422	22
Mound 1	43/291	18
Mound 2	32/248	13
Tonga combined	75/539	14

A = number of skeletal elements with lesions;*n* = number of observations.

Table 6.50. *Chi-square p-values of statistical differences of prevalence of lesions in major limb bones from Taumako and Tonga.*

	Mound 1	Mound 2	Tonga Combined
Taumako	0.029	0.008	0.001
Mound 1		ns	

Table 6.51. *Prevalence of lesions in crania from Taumako and Tonga.*

Group	Subadult		Adult	
	A/n	Per cent	A/n	Per cent
Taumako	46/254	18	261/1168	22
Mound 1	30/146	21	13/145	9
Mound 2	13/49	27	19/199	10
Tonga Combined	43/195	22	32/344	9

A = number of skeletal elements with lesions;*n* = number of observations.

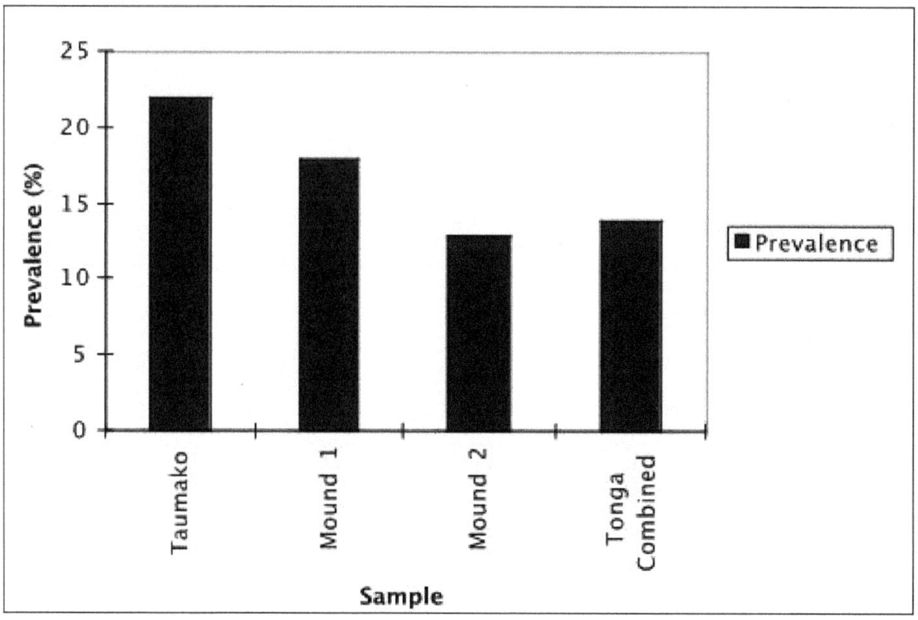

Figure 6.32. *General prevalence of subadult and adult limb bones from Taumako and Tonga.*

Table 6.52. *Chi-square p-values of statistical differences of lesion prevalence in the major limb bones of subadults between Tonga and Taumako.*

	Taumako
Mound1	ns
Mound 2	0.0247
Tonga Combined	0.0001

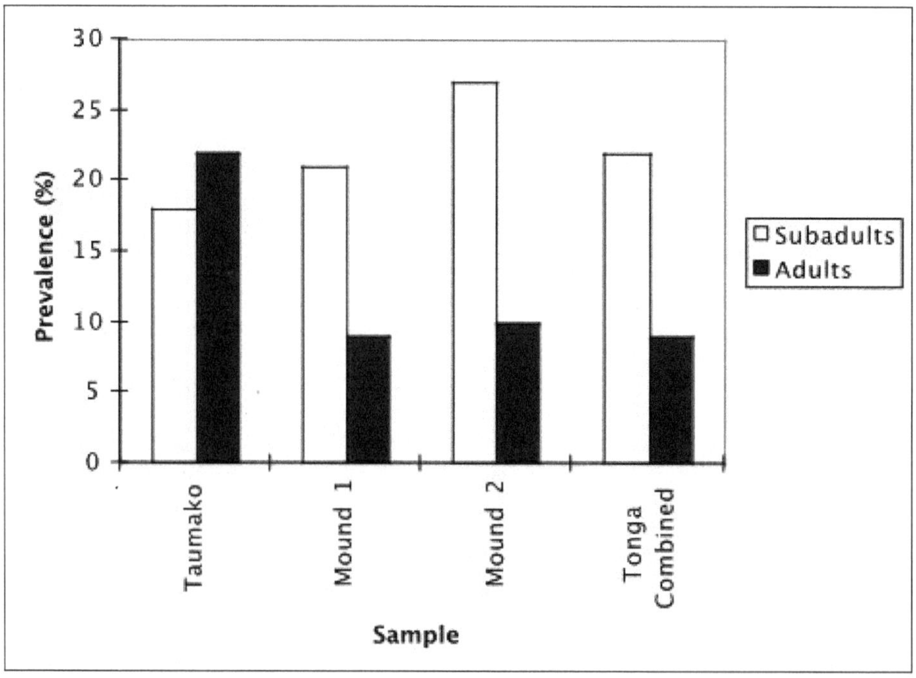

Figure 6.33. *Prevalence of lesions of subadult and adult skeletal elements from Taumako and Tonga.*

Table 6.53. *Chi-square p-values of statistical differences of general prevalence of lesions of adult skeletal elements between Taumako and Tonga.*

	Mound 1	**Mound 2**	**Tonga Combined**
Taumako	0.001	0.000	0.000
Mound 1		ns	

Pattern of skeletal involvement

In all samples the involvement of the tibia was the highest compared to all other bones. In order to aid in the diagnosis of the underlying condition causing the observed lesions, the pattern of involvement between upper and lower limbs of subadults and adults were also assessed. Figure 6.34 shows that more subadult upper limb bones were affected in the Tongan mounds than at Taumako and more lower limb bones were affected at Mound 2 compared to Mound 1 and Taumako. Chi-square statistical analysis showed there were no significant differences between the samples in this pattern of limb involvement.

The Taumako adults had generally greater involvement of both limbs than both the Tongan mounds (Figure 6.35). This result fits with the pattern of a significantly greater amount of lesions at Taumako. However, it is useful to demonstrate that this pattern results from the involvement of both limbs and not a single limb. The statistical values for these differences are presented in Table 6.54.

Lesion status

The status of lesions was similar between Taumako and Tonga. The lesions of subadult postcranial bones were predominantly active in each sample while adult skeletal elements were mixed active and remodelled. Only Taumako and Mound 1 had evidence of gummatous lesions in adult bones. The proportion of limb bones affected with this specific lesion was similar in Taumako (4 per cent n = 44/1168) and Mound 1, Tonga (3 per cent n = 4/145). No gummatous lesions were observed at Mound 2. The small samples from Tonga negate any more detailed comparisons of lesion status.

Resorptive lesions

Resorptive lesions of the vertebrae were observed at both Tonga (Mound 1 = 67 per cent; Mound 2 = 46 per cent) and Taumako (89 per cent). A qualitative analysis of all samples presented in the previous section also revealed a number of individuals with resorptive lesions of the appendicular skeleton. A differential diagnosis of these lesions will be addressed in chapter seven.

6.6 Conclusions

This chapter has contributed to the third objective of this thesis by providing data which will be used to assess the relative levels of disease between the samples. The skeletal element analysis found a similar proportion of proliferative lesions, indicative of infectious disease, at Taumako and Mound 1, while significantly more lesions were observed between Taumako and Mound 2. When the Tongan data were combined, Taumako was shown to have significantly more lesions of adult crania and of all postcranial elements. Significantly more subadult postcranial limb bones were affected at the Tongan mounds than Taumako, while a statistically significant higher number of adult limbs were affected at Taumako compared to the Tongan mounds. Of the bones affected, a greater proportion of lower limb bones including the tibiae were involved in all samples. Gummatous lesions were observed only in adolescents and adults and only at Taumako and Mound 1. The severity of cranial and postcranial lesions was greater at Taumako than Tonga. The implications of these results for the third aim of this study will be addressed in chapter seven.

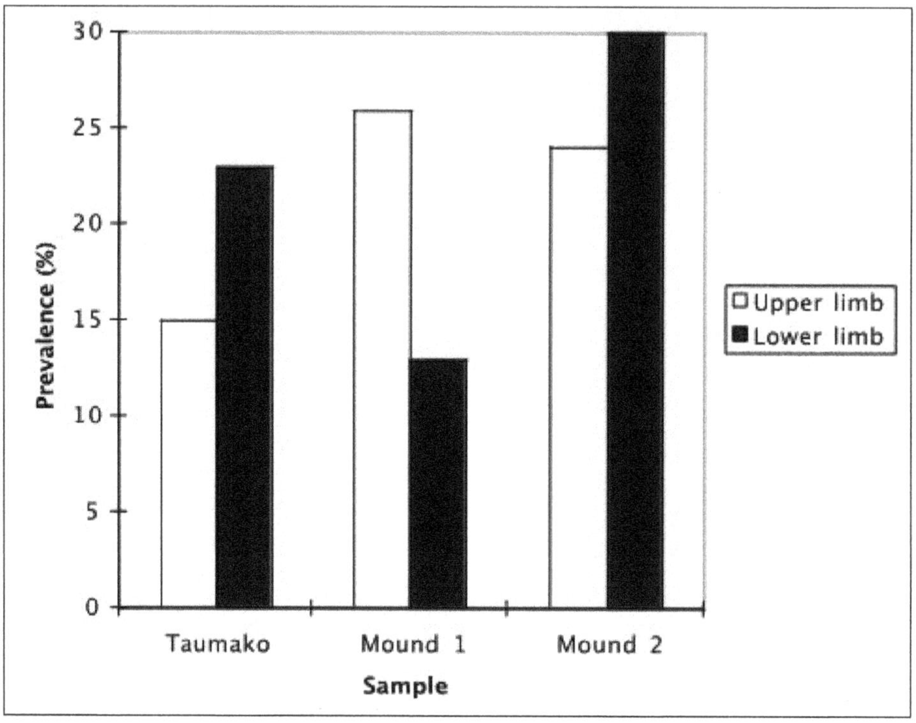

Figure 6.34. *Pattern of involvement of subadult upper and lower limb bones from Taumako and Tonga.*

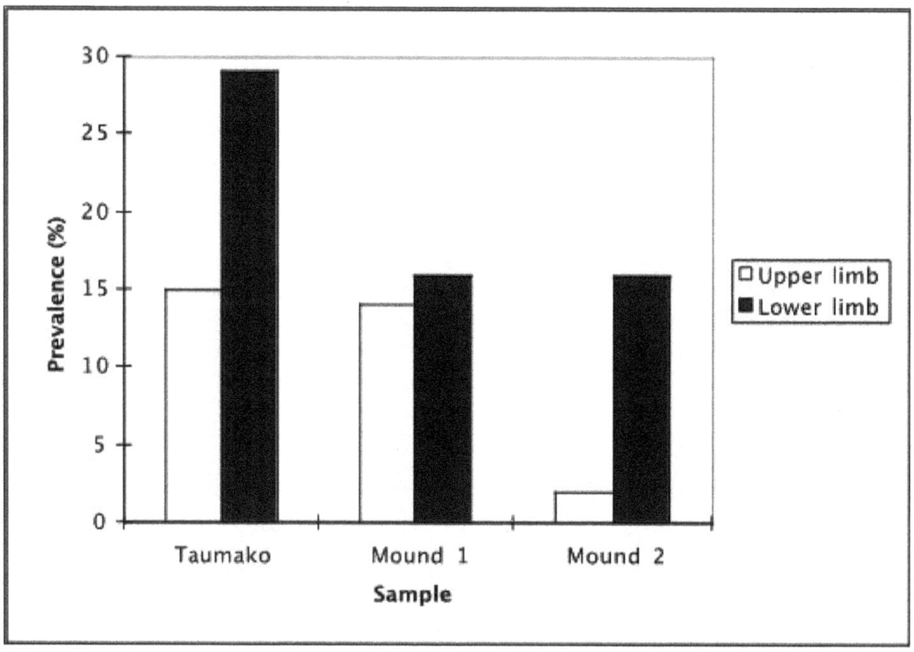

Figure 6.35. *Pattern of involvement of adult upper and lower limb bones from Taumako and Tonga.*

Table 6.54. *Chi-square p-values of statistical differences in the pattern of involvement of adult upper and lower limb bones between Taumako and Tonga.*

	Mound 1	Mound 2
Taumako upper limb	0.013	0.001
Taumako lower limb	ns	0.030

Chapter 7

Discussion and Conclusions

The skeletal indicators of stress and disease have been analysed in chapters three to six of this thesis. These chapters have directly addressed each of the research objectives as outlined in chapter one.

The first purpose of the present chapter is to summarise the results of the preceding four chapters. The second purpose is to interpret the results of this research. The interpretation requires the synthesis of complex and inter-related indicators of stress and disease. In order to adequately address the complexities, two different levels of interpretation are used. Firstly, at the *population level*, the separate parameters are discussed as relative measures of stress and disease within and between the samples. This compares the evidence from different parameters of health and disease within the populations and begins to explain the reasons for the differences observed between Taumako and Tonga.

The second or *individual level* of interpretation requires the examination of affected individuals with evidence of co-existent indicators of stress and disease. This level of interpretation is carried out in order to identify any predisposing factors in the cause of disease and/or mortality in affected individuals.

The discussions of population and individual levels of stress and concurrent lesions are inter-related. Therefore, these two levels of interpretation may seem circular and repetitious. However, the discussion of health and disease at the population level is based on separate sets of data, often representing different individuals; while that of the individual level provides an illustration of the interaction of disease processes in affected individuals. Therefore, the analysis of relative measures of health and disease at a population level may not represent true differences in health between the samples. Furthermore, the population level interpretation of these parameters is based on larger but unrelated samples; while the individual level would seem more valid but is based on smaller samples of individuals. While this is not ideal, it is a limitation inherent in any comparative study of health and disease based on skeletal samples.

The third purpose of this chapter is to discuss the possible causes of the different levels of stress between Taumako and Tonga in relation to diet and nutrition in a Pacific context. The fourth purpose of this chapter is to propose a diagnosis for the cause of skeletal lesions observed in the samples and how the specific diseases may have influenced the expression of the relative parameters of health and disease discussed in the first three sections of the chapter. The final purpose of this chapter is to directly address the aims of this research in light of the interpretations discussed above.

7.1 Results Summary

Firstly, the differences and similarities between the Tongan mounds must be considered, because to directly address the aims of the thesis it would be ideal to ultimately compare the complete Tongan sample with Taumako.

The mortality rates between the mounds show some differences which are important to consider. Differences in infant mortality between the mounds may suggest that infants were more vulnerable to death at Mound 1 or it may be a reflection of sampling differences.

In the adults, a similar pattern of higher mortality among Young adults compared to older adults was found at both mounds. Also, more females than males were identified at both mounds. This suggests that females died younger than males at Tonga. So, similar patterns of adult mortality were found between the mounds while subadult mortality was different. Considering the sampling issues associated with Mound 2, it is reasonable to argue that the differences in subadult mortality between these mounds might be more a reflection of sampling differences than real differences in mortality rates.

As shown in chapter four, similar patterns of LEH prevalence in the permanent teeth were found at both mounds. These similarities in the prevalence and pattern of LEH in the permanent teeth suggest that the environmental pressures which caused the development of LEH were similar at both sites. A difference was observed in the frequencies of LEH and circular caries in the deciduous teeth. Both of these parameters of prenatal stress were observed in very low frequencies which probably means that the differences in prevalence were due to sampling issues.

The mean stature of females from Mound 1 was 7cm taller than females from Mound 2. Males were a similar height at both mounds. The spread of the difference in stature was greater in Mound 2 females than Mound 1 females and more in Mound 1 males than Mound 2 males. These differences in adult stature might reflect different levels of childhood stress between the mounds, however considering the very small sample of Mound 1 females, the differences are probably the result of sample size.

Overall, a higher prevalence of cribra orbitalia in subadults and adults was found at Mound 2 compared to Mound 1. This difference was not statistically significant. Although fewer crania were present at Mound 2, all ages except infants were affected at this mound while only young children and a single Young adult were affected at Mound 1. These differences in prevalence of cribra orbitalia suggest that environmental circumstances predisposing children to iron-deficiency anaemia were more severe at Mound 2 than Mound 1. Despite these differences in prevalence rates of cribra orbitalia the severity of lesions observed were similar.

When examining proliferative lesions of the crania a higher overall prevalence of lesions was found at Mound 1. This may be explained by the fact that no adult crania were affected at Mound 2. A higher proportion of subadult crania were affected at Mound 2, but this constitutes only two individuals. The type of cranial lesions in subadults were similar in both mounds which suggests a common aetiology. In the postcranial material, similar proportions of skeletal elements were affected at Mounds 1 and 2. The proportion of affected upper limb to lower limbs bones were similar at Mound 1 while a significantly higher number of lower limb bones were affected at Mound 2. This difference in limb involvement is probably be a reflection of the more frequent involvement of subadult upper limb bones at Mound 1, or it could suggest different aetiology for the lesions at each mound.

The cranial and postcranial lesions were not severe with discrete gummatous lesions observed only at Mound 1. The differences in the involvement of crania between the mounds is probably a symptom of the poor recovery of crania at Mound 2. Many adults from Mound 2 were represented only by lower limb bones which may have skewed the proportion of upper and lower limb bones involved. Despite these sampling issues the pattern of proliferative lesions was similar at both mounds.

While some differences in the levels of evidence of stress and disease were noted between the mounds, most of these might be best explained by sampling issues. The overall higher prevalence of cribra orbitalia at Mound 2, although this difference was not statistically significant, is perplexing and may represent a real environmental difference which affected health. Despite this, the indicators of LEH and proliferative lesions suggest that the levels of stress and disease were largely similar at both mounds. Therefore, it is considered valid to compare the Tongan mounds with Taumako as a single sample in subsequent discussion.

7.1.1 Taumako compared with Tonga

Table 7.1 summarises the results from chapters three to six and shows whether the results found higher, similar, or lower prevalences of the stress and disease indicators and higher mortality at Taumako compared to Tonga. The data from this table are also used when discussing the relative measures of stress and disease at the population level in the following section.

As discussed in chapter three the mortality rates of Taumako and Tonga were similar in subadults, particularly when the Tongan data were combined. A difference in mortality between the sexes was observed between Taumako and Tonga where more males than females were identified at Taumako. Also at Taumako, more males lived longer than females while the opposite was found at Tonga. Before 20 years of age females were more vulnerable to death at both Taumako and Tonga. This mortality pattern among the sexes suggests that Taumako males were more vulnerable to early death than females at all ages, while at Tonga the opposite was indicated.

The results of LEH analyses show consistently higher prevalences at Taumako indicating more severe levels of childhood stress. The significantly higher number of circular caries at Taumako compared to Mound 1 suggests that stress levels at Taumako began during prenatal life while prenatal stress at Tonga was not a serious risk to health. The differences in LEH of the deciduous teeth would seem to support this. The statures of Taumako and Tongan males suggest that childhood stress did not affect the genetic potential for growth at Taumako; while Taumako females were shorter than those from Mound 1. This difference is probably a reflection of the small sample of females from Mound 1 more than a real difference in height. However, the greater range in heights of Taumako adults suggests that more individuals were susceptible to perturbations in growth during childhood than at Tonga. This assumes that the genetic potential for growth was similar between Taumako and Tonga, which given the 'mixed' population of Taumako, may not be the case.

The evidence of anaemia was consistently higher at Taumako compared to Tonga. The prevalence of proliferative lesions in total number of skeletal elements was significantly higher at Taumako compared to Tonga, except in the subadult limbs where the Tongan mounds had significantly more lesions than Taumako. Some results showed similar prevalences between Taumako and Tonga, particularly in the prevalence of cranial lesions in subadults. However, adult crania at Taumako were significantly more affected compared to the combined Tongan data. This similarity in cranial involvement of subadults is not surprising considering the anomalous result of the subadult limbs, because assuming a common aetiology for the lesions in Tongan subadults, it would be expected that the prevalence of subadult cranial lesions would be similar or higher than at Taumako.

These results show significantly higher levels of disease and stress at Taumako in all parameters except pathological lesions in subadult limbs and crania. Evidence of anaemia was higher at Taumako but this was not significant. These higher levels of skeletal and dental lesions at Taumako sug-

Table 7.1. *Summary results of relative prevalence of indicators of stress and disease at Taumako compared to Tonga.*

Indicator	Unit	Mound 1 Tonga	Mound 2 Tonga	Tonga Combined
Infant Mortality	Infants			Similar
Child Mortality	Children			Similar
Young Mortality	Young Adults			Similar
Mid Mortality	Mid Adults			Similar
Old Mortality	Old Adults			Higher
LEH	Deciduous LEH	-	Similar	-
	Circular Caries	Higher*	-	-
	Tooth	Higher	Higher	Higher*
	Individual	Higher	Higher	Higher*
	Subadult Permanent	Higher	Higher	Higher
	Adult	Higher*	Higher*	Higher*
Male Stature		Similar	Similar	
Female Stature		Shorter	Similar	
Cribra Orbitalia	Total	Higher*	Similar	Higher
	Adult	Higher	Higher	Higher
	Subadult	Higher	Higher	Higher
Skeletal Lesions Cranial	All Crania	Similar	Higher*	Similar
	Subadult Crania	Similar	Similar	Similar
	Adult Crania	Similar	Higher	Higher*
Skeletal Lesions Postcranial	All Elements	Similar	Higher*	Higher*
	All Limbs	Higher*	Higher*	Higher*
	Subadult Limbs	Similar	Lower*	Lower*
	Adult Limbs	Higher*	Higher*	Higher*

* = The difference between Taumako and Tonga was of statistical significance.

gests that levels of stress and disease were more severe at Taumako. It also suggests that the aetiology of lesions may be different at Taumako than at Tonga, particularly with regard to subadult pathology.

7.2 Population Level Indicators of Stress and Disease

As discussed above, the purpose of this section is to examine the parameters of relative levels of stress and disease at a population level. First these parameters will be considered within the separate populations and then Taumako and Tonga are compared.

The measures of growth and infection aid in reflecting conditions present at death but also of stress events during life (Goodman, 1993). High frequencies of these indicators provide insights into the efficiency of environmental exploitation, whether nutrition was adequate, and whether cultural practices such as weaning and food proscriptions may have negatively influenced the health of the population. They might also provide some indication of relative vulnerability to death and they may provide some information on the impact of social inequality in stratified societies. These indicators are combined in this section because if they are considered only in isolation then the full picture of relative health between the two Pacific Island regions would not be complete.

An examination of multiple indicators of disease has been carried out in other studies of health and disease in prehistoric skeletal remains. For example, an examination of a large skeletal series of infants and children from the prehistoric Libben site from Ohio found a significant correlation between porotic hyperostosis and periosteal reactions in infants from 6 to 24 months old (Mensforth et al., 1978). Mensforth et al. (1978) interpreted this association as comparable to modern frequencies of iron-deficiency anaemia in infants and children and correlated this with the age-specific distribution of weanling diarrhoea. Similar patterns of concurrent subperiosteal lesions and cribra orbitalia have been noted in other studies (e.g. Lallo et al. 1977).

Another study of infant and child health in a prehistoric Amerindian population found a significant correlation between circular caries and porotic hyperostosis and also circular caries and endocranial new bone (Cook and Buikstra, 1979). The association with endocranial lesions was interpreted based on the study of Sweeney et al. (1969). The study found that children with dental lesions were more likely to suffer fatal infections in the first few weeks of life than children without dental lesions (Sweeney et al., 1969). In a Pacific context, Stodder (1997) found similar age-specific patterns of LEH and cribra orbitalia in a study of dental defects in infants and children from prehistoric Guam. These age-specific patterns were interpreted by Stodder (1997) as indicating a common aetiology of both indicators of stress. She also found a significant association between LEH and proliferative lesions of the skeleton.

The researchers mentioned above all interpreted the proliferative cranial and postcranial lesions of subadults as evidence of concurrent infection. For example, Stodder (1997) concluded that all proliferative lesions of the subadults were due to infection, and some were due to yaws. Similarly, Mensforth et al. (1978) and Lallo et al. (1977) attribute the presence of subperiosteal new bone to an infectious process. As discussed below, this may not always be valid when considering subadult material. Therefore, the following discussion of relative measures of stress indicators does not assume a solely infectious origin of the proliferative lesions in young children and infants. In adults the development of proliferative lesions are interpreted as more likely to be the result of an infectious process.

7.2.1 Taumako

A lower mortality rate in infants compared to older children was correlated with a low frequency of circular caries, moderate levels of cribra orbitalia and high levels of endocranial and postcranial lesions. As discussed in chapter five, iron-deficiency anaemia is not usually known to develop in the infant until after 6 months of age. Therefore it would not be expected to find high levels of cribra orbitalia until later in infancy and early childhood.

It is generally accepted that newborn infants are protected from microbial infection by maternal antibodies imparted by way of breastmilk. Despite this, the newborn infant in tropical environments is vulnerable to attack from certain pathogens which can lead to conditions such as respiratory tract infections, general sepsis of the newborn and meningitis, caused by remote infections such as pneumonia (Jelliffe, 1970). These conditions, as well as specific endemic disease may have contributed to the high levels of postcranial and endocranial lesions observed in the Taumako infants. The universal involvement of the endocranial vaults in infants may correlate with the premature death of these infants. The underlying cause of the cranial and postcranial lesions may also be related to the nutritional stresses associated with weaning.

In the young children from Taumako, very high levels of circular caries were observed. This is correlated to a higher mortality compared to infants and higher levels of cribra orbitalia and postcranial lesions. The levels of circular caries in this age-group might suggest that these children were fed weaning foods, initiating the development of a carious lesion over the site of defective enamel. This age-specific pattern may infer that weaning foods were introduced between one and five years of age rather than during infancy. It may also suggest that the dental defect was associated with increased risk of death in this age-group. This risk of death may not be directly associated with the presence of the dental defect but a reflection of the cumulative effects of prenatal stress and increased risk of intestinal infections associated with weaning. Weanling diarrhoea may have been a contributing factor in the morbidity and mortality patterns observed in this age-category at Taumako. The presence of moderate levels of LEH in the permanent teeth of these young children suggests a chronicity of stress from birth to death.

Mild malnutrition can predispose children to repeated and chronic diarrhoea episodes, which in turn leaves the child more vulnerable to other infectious disease. This vulnerability to other infectious disease is reflected in the highest levels of postcranial lesions recorded in this sample. Another factor which makes this age-group more vulnerable to infectious disease is that most will be independently mobile as toddlers. A toddlers ability to interact with other members of the community would dramatically increase the opportunity of transmission of diseases endemic in the community.

The lower mortality rate of older children (6-15 years old) suggests that some children were less vulnerable to the stresses of infancy and early childhood than those which entered the mortality sample at a younger age. However, the high levels of cribra orbitalia and LEH in these children suggests that some finally succumbed to repeated episodes of ill-health as younger children. Conversely, high levels of these indicators may also reflect a subsample of children strong enough to combat acute episodes of ill-health as young children but died of causes unrelated to the condition which caused the dental defects. If the presence of proliferative lesions in subadults is interpreted as the result of an infectious process, then the low to moderate levels of lesions in these older children may indicate a period of latency of infection. The 16-20 year olds seem less likely to have suffered from iron-deficiency anaemia as young children which may have contributed to their survival until this age. The underlying condition which resulted in the high levels of LEH in this age-group may have contributed to them succumbing to infection, as reflected in increased levels of proliferative lesions. Despite this, the low mortality rate of this age group suggests that late adolescence was not an especially vulnerable time at Taumako.

The moderate to high levels of LEH and cribra orbitalia in all adult age groups suggests these individuals were able to survive the stresses of childhood but succumbed to death as adults. The higher levels of proliferative lesions in young adults, the higher mortality and levels of LEH in this age-group could be interpreted as the cumulative effects of chronic childhood disease contributing to premature death. When considering the number of dental defects in the permanent teeth of affected individuals this pattern continues. For both the subadults and adults at Taumako more individuals suffered periods of multiple stress than single episodes. This indicates more repeated episodes of stress during childhood in this sample. However, the subadults had a higher proportion of multiple defects indicating that multiple episodes of stress during childhood increased the risk of death before reproduction was possible.

Repeated episodes of weanling diarrhoea, respiratory infections and seasonal malnutrition could have contributed to this pattern of enamel defects at Taumako. However, as the epidemiological and clinical literature concerning *P. vivax* showed, the chronic nature of infection with this malaria species could also have contributed to this pattern of more multiple than single defects in this population.

An analysis of age at occurrence of defect development at the intertooth level of Taumako showed that subadults suffered from episodes of stress earlier in childhood than those individuals who survived into adulthood. This pattern is carried through in the individual analysis of mean age at occurrence of defect development. The data show that the age at occurrence increases with age at death. This suggests that if a period of stress occurred prior to three years of age then death during early childhood was more likely. A comparison between the sexes in mean age at occurrence revealed an interesting pattern where an early age at death in males was correlated with early age of occurrence of defect development. This pattern fits with other studies of the greater vulnerability of males to the development of dental defects. One study found that more male boys had hypoplastic defects of the deciduous teeth than females. This was interpreted as a result of greater demands for nutrition due to heavier birth weights (Infante and Gillespie, 1974). Similarly, the immune response to infection is thought to be less effective in males than females (Stini, 1969; Ortner, 1998). Therefore, the results of this study might suggest that the Taumako males were more vulnerable to the stress events of early childhood than females and predisposed them to early death. The fact that the age of occurrence is constant in females might suggest that low life expectancy was not highly correlated to childhood stress in these individuals. The wider range of heights in males may be a reflection of a greater impact of childhood stress on males than females as discussed below. This is complementary to evidence of greater vulnerability to early death with younger age at occurrence of defect.

7.2.2 *Tonga*

In the subadults only the 1–5 year olds had evidence of LEH and cribra orbitalia. Similar mortality rates between the infants and young children, with no evidence of growth disturbances or iron-deficiency anaemia suggests these infants died of acute stress before the development of lesions. Conversely, it is possible that levels of stress during infancy were not severe enough to cause the development of lesions. The young children with lesions may be individuals who survived the stresses of infancy but succumbed to pressures related to weaning. The individual analysis did show that some infants developed postcranial skeletal lesions while the older infants had more diffuse postcranial subperiosteal lesions with co-existent endocranial new bone formation. This suggests a greater severity of the causal condition with age.

In adults, low levels of LEH and cribra orbitalia in the young adults suggests these individuals were able to survive the stresses of childhood, but as a consequence, may have been more vulnerable to premature death. Indicators of childhood stress were higher in the mid and old adults than in younger. This suggests these indicators were not highly correlated with premature death. It is also possible that the cause of death may have been unrelated to these lesions.

7.2.3 Comparison between Taumako and Tonga

As discussed in chapter three, levels of subadult mortality were similar between the samples. This pattern suggests that the pressures influencing mortality in children were similar between these two Pacific Island regions. The threat to life appears to have been most extreme between one and five years of age. This is consistent with the age of weaning when exposure to pathogens and seasonal food shortages could lead to chronic illness and eventual death. So, like many regions of the world, the Pacific Island early childhood experience was a time of periodic illness and a particular vulnerability to death.

Where differences between the samples begin to become apparent is when the levels of the expression of childhood stress are compared. The very high levels of circular caries at Taumako and their virtual absence in Tongan children suggests that stress severe enough to cause significant alteration to the deciduous teeth affected Taumako children from prenatal life. Similarly, the higher levels of LEH in the permanent teeth suggest that the pressures experienced by the Taumako population were more extreme and more prolonged than at Tonga.

The higher prevalence of indicators of growth disruption in the Taumako sample supports one of the expectations of this thesis, but the tall stature of the adults does not. However, adult stature is a measure of the growth of individuals who have survived childhood stress and who may have had adequate catch-up growth to achieve their genetic potential for height. The wider range of heights in Taumako males may be a reflection of a greater impact of childhood stress on males than females as discussed above.

A higher prevalence of cribra orbitalia at Taumako suggests that iron-deficiency anaemia was a greater threat to child health than at Tonga. The underlying cause of iron-deficiency anaemia at both these sites may be related to the cause of LEH as both of these represent childhood conditions. If this were the case then the different levels of these stress indicators supports the premise of more severe childhood disease environment at Taumako. Similarly, the presence of infants at Taumako with cribra orbitalia, while only children over one year were affected at Tonga, is complementary with the evidence of more prenatal defects at Taumako. These inter-related results suggest that exposure to disease and the development of anaemia occurred earlier at Taumako and with greater impact than at Tonga. The statistically higher prevalence of proliferative lesions in adults at Taumako would also suggest that the disease experience continued throughout life with more virulence at Taumako than Tonga.

The analysis of proliferative lesions in the skeletal elements of subadults showed more were affected at the Tongan mounds than at Taumako. This pattern is anomalous to the consistently higher prevalence of other indicators of stress and disease at Taumako. This may reflect more a severe disease environment affecting Tongan children, or it may suggest a different aetiology for the lesions. The implications of this are discussed in the context of the differential diagnosis in detail below.

7.3 Individual Level Indicators

As discussed above, the sub-samples of individuals with multiple indicators of childhood stress provide a more immediate illustration of the interaction between different disease processes in prehistory compared to assessing relative measures of health in unrelated sub-samples of individuals. This discussion is based on affected individuals from chapters four and six.

7.3.1 Taumako

Table 7.2 presents the subadult individuals with all variables testable for multiple indicators of stress and disease. Of the eight individuals with circular caries, all had at least one other indicator of stress and disease. Subjectively, it would seem that individuals with prenatal physiological insults, reflected in the deciduous teeth, were more likely to develop subsequent conditions resulting in the apposition of subperiosteal new bone of the limbs and crania. The association between proliferative lesions and dental defects may suggest a common cause for both lesions. If this were the case then the disease which affected these individuals may have been present during prenatal life or predisposed them to neonatal infection.

Overall 67 per cent ($n = 8/12$) of those with postcranial lesions also had one other indicator of stress. When subadults with postcranial lesions were assessed for concurrent dental and cranial lesions, the association between these two parameters is strongest in the children two years of age and younger. Except for Burial 10, the association between postcranial lesions and other indicators of stress decreased after two years of age. Of the individuals with cranial lesions and cribra orbitalia, equal numbers had co-existent circular caries or postcranial lesions. Again, the association between postcranial and cranial lesions is strongest in children two years of age and younger.

A lower association between postcranial lesions and dental defects may suggest the development of a condition or transmission of infection after birth which was unrelated to the cause of the dental defects. These data suggest that whatever the cause of circular caries in this population, affected individuals were predisposed to later infection; while children without circular caries had skeletal lesions even in the absence of existing conditions. This pattern suggests a highly endemic disease environment at Taumako and that exposure to infection may have occurred soon after birth in some individuals.

In adults, the association between postcranial lesions and childhood stress in the form of LEH and cribra orbitalia

Table 7.2. *Subadult individuals with circular caries and the association with other indicators of childhood stress: Taumako.*

Burial	Age	Circular caries	Postcranial Lesions	Cranial Lesions	Cribra Orbitalia
125	0	-	x	-	-
124	0.75	x	x	x	-
130	1	x	x	-	-
50	1.25	x	x	-	x
129	1.25	x	x	x	x
107	1.5	x	x	x	-
92	2	-	x	x	x
158	2	x	x	-	-
93	2	x	-	x	x
104.2	4	-	x	-	-
122.4	6	-	x	-	-
10	6.5	x	x	-	x
106	12	-	x	-	-
n	**13**	**8**	**12**	**5**	**5**

x = parameter of stress present.

was higher than for cranial lesions (Table 7.3). This suggests the development of cranial lesions was the result of a different condition than that responsible for LEH and cribra orbitalia. It might also suggest that the cranial lesions developed later in life. When assessing adult individuals with postcranial lesions, more young adults had co-existent LEH than older adults (Table 7.4). Overall, these data show that LEH was not as highly correlated with the development of subsequent postcranial lesions in adults, as circular caries in deciduous teeth was in children. However, the higher association between LEH and other indicators in the Young adults may illustrate an interaction between growth disruption and predisposition to later infection which increased frailty and the chances of early death. The dental defects in these Young adults may also be an indication of the time of initial infection during childhood of the disease which caused the skeletal lesions.

7.3.2 Tonga

One subadult had LEH of the deciduous teeth. This infant was 9 months old at death and had concurrent proliferative lesions of the cranial and postcranial skeleton. Of the two subadults with circular caries, one was 1 year old at death with co-existent cranial and postcranial proliferative lesions. This child also had cribra orbitalia. The other child with circular caries had concurrent proliferative postcranial lesions only. The association of proliferative lesions with defects of the deciduous teeth would suggest a common cause for both of these indicators of stress. With so few individuals to consider it is recognised that this interpretation is subjective.

Four subadults had lesions of permanent dentition. Of these only one had co-existing lesions of the postcranial skeleton with circular caries. These data, while subjective, do not demonstrate a high correlation between LEH in permanent teeth and other lesions. This suggests that whatever caused the growth disruption did not necessarily predispose individuals to infectious disease and other indicators of childhood stress in this population.

Table 7.5 presents the subadult individuals with multiple indicators of stress and disease. Only four out of nine (44 per cent) individuals had another indicator. This suggests that the condition which caused the postcranial lesions occurred in the absence of predisposing stress reflected in the dental lesions and cribra orbitalia.

Of the adults from Tonga, sixteen had postcranial lesions (Table 7.6). No adults had concurrent LEH. Of the three Young adults with lesions, all had concurrent cribra orbitalia.

The preceding sections of this chapter have shown that each parameter examined found higher and more severe levels of stress and disease at Taumako compared to Tonga. The examination of these combined parameters support these conclusions.

At the population level, higher prevalences of prenatal stress at Taumako were found to correlate with higher proportions of cribra orbitalia and proliferative lesions in younger infants and children compared to Tonga. Similarly, higher levels of LEH in the permanent teeth of subadults and adults suggests more prolonged periods of childhood stress at Taumako compared to Tonga. The presence of multiple dental defects in adults from Taumako and their absence in adults from Tonga, further support the premise of repeated and prolonged childhood stress at Taumako. The levels of Tongan childhood stress seemed to be less severe than Taumako and if these occurred were acute rather than chronic.

At the individual level a strong association was found between all indicators of stress and disease in subadults from Taumako, especially those two years of age and younger. This association was not found in the Tongan subadults.

Table 7.3. *Adult individuals with LEH of the permanent teeth in association with other indicators of childhood stress and disease: Taumako.*

Age	LEH	PC	Per cent	Cranial	Per cent	Cribra	Per cent
Young	21	10	48	6	29	9	43
Mid	12	3	25	-		2	17
Old	4	-		-		2	50
Total	**37**	**13**	**35**	**6**	**16**	**13**	**35**

PC = postcranial lesions; Cranial = cranial lesions; Cribra = cribra orbitalia.

Table 7.4. *Adult individuals with postcranial lesions and the association with indicators of childhood stress and disease: Taumako.*

Age	PC	LEH	Per cent	Cranial	Per cent	Cribra	Per cent
Young	16	11	69	6	38	6	38
Mid	7	2	29	3	43	3	43
Old	4	2	50	0	0	0	0
Total	**27**	**15**	**56**	**9**	**33**	**9**	**33**

PC = postcranial lesions; Cranial = cranial lesions; Cribra = cribra orbitalia.

Table 7.5. *Subadults with postcranial lesions and the association with other indicators of childhood stress: Tonga.*

Burial	Age	Postcranial Lesions	Dental Lesions	Cranial Lesions	Cribra Orbitalia
5	0.25	x	-	-	-
29.3	0.5	x	-	-	-
1.2	0.75	x	x	x	-
3	1	x	-	x	x
7	1	x	x	x	x
17	1	x	-	-	-
1.1	1.5	x	-	x	-
15	1.5	x	-	-	-
36	2	x	-	-	-
Total	**9**	**9**	**2**	**4**	**2**

PC = postcranial lesions; Cranial = cranial lesions; Cribra = cribra orbitalia.
x = parameter of stress present.

Table 7.6. *Adults with postcranial lesions and the association with other indicators of childhood stress and disease: Tonga.*

Age	PC	LEH	Per cent	Cranial	Per cent	Cribra	Per cent
Young	7	0	-	2	29	3	43
Mid	1	0	-	0	-	0	-
Old	-	-	-	-	-	-	-
Adult	8	0	-	0	-	0	-
Total	16	0	-	2	13	3	19

Per cent = postcranial lesions; Cranial = cranial lesions; Cribra = cribra orbitalia.

This pattern suggests that the disease environment was more virulent at Taumako compared to Tonga. In adults from Taumako the association between LEH and other indicators was strongest in the Young adults which support the population level data of Young adults with LEH as more vulnerable to death. This level of interpretation also showed they seemed more vulnerable to lesion development of the postcranial skeleton with co-existent LEH. This pattern was not found in the Tongan adults.

In conclusion, the examination of relative measures of stress and disease at the population level found an association between mortality and LEH at Taumako; while this was not so strong at Tonga. At the individual level, the evidence is suggestive of a highly virulent disease environment at Taumako; while the lower association between LEH and skeletal lesions at Tonga does not. It would seem therefore that the individual level interpretation has supported the conclusions of the population level discussion.

7.4 Nutrition and Growth

Based on present evidence it can be concluded that the childhood experience of Taumako was not ideal for growth or survival. The following discussion will consider the environmental and cultural factors which may have influenced the different levels of LEH and cribra orbitalia between the populations within a Pacific Island context.

Within a broader Pacific context the level of LEH in Taumako adults (21 per cent) is not extreme and in the Tongan samples remarkably low (5 per cent for Mound 1 and 8 per cent for Mound 2). In a study of LEH presence in the canines of adult individuals from several Pacific Islands samples, Evans (1987) presents results that are comparable or higher than those at Taumako. For instance sites in malarious zones such as Nebira (NG) had 100 per cent prevalence of LEH, Eriama (NG) had 25 per cent and a site in Vanuatu had 33 per cent of adult teeth affected. A similar prevalence was found from the Lakeba site in non-malarial Fiji with 25 per cent of adult teeth affected (Evans, 1987).

In a study of non-specific indicators of childhood stress in Guam, a malaria-free island in prehistory, Stodder (1997) found 12 per cent of individuals had LEH of the deciduous teeth. This is considerably higher than the frequencies found in any of the samples of this study. These comparative results suggest that the level of LEH at Taumako could not be interpreted purely as the result of repeated episodes of infection with malaria. Other factors such as maternal and child nutrition may have been equally important. These factors are discussed below.

7.4.1 Diet and nutrition

The importance of adequate nutrition during infancy and childhood for child growth cannot be overstated and will be discussed below to examine whether nutrition patterns in prehistory may have contributed to the higher prevalence of LEH and circular caries during childhood at Taumako than at Tonga. Given the rich resources of Tongatapu and feeding strategies so advantageous to growth it would not be expected that high levels of LEH would be found in Tongan skeletal samples. Compared to Taumako, the present study confirms this expectation. However, some evidence of growth disruption was found in the Tongan skeletal material. Differences in access to resources among social classes may explain the variation of LEH in the permanent teeth of subadults and adults. As outlined in chapter four, a wide range in heights in some of the Tongan adults may reflect inadequate nutrition during growth.

In the context of Tongan social structure this variation may be explained by differential access to resources. As explained in chapter two, the Mounds at 'Atele were most probably communal burial sites used by local commoners. There was nothing to indicate these mounds were used by persons of high status. Therefore, it is reasonable to suppose that during periods of environmental or social strain some evidence of inadequate nutrition would be reflected in the skeletal material of these people.

In Polynesian islands with a highly stratified social structure, such as Tonga, obesity was considered a mark of the person's status. In modern Tonga protein-rich foods are available only to persons of high status, while those with lower incomes are forced to consume large quantities of root vegetables such as taro in order to fulfil the body's requirements for protein (Prior, 1976). This modern practice was probably followed in prehistory, where chiefly persons also controlled the use of land and access to resources. Therefore, it is important to remember that while the environment of Tongatapu provided a plentiful subsistence base, adequate nutrition may not have been available to all members of the community. Certainly, in times of drought or environmental disaster persons of common rank may have had less access to available resources.

Compared to the high levels of circular caries in some other Polynesian communities, it is quite surprising that such low levels of this dental defect were found at Tonga. However, when considering the context in which these defects were observed in Polynesia this may not be so unusual. For example, high levels of this defect were observed on Pukapuka where environmental constraints on subsistence were extreme (Prior, 1976). These constraints affected maternal health considerably (Davies, 1958) and probably contributed to the high level of circular caries. In normal circumstances maternal nutrition was probably adequate in Tonga.

The traditional feeding habits of the Taumako people are not known. However, it is known that the modern Polynesian-speaking inhabitants of the Solomon islands rely on a staple diet of breadfruit, nuts and coconuts (Willmott, 1969a). It is also known that the people of Taumako were accomplished exploiters of marine resources (Parsonson, 1966). Therefore, the diet of these people should not have been restricted by ecological constraints such as on Pukapuka, but high levels of prenatal and childhood stress were found in this study.

If it is assumed that the Taumako people adopted the food proscriptions of the neighbouring Melanesian islands it may go some way to explaining the high level of LEH in this sample. We know that a considerable amount of contact occurred between Taumako and Santa Cruz, and also that the Taumako community included some individuals captured from neighbouring islands. Therefore, it is reasonable to argue that a Polynesian speaking population living in a highly malarious zone might adopt infant feeding strategies which would aid in the survival of their children. The pattern of circular caries, in part indicating maternal under-nutrition, may also be a reflection of prohibitions affecting pregnant and lactating mothers. Moderate levels of LEH in subadult permanent teeth and high to moderate levels of LEH in adults support the argument that chronic under-nutrition from foetal life through to around five years of age could cause this pattern of growth disruption. It is concluded that differences in nutrition could be a major contributing factor for the different levels of prenatal and childhood stress found in this study.

7.4.2 Nutrition and infection

Throughout this thesis it has been argued that nutrition and infection are inextricably linked as contributing factors of non-specific indicators of growth disruption in skeletal and dental material. It has been argued that malaria was probably endemic on Taumako in prehistory and should therefore be considered as a contributing factor in the high levels of prenatal and childhood growth disruption at Taumako. The mechanism by which malaria may have contributed to this observed pattern is complex. It is proposed here that while malaria may not have directly caused high levels of maternal death at Taumako it may have contributed to the presence of prenatal dental defects in the foetus. If the Taumako people adhered to feeding proscriptions during pregnancy, then maternal under-nutrition may also have contributed to the pattern of deciduous dental defects. Therefore, it could be argued that malaria was also an indirect contributor to any ill-effects caused by this feeding pattern in prehistory.

Weanling diarrhoea probably affected the Tongan infants and contributed to the development of LEH in permanent teeth. However, with a child feeding pattern which would ameliorate the effects of the diarrhoeal attacks it is expected that the long-term affects of this condition would be less severe at Tonga than Taumako. The results of this study are consistent with this interpretation.

7.5 Anaemia

It was concluded that anaemic episodes suffered by the children of Taumako were more severe and prolonged than at Tonga. The involvement of infants at Taumako but not in the Tongan population suggests that the cycle of anaemia began at a younger age at Taumako and may be related to maternal iron depletion. This pattern of infant involvement is complimentary to the evidence of prenatal stress in the dentition at Taumako.

Other palaeopathological studies have found an association of skeletal evidence of endemic infection and prenatal under-nutrition with high prevalences of cribra orbitalia (Cook and Buikstra, 1979; Mensforth, 1985). The data from Taumako is in accordance with the conclusions of these other studies.

Within a broader context the prevalence of skeletal manifestations of anaemia at Taumako is not exceptionally high. In the non-malarious prehistoric Americas a prevalence of 38 per cent was reported for subadults from Channel Island off the south coast of California (Walker, 1986). The highest frequency in the Californian study was found in infants under one year of age. Similarly, a prevalence of 38 per cent was reported in subadults from four Amerindian sites (Lallo et al., 1977).

Other Pacific Island samples from non-malarious areas have reported a prevalence higher than Taumako. For example Stodder (1997) found 54 per cent of subadults from a site in Guam had cribra orbitalia. This is similar to a prevalence of 50 per cent from the site of Nebira, an endemic malarious zone in NG (Pietrusewsky, 1976), and a prevalence of 77 per cent of cribra orbitalia in subadults from a malarious region of Thailand (Tayles, 1996). Evidence of cribra orbitalia from eastern Polynesian sites have shown lower frequencies than Taumako and Mound 2 at Tonga. For example, a number of sites from prehistoric Hawaii yielded an overall prevalence of cribra orbitalia of 20 per cent (Pietrusewsky, 1994), while only two individuals from a skeletal series of 42 (5 per cent) individuals from the Marquesas Islands have cribra orbitalia lesions (Pietrusewsky, 1976).

Within this broader global context the prevalence of cribra orbitalia at Taumako is not anomalous. However, it is higher than the Tongan mounds which lends some support to malaria as a contributing factor in the aetiology of anaemia in this population.

7.5.1 Weaning and iron

The iron content in the diet of the Taumako people was probably similar to that of the Tongans. Parsonson (1966) describes the modern Taumako people as very successful at exploiting the resources of the sea. Therefore, the development of anaemia from a dietary source alone would seem unusual with such a rich source of dietary iron. However, large quantities of coconuts were consumed in the Polynesian speaking islands of Melanesia (Willmott, 1969a). The inhibitory properties of coconut milk may have counteracted the benefits of a rich marine subsistence base at Taumako. A further factor to consider is that young children may have been denied access to foods rich in iron, such as fish, green leaves and pork, while Tongan children probably had free access to these foods when they were available. Also, prolonged lactation, which may have been adhered to at Taumako, reduces the iron content in breastmilk and contributes to anaemia in children (Lepowsky, 1987).

If access to iron rich foods was restricted in children on Taumako, this may be a contributing factor in the difference in prevalence of cribra orbitalia between Taumako and Tonga. The greater consumption of coconuts in the Polynesian outliers may also have been an exacerbating factor in the development of anaemia in this population.

As discussed above, no prohibitions on eating protein and iron rich food were imposed on children in Polynesia. Maternal iron stores were probably adequate in Tonga which aided foetal development. These women would also have been able to supplement their diet with marine resources higher in iron than the staple vegetable foods. Nutrition surveys discussed in chapter two have shown that Polynesian infants were solely breastfed for no longer than six months after which time dietary sources of iron would maintain adequate requirements for optimal growth during infancy. The fact that no infants under one year of age had evidence of anaemia in the Tongan samples suggests they were supplied with adequate iron during this time of development. Factors which may have contributed to the levels of anaemia in this population were the inhibiting properties of foods such as taro leaves and coconut milk. Taro leaves act as inhibitors to iron absorption but they are also high in oxalic acid (Pollock, 1992). This acid may have irritated the immature digestive tract of weaning children and exacerbated any existing diarrhoea episode which in turn inhibited iron absorption.

Probably the most significant factor in a dietary cause for anaemia in prehistoric Tonga would have been the unpredictable availability of adequate quantities of food. The data show that cribra orbitalia began after one year of age in the Tonga mounds which supports a conclusion that the most vulnerable time for children in this region of the Pacific Islands was the weaning period. The lack of severe lesions in any of the affected individuals and relatively low prevalence compared to Taumako also supports the conclusion that anaemia was not an important threat to health in this population.

7.5.2 *Infection and anaemia*

As stated above for growth disturbance in the teeth, malaria does not directly affect the skeleton but it probably contributed to the clinical severity of anaemia during prehistory. Malaria has also been implicated as leading cause of anaemia in children in a number of western Pacific Islands (Bailey, 1966; Willmott, 1969a; Kariks and Woodfield, 1972; Bowden et al., 1985). This disease is therefore strongly implicated as a contributing causal factor in the skeletal manifestations of anaemia at Taumako. Willmott (1969, 29) summarises the relationship between anaemia and malaria in the Solomon islands thus:

> It is not easy to separate... the effects of malaria and malnutrition since they tend to interact and play a combined part in reducing the health of the population.

Hookworm may also have been present in both locations. The anaemia associated with hookworm infection is more severe when the individual is malnourished, particularly if iron stores are low (Jelliffe, 1970) and the effects of malaria exacerbate the effects of hookworm and nutritional deficiency (Willmott, 1969a).

The causes of anaemia in the context of these two prehistoric populations are probably multi-factorial with the combined factors of nutrition and infection equally responsible. The degree to which each of these factors contributed to the development of anaemia in affected individuals is impossible to gauge. However, as with the difference in dental defects, two factors stand out; these are the presence of malaria at Taumako and possible food restrictions which may have exacerbated the expression of anaemia. From the discussion of LEH and anaemia above it would seem that diet, endemic disease, and non-specific childhood infection and malaria all converged to leave the infants and young children of Taumako more vulnerable to the development of dental defects and a skeletal response to anaemia than at Tonga. These factors may have predisposed many to an early death.

7.6 Differential Diagnosis of Skeletal Lesions

The previous sections have discussed proliferative lesions as a measure of relative health between the samples. However, a differential diagnosis of the skeletal lesions is now important to assess the impact of specific infectious disease between the populations. Therefore, by offering a diagnosis of the diseases causing the lesions the non-specific parameters of health can be integrated within a context of the disease environment specific to each population.

Table 7.7 summarises the distinctive characteristics of the disease in the differential diagnosis, as reviewed in chapter two.

7.6.1 *Taumako skeletal lesions*

As chapter six shows, postcranial skeletal pathology was found in over half of the individuals in this sample from birth through to old age. This suggests a highly endemic disease environment at Taumako in prehistory. The prevalence of 53 per cent in individuals at all ages is also higher than Houghton's (1996) estimate for the prevalence of yaws in the Taumako sample. However, Houghton (1996) did not report any skeletal pathology in children under 13 years of age which probably accounts for some of the difference between these two studies. The Houghton study also assumes that all evidence of skeletal lesions was a response to yaws, which may not be the case.

Overall, 53 per cent of the subadults had lesions of the postcranial skeleton. This is high compared to other studies from the Pacific Islands. For example, Stodder (1997)

Table 7.7. *Characteristic features of each disease included in the differential diagnosis of skeletal lesions.*

Disease	Type of Bone Change	Bone Part Affected	Common Bones Involved	Distinctive Lesions
Osteomyelitis	Destructive and reparative		Tubular bones-any	Sequestra, involucra, cloaca
Infant		Metaphysis	Tubular bones of lower limb- involvement of multiple bones more common than children	
Child		Epiphysis	Tubular bones of lower limb	
Adult		Subchondral	Spine, pelvis and hands and feet	
Treponemal Disease	Predominantly osteoblastic (secondary). Lytic foci form in new bone (tertiary)	Diaphysis	Tibia predominates; and all other bones of lower limb. More distal bones of limbs. Cranial and facial bones	Multiple bones affected symmetrically with diffuse osteoblastic changes. Anterior tibial bowing. Gumma, caries sicca
Yaws	More destructive to face	Diaphysis	More destructive lesions of face and palate than syphilis	
Syphilis	More destructive to cranial vault	Diaphysis and subchondral	More destructive changes to cranial vault than yaws	
Congenital Syphilis	Early destructive; later reparative new bone	Early metaphysis; later diaphysis	Same as other treponemal conditions	Same as other treponemal conditions. 'Wimberger's sign', pathognomonic changes to permanent dentition
Leprosy	Resorptive, some osteoblastic restricted to tibia and fibula	Extremities- sometimes diaphysis		Facies leprosa syndrome. Resorption of extremities of small bones of hands and feet
Tuberculosis	Solely resorptive. Some reports of osteoblastic production on pleural aspect of ribs	Subchondral in limb bones. Subchondral and body of vertebrae-depends on pathway.		Pott's disease of spine. Joint destruction

found evidence of skeletal lesions in only 10 per cent of the subadults from a site in Guam.

No significant differences were found between the sexes which suggests a disease environment that was neither ameliorated nor worsened by sex specific biological or cultural factors.

Differential diagnosis of adult proliferative lesions

The prevalence of proliferative lesions in the Taumako adults was high with 53 per cent of individuals affected in the postcranial bones and 17 per cent of individuals with cranial lesions. The predominant type of lesion was purely proliferative characterised by new bone formation, but some had resorptive foci within the periosteal new bone, characteristic of gumma. In the postcranial skeleton, the diaphyses were primarily affected, except where gummatous destruction encroached on the joint space.

The pattern of affected skeletal elements showed a significantly higher involvement of multiple bones, including the tibiae which were affected significantly more than any other bone except the fibula. As illustrated in Figure 7.1. all postcranial limb bones were involved to some degree and lesions of the tibia and other lower limb bones were most frequently observed. When the number of affected lower limb bones were combined, the involvement of this limb was significantly higher than the upper limb. Lesions of the hands, feet scapulae and pelvis were only occasionally observed.

Based on the features detailed in Table 7.7 a diagnosis of osteomyelitis can be ruled out. This is primarily because no pathognomonic lesions such as sequestrae and cloacae were observed in this sample. Also osteomyelitis in adults rarely affects multiple bones. Tuberculosis is tentatively excluded as the cause of the proliferative lesions of the diaphyses. This is because tuberculosis primarily causes resorptive lesions of the vertebral column or joints of the appendicular skeleton. However, some individuals had mixed proliferative and resorptive lesions of the appendicular joints. Therefore, the possibility of tuberculosis contributing to the pathology of some individuals cannot be entirely ruled out.

The most likely condition affecting this group of adults from Taumako is one of the treponemal syndromes. The pattern of skeletal elements affected and the degree to which this infection has affected the population is strikingly similar to palaeopathological accounts of treponemal disease in other regions of the world (Powell, 1988; Ortner, 2003). This pattern also fits the criteria of treponemal diagnosis suggested by research on other Pacific Island samples (Stodder and Martin, 1992; Trembly, 1996).

No other infectious disease is known to elicit such a prolific apposition of new bone or to produce characteristic gummatous lesions of the bone tissue. The presence of lesions pathognomonic of treponemal disease such as gummatous lesions of the limbs and cranial vault and anterior tibial bowing support the diagnosis of a treponemal condition at Taumako in prehistory. Therefore, based on the pattern of involvement and the types of bone changes, a diagnosis of treponemal disease is suggested as the disease responsible for the majority of the lesions found in the Taumako adults.

Powell (1988) argued that it would be epidemiologically unrealistic to attribute all proliferative lesions of multiple bones to a treponemal condition. It is possible that some of the lesions observed in this sample are the result of other infections. For instance, leprosy cannot be ruled out in individuals with only the tibia involved and some individuals had proliferative lesions of the nasal bones indicative of early rhino-maxillary changes in leprosy (Ortner, 1996). However, the absence of pathognomonic *facies lesprosa* lesions suggests that leprosy was not the predominant infection in this population.

The pattern of skeletal involvement in each affected individual was used to estimate the proportion with a systemic infection. As shown in Table 6.8, 8 per cent of individuals were affected in only the tibia and 18 per cent were affected in bones other than the tibia. This pattern of involvement does not fit that of a systemic infection where multiple bones are usually involved (Ortner et al., 1992). The lesions of these two groups of individuals were probably the result of local trauma to the affected bone or some other non-specific infectious aetiology. These lesions were discrete and localised suggestive of a non-systemic infectious process, possibly from ulcerous infection of the soft tissue. Therefore, these lesions may represent an infectious process of less clinical significance than a systemic disease. Although, they may also be indicative of the early stages of a systemic infection or provide a pathway for the development of systemic disease in immune compromised individuals.

Trembly (1996) argued that only individuals with pathognomonic gummatous lesions should be diagnosed as unequivocal treponematosis in a skeletal series, while those with periostitis of the tibia and three other limb bones could be diagnosed as possible treponematosis (Stodder and Martin, 1992; Trembly, 1996).

These criteria have been applied below in order to differentiate those individuals with treponemal disease and those with non-specific infection. In the Taumako adults eight of the 38 affected individuals had gummatous lesions of the postcranial skeleton, supporting a diagnosis of probable treponematosis in at least these individuals. Three individuals with *caries sicca*-type lesions of the crania but no postcranial lesions are included in the probable group. The other nine adults with cranial lesions pathognomonic of treponemal disease did not have gummatous postcranial lesions but are also included in the probable group. In total, 49 per cent ($n = 20/41$) of affected individuals are diagnosed with 'probable treponemal disease'. If Trembly's criteria is followed, nearly one third ($n = 11/41$) of the individuals with periostitis of multiple limbs would be diagnosed as possible treponemal disease. Ten (24 per cent) of the individuals with proliferative lesions at Taumako were diagnosed as probably non-specific infection. Despite the fact that nearly

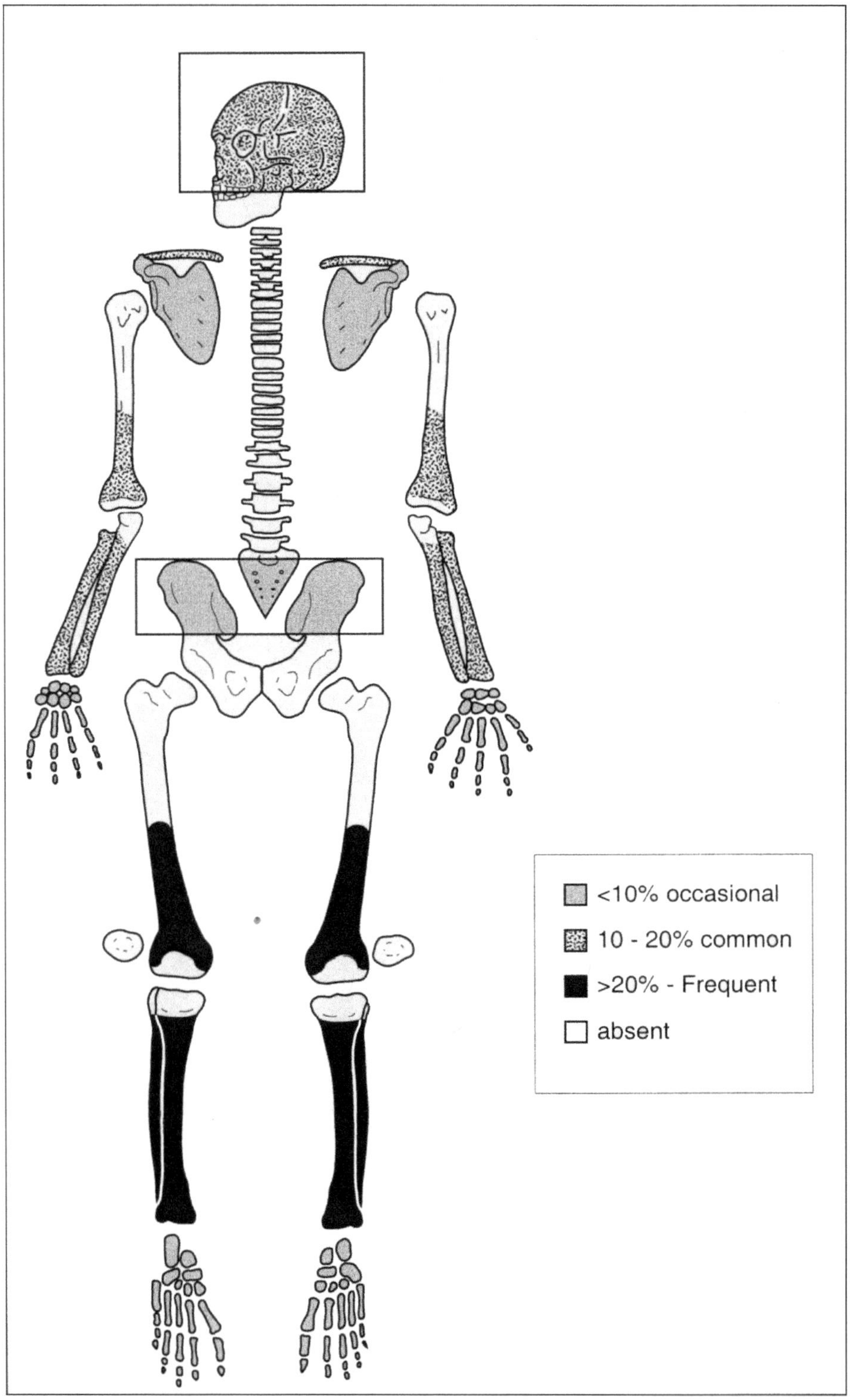

Figure 7.1. *Pattern of skeletal elements affected with proliferative lesions in Taumako adults.*

one third of the adults did not have pathognomonic lesions of the postcranial skeleton, the nature of the lesions and the pattern of skeletal elements involved would lend support to a conclusion that the predominant infection at Taumako was some form of treponemal disease.

Subadult skeletal pathology

The subadult pattern of skeletal involvement at Taumako was that of universal active proliferative lesions of the diaphyses. No lesions were characteristic of osteomyelitis or tuberculosis. The pattern of skeletal involvement in affected subadults was similar to the pattern observed in adults (Figure 7.2).

The tibia and other lower limb bones were significantly more affected than upper limb bones. Of the affected subadults, 56 per cent were affected in multiple bones including the tibia. This proportion is similar to the adult pattern. Based on this evidence the cause for the skeletal lesions in subadults from Taumako was probably a systemic infection. The prevalence of probable treponemal disease in adults and the *caries sicca* changes of the crania in affected adolescents would support a diagnosis of treponemal disease for most of the lesions in the subadults.

Treponemal disease is the most likely aetiology of these lesions, although congenital syphilis cannot be ruled out as an aetiology for the lesions in some subadults. The presence of circular caries with a high correlation of postcranial lesions in affected children suggests a transplacental transmission of infection. However, the absence of classic trophic changes of early congenital syphilis and pathognomonic changes to permanent dentition preclude an automatic diagnosis of this disease.

The complexities of diagnosing proliferative lesions in subadult skeletal material, especially in the Pacific Islands, has been addressed in a recent publication (Buckley, 2000b). This paper suggests that the possibility of metabolic disease, such as scurvy and traumatic injury, contributing to lesion development in subadults cannot be ruled out. For example, in the Taumako sample, six of the sixteen affected subadults were affected either in multiple bones other than the tibia or only in the tibia. The three children who were affected in bones other than the tibia were under five years of age at death. This does not fit with the expected pattern of skeletal involvement in treponemal disease and may suggest a non-specific aetiology other than infection, such as metabolic disease. In support of a non-infectious aetiology for some of the lesions, gumma were not observed in any subadults under 16 years of age in this population.

However, this non-specific pattern of skeletal involvement in some subadults from Taumako may also indicate early infection before other bones became involved. This is supported by the greater involvement of the femur in adults compared to subadults, possibly indicating that this more proximal bone does not become affected until later in the secondary and tertiary stages of the disease. Most of the lesions were active at death in subadults but some evidence of remodelling was seen in the 1–5 year olds and 16–20 year olds. This suggests that the disease which caused the lesions was not wholly responsible for the cause of death in these children.

The cranial pathology of younger subadults (< 10yrs) is primarily endocranial, although two of these have ectocranial new bone and porosity of the facial bones, including the sphenoid. These changes to the facial bones fit with recent published accounts of cranial changes in scurvy (Ortner and Ericksen, 1997). However, increased vascularity leading to porosity of these sites and new bone apposition could also be the result of general inflammation due to a systemic infection. With the presence of lesions indicative of possible scurvy and proliferative lesions of the limbs either from infection or subperiosteal haemorrhage associated with scurvy, an infectious origin other than yaws and/or a nutritional disorder such as scurvy cannot be ruled out in some cases.

Of the six subadults under five years of age with cranial lesions, five were under one year old at death. The lesions were predominantly on the endocranial aspect of the vault and the facial bones. This significantly higher prevalence of cranial lesions in infants compared to adults might suggest that these lesions contributed to mortality in infants. The literature concerning the aetiology of endocranial new bone is unclear. However, Cook and Buikstra (1979) and Mensforth (1978) both suggest these lesions are caused by chronic meningitis in the first few months of life. The aetiology of these cranial lesions deserves rigorous discussion in the anthropological literature, because this type of lesion is ubiquitous among infants and young children from diverse geographic and temporal backgrounds. The fact that this type of lesion was found in infants and young children might suggest the underlying cause of the lesions was more lethal than those of the ectocranial vault in the older children and adults. However, remodelling of endocranial lesions in some of these young children suggests the process causing the lesions was alleviated some time before death. The other studies mentioned above do not differentiate between active and remodelled endocranial lesions so may not have appreciated the complexities of this type of lesion. It is possible that the presence of endocranial lesions in infants attests to a congenital transmission of infection in this population. This is supported by the high prevalence of prenatal defects in subadults, however the endocranial lesions may not be of the same aetiology as the postcranial or dental lesions.

Differentiation between the treponemal syndromes: Yaws or syphilis at Taumako?

As discussed in chapter two, differentiating between the treponemes is difficult based on skeletal evidence alone, but because it is important to understand the biological cost of the treponemal disease present at Taumako it would be ideal to make a decision as to whether the treponemal disease was yaws or syphilis. Firstly, based on archaeological evidence the burial mound at Taumako was in use during

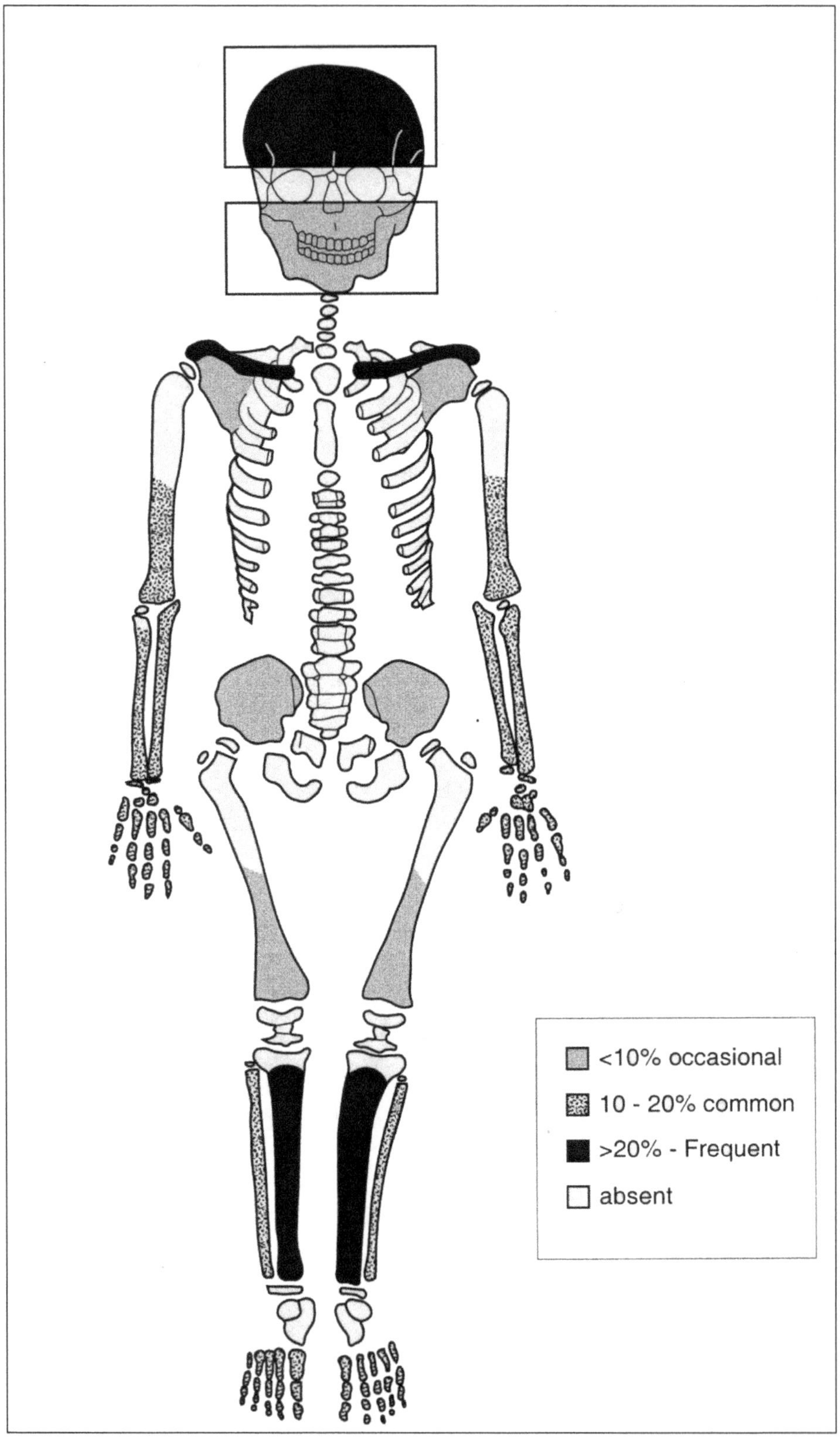

Figure 7.2. *Pattern of skeletal elements affected with proliferative lesions in Taumako subadults.*

the prehistoric period. On the whole dates from Taumako would support a pre-European non-venereal pathogen, although the later period of use of the mound does suggest that contact with European pathogens was possible.

As reviewed in chapter two, yaws is considered to be a childhood condition while venereal syphilis is generally transmitted in adults through sexual contact. The high prevalence of lesions in subadult material further supports a diagnosis of yaws rather than syphilis. Furthermore, no sex differences in skeletal involvement were found at Taumako. Yaws is the only treponemal disease which has a 1:1 ratio of males to females affected, while syphilis has a 3:1 ratio (Ortner, 1998). Finally, periostitis of particularly the tibia as seen at Taumako is known to be more florid and more frequent in yaws than syphilis (Aufderheide and Rodriguez-Martin, 1998).

These comparisons of the differences between yaws and syphilis would support a diagnosis of yaws and not venereal syphilis at Taumako in prehistory.

The palaeoepidemiology of disease at Taumako

The analyses of the status and severity of the lesions in the Taumako sample lend some clues to the impact of infectious disease at Taumako. This includes data of postcranial lesions in all affected individuals. The frequency of skeletal involvement fits with the expression of yaws in the Pacific Islands although the prevalence is higher at Taumako than other sites. Heathcote et al. (1998) tabulated the frequencies of 'presumptive yaws' in several large skeletal series from Guam. These frequencies range from 8 per cent to 20 per cent of adult individuals which is considerably lower than the prevalence of 49 per cent of probable yaws cases reported in this study. In subadults from Latte Period sites in Guam, 38 per cent of those with lesions were diagnosed by Stodder (1997) as yaws, while 6 per cent of the total population had lesions suggestive of yaws. The difference in prevalence between studies in Guam and the present study may have been influenced by the method chosen to estimate disease prevalence.

It is interesting that Stodder (1997) defines a prevalence of yaws ranging from 19 to 25 per cent in Latte period sites of Tumon Bay in Guam as hyperendemic. This would suggest then that the frequency found at Taumako reflects a more 'holoendemic' state of the disease. Based on this evidence it is probably reasonable to argue that with over half of the Taumako sample of individuals with skeletal lesions, most of the living population bore active and healed cutaneous lesions of the disease; 'every child would have seen its fate in the adults around, and secondary infection of the soft tissue lesions must have been a common source of debility and death (Houghton 1996, 215)'.

These lower prevalences of yaws in other Pacific Island sites would seem to set Taumako apart but the pattern of pathology at Taumako is not so dramatic when compared to studies from the Americas. For example, Cook (1976: cited in Powell 1988) reported a prevalence of 50 per cent of individuals affected with what she described as an endemic treponemal infection.

Based on skeletal evidence from Guam, Stodder (1997) suggested that transmission of the disease occurred in the first year of life and would have increased in toddlers and older children, with new cases of the disease declining after adolescence. The infectious lesions at Taumako follow a similar pattern. Firstly, a prevalence of 60 per cent of infants with proliferative lesions of the postcranial skeleton suggests that transmission of the disease occurred very early in life. This is because only the secondary stage of treponemal disease will produce changes in the bone tissue. The presence of two new-born infants with proliferative changes of the skeleton might indicate a congenital transmission of disease. While this possibility would then suggest that venereal syphilis was the treponemal disease responsible it should not be automatically assumed. Congenital transmission of the non-venereal forms of treponematosis are rare but not impossible (Parramore, 1970) and it is possible that in the holoendemic environment of Taumako, transplacental transmission of the disease occurred in some cases. However, the presence of neonates with skeletal lesions might reflect extremely rapid development of skeletal lesions after initial infection with yaws.

The highest prevalence of osteoblastic lesions in young children (1–5yrs) is similar to the peak prevalence of lesions noted in the Cook (1976) study. This high prevalence may also be a reflection of the greater proportion of individuals dying between one and five years. It is also possible that the cause of the skeletal lesions was partly responsible for the early death of affected children. The correlation between yaws and mortality is usually considered to be very low (Powell, 1992), however, this conclusion has been based on clinical studies in Africa and the Americas. The transmission of yaws in Africa did not occur as early in childhood as in the Pacific Islands (Hackett, 1946) where infection could occur as young as three months of age. It has been suggested that in a Pacific context, yaws was a significant contributor to early death in many young children (May, 1958). Therefore, given the association of dental defects and cribra orbitalia suggestive of chronic non-specific infection and under-nutrition at Taumako, it is reasonable that in this context yaws may have contributed to the mortality pattern seen in the sample. The lowest proportion of skeletal lesions were found in older children (6–15yrs). This may indicate a common period of disease latency for this population. While it seems counter-intuitive that active lesions in this age group would reflect a quiescent stage of treponemal disease, Hackett (1946) noted the presence of active bone lesions in children with no cutaneous sign of the disease.

After the latency period, some individuals may have experienced a spontaneous remission, where the immune system was able to overcome the disease, while some individuals would have progressed directly to the tertiary stage (Hackett, 1946). The skeletal evidence suggests that tertiary yaws did not develop until adolescence when gummatous lesions were first observed. If the epidemiology of the disease can be extrapolated from the skeletal data, it would seem that latency of the disease continues until after 20 years of age. The prevalence of skeletal lesions in adults is highest in the young adults and decreases with age, suggesting remission of the disease in some individuals. Although yaws was probably not the direct cause of death in these Young adults, it may have predisposed them to other more fatal infections. The presence of remodelled lesions with concurrent active lesions during adolescence and adult years indicates that this disease was capable of remaining throughout life. Some adults had evidence of wholly remodelled lesions suggesting respite from infection.

It is recognised that a skeletal population cannot directly reflect the disease pattern of a living population. However, the evidence of infectious disease presented in this study correlates closely with clinical surveys of yaws in the Pacific Islands. Except for the extreme prevalence, the results reported here fit with other palaeopathological studies of treponemal disease in the Pacific Islands.

7.6.2 *Tongan skeletal pathology*

The analysis of skeletal lesions from the Tongan mounds provided some information of the nature of pathology at these sites. Most types of bones were affected in varying degrees at both mounds which indicates a systemic infection or metabolic condition affecting multiple bones. More subadult bones were affected than adult bones. This suggests a disease that affects individuals early in life and the active status of all lesions in subadult bones may suggest that the underlying condition may have contributed to early death in childhood.

The pattern of skeletal involvement was different between the sites. The upper and lower limbs were similarly affected at Mound 1, while at Mound 2 a significantly higher involvement of the lower limb was found.

Differential diagnosis of adult proliferative lesions

Because only skeletal element data are available for Tonga, in the context of differential diagnosis, it is not possible to treat the data in the same detailed manner as the Taumako sample. The type of lesions observed in the Tongan mounds were similar to Taumako although significantly less prevalent in cranial and postcranial material. Lesions of the postcranial material were primarily osteoblastic, affecting the diaphyses of bones and with greater involvement of the tibia.

The pattern of skeletal involvement within individuals could not be quantitatively analysed but the descriptive tables of adults in chapter six shows the involvement of multiple bones in some individuals. No lesions could be defined as pathognomonic of osteomyelitis in either of the Tongan mounds. The proliferative nature of the lesions with universal involvement of the diaphyses would also exclude tuberculosis.

As shown in Table 7.7, leprosy can produce osteoblastic lesions of the diaphysis, particularly the tibia and fibula. However in the absence of any pathognomonic *facies leprosa* lesions, this disease could only be tentatively suggested as a cause of some lesions in the adults.

Figures 7.3 and 7.4 show a pattern of skeletal involvement similar to Taumako, although the skeletal involvement is not so widespread. Based on this pattern and the presence of gummatous postcranial lesions and *caries sicca* cranial changes in three adults from Mound 1, it is argued that some form of treponematosis, probably yaws, was present in Tonga in prehistory. However, a diagnosis of probable yaws could only be suggested for those few individuals with gummatous lesions. That yaws contributed to the development of other lesions is also possible but it is more likely that a non-specific infection of less clinical importance, such as tropical ulcer, or other non-infectious aetiology caused the majority of the postcranial lesions in the Tongan adults.

Subadults

Figure 7.5 and 7.6 illustrate the pattern of skeletal involvement in the Tongan subadults. The involvement of subadult skeletal elements is more widespread than in adult, with upper limb bones more frequently involved. This may suggest a different aetiology for skeletal lesions in subadults compared to adults. As mentioned above, a differential diagnosis of skeletal lesions in the subadults from the Tongan mounds has been addressed in a recent publication (Buckley 2000b). This article concluded that the most likely cause for the lesions in Tongan children was the result of a synergistic relationship between infection, probably yaws, weanling diarrhoea, and metabolic disease. This conclusion was based on the pattern and type of skeletal involvement and on clinical and nutritional literature reviewed in this thesis. It was also argued that in this context a traumatic origin could not be ruled out as contributing to the development of some of these lesions. Based on the skeletal evidence and literature reviewed it was also argued that it is more pragmatic to consider multiple causes for skeletal pathology in subadults than a single disease. In the light of the non-specific nature of the adult pattern of pathology in these samples, this conclusion is further supported.

7.6.3 *Comparison of skeletal pathology between Taumako and Tonga*

At Taumako subadults and adults were affected with similar proportions of cranial and postcranial lesions suggesting a highly endemic disease affecting a high proportion of the population from an early age. The significantly lower involvement of all postcranial bones in adults from the Tongan mounds suggests that the level of disease transmission was lower than at Taumako with a less virulent disease environment. This pattern may also be a reflection of a different host response to infection in the two samples, as discussed in relation to nutrition above.

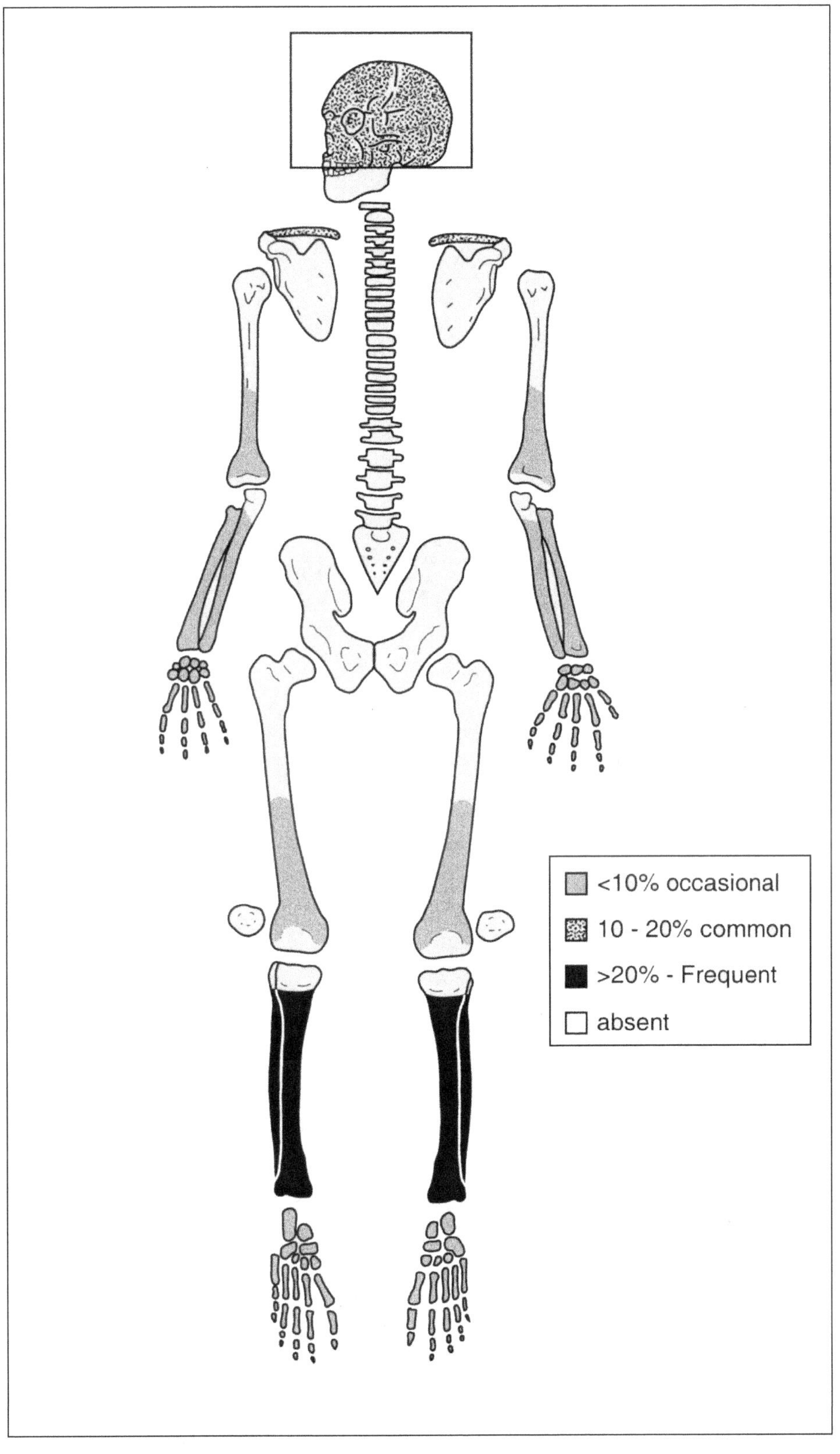

Figure 7.3. *Pattern of skeletal elements affected with proliferative lesions in adults from Mound 1, Tonga.*

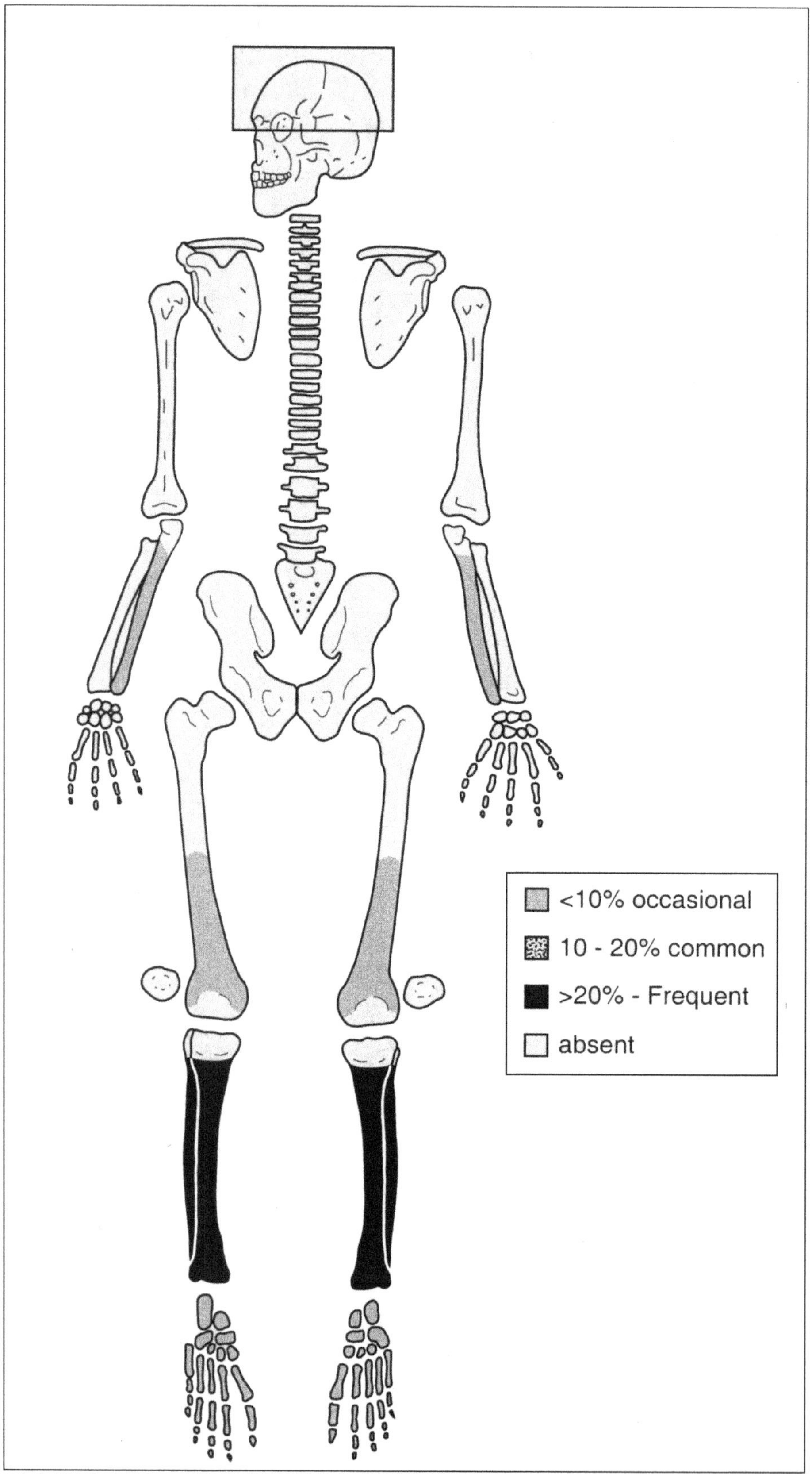

Figure 7.4. *Pattern of skeletal elements affected with proliferative lesions in adults from Mound 2, Tonga.*

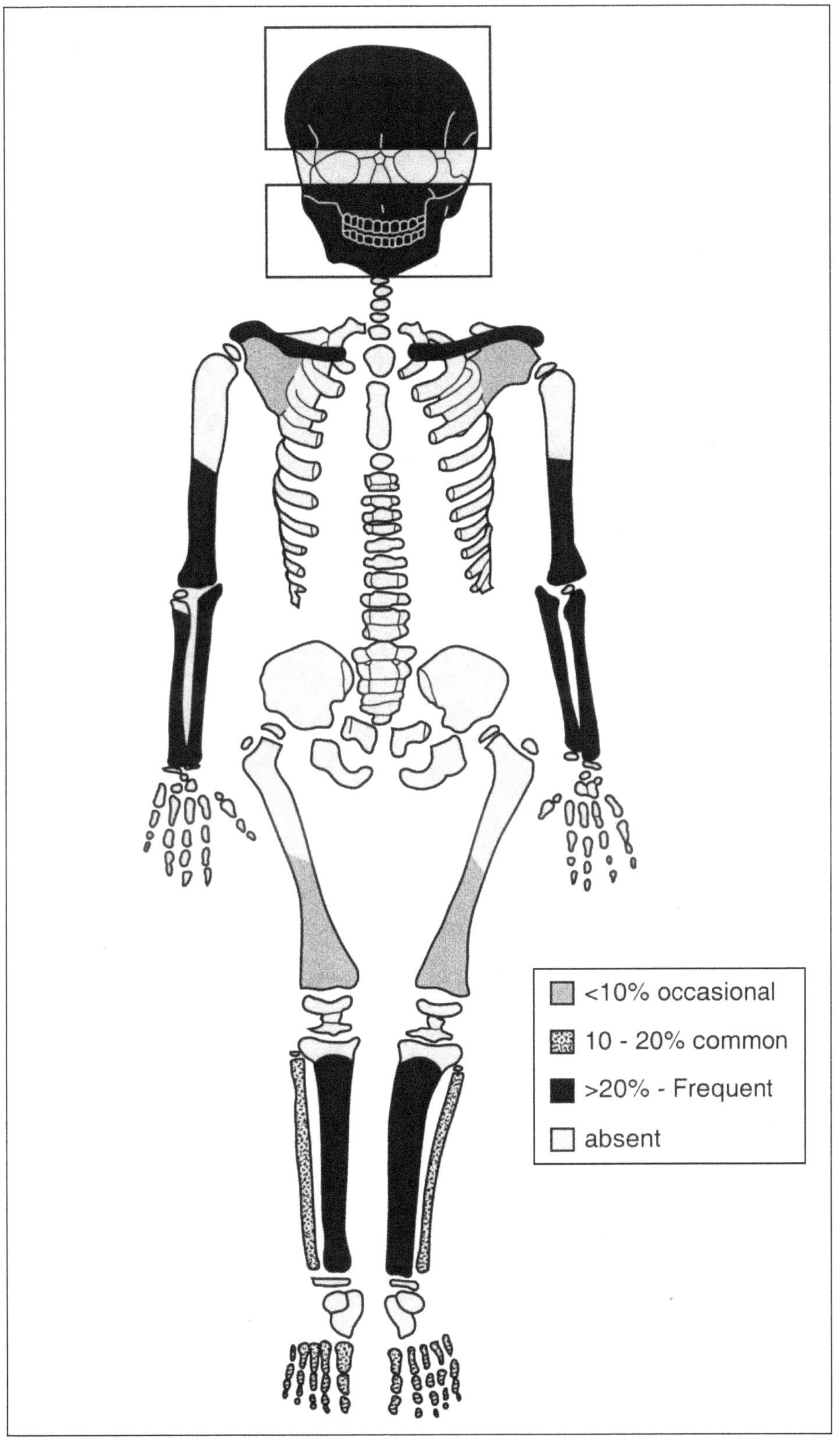

Figure 7.5. *Pattern of skeletal elements affected with proliferative lesions in subadults from Mound 1, Tonga.*

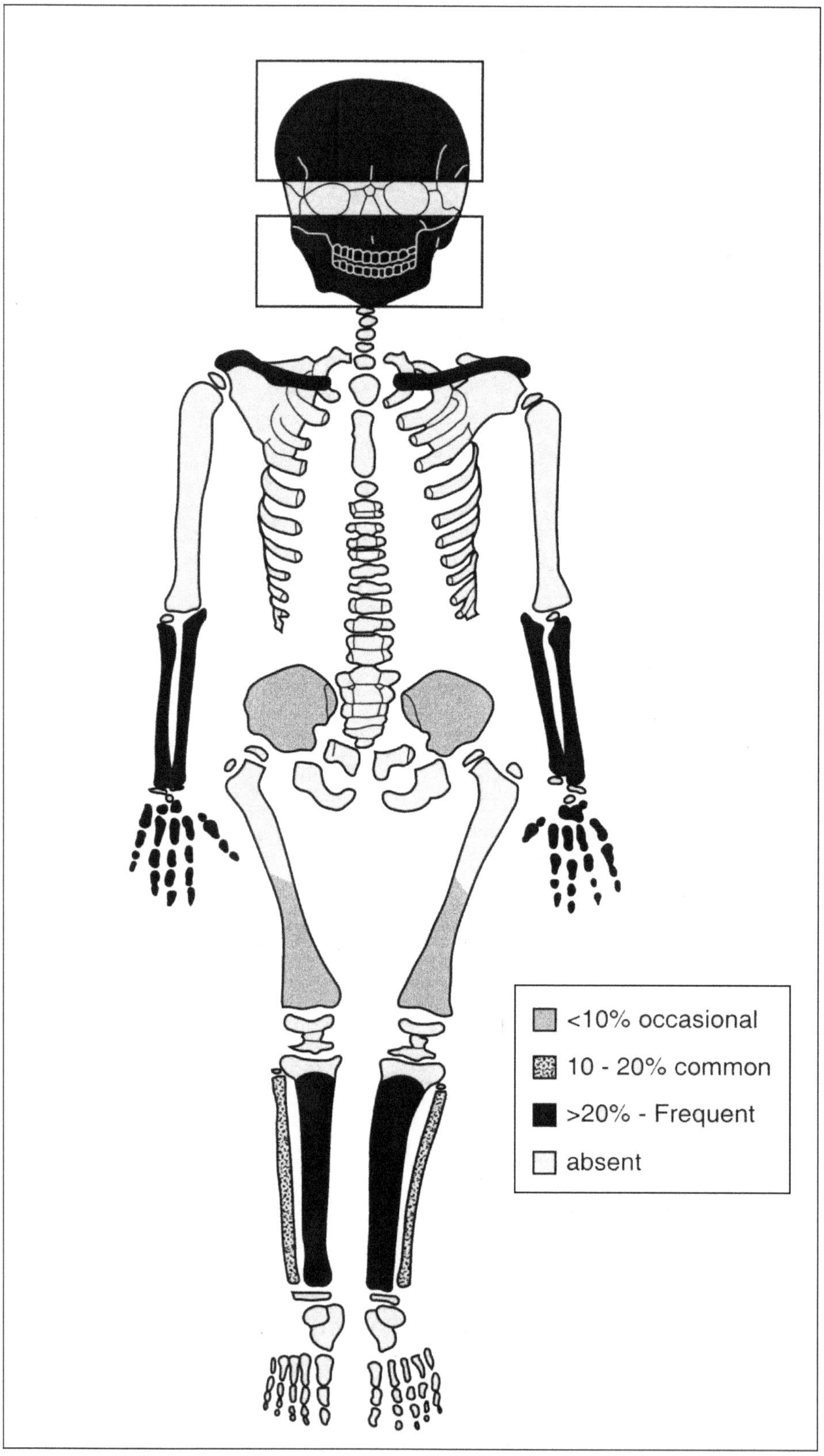

Figure 7.6. *Pattern of skeletal elements affected with proliferative lesions in subadults from Mound 2, Tonga.*

While the status and severity of pathology in Tongan adults could not be quantified, a more subjective discussion of the type of lesions observed at Taumako compared to Tonga is considered useful to further illustrate the greater severity of pathology at Taumako. At Taumako lesions were more severe than at Tonga. As shown in chapter six many of the affected individuals had lesions which had progressed to grades 3-4 of bone apposition while at Tonga no adults had lesions of this severity. Gumma were common in the postcranial bones of Taumako adults and were usually observed in clusters rather than in isolation. Only one individual had gummatous lesions at Tonga and these were discrete foci of destruction. Similarly, the *caries sicca* changes of the cranium progressed to grade 5 and over in nearly half of the affected adults at Taumako while no lesions over grade 5 were observed in the three Tongan adults with *caries sicca* lesions. Lesions of the facial bones characteristic of *gangosa* were observed only at Taumako. In terms of differences in types of lesions between the two samples, this supports the evidence from the quantitative data that the expression of disease was more severe at Taumako. The progression of lesions in the postcranial skeleton to grades 3 and 4 in some adults suggests a disease process which was of a long duration. The continual inflammation and pain associated with these lesions, and those less severe, would have made life uncomfortable and possibly predisposed these individuals to acute, more lethal, infectious disease. The skeletal evidence would suggest that the clinical effects of infectious disease were not so severe at Tonga.

Certain lesions recorded in the Taumako skeletal element analysis attest to morphological changes of the limbs, probably as a result of yaws infection. These were anterior diaphyseal bowing of the tibia, or sabre shin, considered pathognomonic of treponemal disease, and anterior bowing of the distal humerus. The latter morphological change has not been previously described in the literature but was observed in association with characteristic treponemal lesions of other postcranial elements in affected individuals. This suggests that bowing of the humerus is associated with treponemal disease. This change is associated with long term apposition of new bone on the posterior surface of the distal humerus shaft. The process in the tibia is similar but the apposition of bone occurs on the anterior and medial aspects.

It is suggested that the bowing of the humerus is due to the destruction of cutaneous tissue by gummatous lesions, leading to contracture of remaining soft tissues and inducing extreme flexion of the elbow joint. Over time the distal diaphysis will become bowed in a response to the pressure created by constant joint flexion. Sengupta (1985) described a clinical case of a Malaysian man who suffered severe contractions of the elbow, wrist and finger joints as a result of gummatous destruction of cutaneous and tendinous tissues. Bowing of the anterior humerus was observed in several individuals from Taumako, in varying stages of severity and in some cases with associated gummatous lesions. The occurrence of this bony change at Taumako, coupled with several cases of *gangosa*, conjure an image of many disfigured individuals. Individuals with gross contracture of limb joints would also be considerably disabled and probably unable to contribute labour to the community in terms of food procurement. This gross disfigurement of some individuals may also have decreased their sexual desirability (Houghton 1996). Therefore, this severe expression of treponemal disease at Taumako may have indirectly affected the population's success at reproduction. Bowing of the tibiae which was frequently observed at Taumako, would have caused considerable discomfort but was probably not as disabling as joint contracture of the elbow because it did not affect the joint space and movement was not impaired. Neither of these morphological changes were observed at Tonga.

While similar proportions of subadults were affected in the crania in both samples, the type of cranial involvement was different between the samples. At Taumako a significantly greater number of young children had endocranial lesions of the vault, while in Tonga endocranial lesions were identified, but they were not the predominant cranial lesion. This is relevant because at Taumako the data suggests that the presence of endocranial lesions in young children may be associated with early death. This association was not found in the Tongan samples where no children under six months of age had these lesions. Therefore, it may be concluded that the disease responsible for the development of endocranial lesions at Taumako affected the infants and young children more severely than at Tonga. This pattern further supports a hypothesis of a more severe disease expression at Taumako.

A possible explanation for the anomalous pattern of postcranial involvement of a higher prevalence in subadults at Tonga is that the Taumako children were dying before lesions could develop, while the Tongan children were less frail and lesions were able to develop before death. This interpretation is also in accordance with a disease of less virulence affecting the Tongan children and may suggest a non-infectious aetiology. As mentioned above, no Tongan infants less than three months of age had lesions of cranial or postcranial elements. They also had no evidence of anaemia prior to 1 year old while Taumako infants had evidence of anaemia and infection. More dental defects of the deciduous teeth at Taumako also suggests the assault of infection and stress began earlier and more severely than at Tonga. So, while the overall prevalence of postcranial lesions was higher at Tonga the more age-specific patterns of pathology suggest the skeletal response to disease occurred earlier at Taumako. Some infants would be expected to die before skeletal lesions could develop and those children who survived infancy may have finally succumbed to re-exposure to disease later in childhood, before the development of lesions. This would also suggest that the disease responsible for the lesions at Taumako was more virulent than at Tonga and supports the possibility of a different aetiology for the causes of the lesions at each site.

This interpretation is consistent with one of the expectations of this thesis that the expression of disease may have been exacerbated by the presence of malaria at Taumako. If this explanation were accepted then it would be expected that the Taumako infant mortality rate would be higher than at Tonga, as a reflection of greater frailty. The osteological paradox may an explanation but as shown in chapter three, the infant and subadult mortality rates were very similar at both sites so the 'paradox' cannot solely explain this difference. Differences in sample preservation between the sites may also explain this anomalous pattern of subadult pathology.

Therefore, it can be concluded that the skeletal evidence supports the expectation that the prevalence of infectious disease was higher at Taumako and probably had a more significant impact on health. However, following the ethos of the osteological paradox, the following questions must be asked: Firstly, does the overall higher level of infectious lesions at Taumako indicate an unhealthy population compared to Tonga or a more *adequate* immune response allowing longer survival and lesion development? The evidence of high levels non-specific stress indicators at Taumako does not support this interpretation and suggests that this population was under continual assault from infectious and non-infectious environmental stressors. Conversely, it could be argued that the expression of yaws was more florid at Taumako because of an *inadequate* immune response. The population's immune response may have been adversely affected by concurrent non-specific infection and iron-deficiency anaemia. Yaws does not directly increase mortality in a population but the *expression* of the disease may have been enhanced. The decreased immune response may have been exacerbated by the presence of malaria. This argument is supported by the fact that the prevalence of yaws at Taumako was much higher than prevalences reported in Guam (Stodder, 1997) where no malaria was present in prehistory.

By default, this line of reasoning suggests that the lower level of infectious lesions in the total Tongan sample indicate better health. This is supported by the very low levels of LEH at these sites. Conversely, the Tongans may have been dying early as a result of infection, before skeletal lesions could develop. The low levels of non-specific indicators of stress do not support this interpretation at Tonga.

Other factors which may have influenced the greater prevalence of lesions at Taumako, besides disease virulence and/or malaria presence, should be mentioned here. One of these is climatic variation between the two regions. Several researchers have noted that the expression of non-venereal treponematosis can be influenced by local temperature and humidity levels (e.g. Parramore 1970). As reviewed in chapter two Tongatapu is cooler and drier than other islands in this group and drought is frequent. The Solomon Islands, and Taumako by extrapolation, are typically hot and humid throughout the year. A hotter, more humid climate has been shown to favour the more severe clinical expression of yaws where more florid lesions develop (May, 1958). Therefore, part of the reason for the difference in disease prevalence between Taumako and Tonga may be climatic.

A further factor to consider is differences in population density between the two regions. It is possible that the concentration of the community on the tiny artificial island of *Tahua* favoured the transmission of yaws in this population. Finally, poor nutrition may have exacerbated the expression of disease at Taumako compared to Tonga. This final factor would seem to be supported by the higher prevalence of LEH and cribra orbitalia at Taumako as discussed above.

7.7 Differential Diagnosis of Resorptive Lesions from Taumako and Tonga

7.7.1 Vertebral lytic lesions

As outlined in chapter six, a number of individuals from the Taumako and Tongan samples had lytic lesions of the vertebrae. The presence of these lesions may have implications for the clinical impact of other diseases discussed above.

Firstly, it must be ascertained whether the lesions are in fact an anomalous expression of normal vertebral anatomy. However, the changes in the affected vertebrae illustrated in Figure 6.20 and 6.21 show that these changes are probably outside the normal continuum of vertebral body variation. It is therefore assumed that these vertebral changes represent an abnormal state.

Causes such as tumours and osteomyelitis can be ruled at both sites out on the basis of the rarity of these conditions and the very high frequencies of individuals with these lesions in these samples. Also osteomyelitis would not be expected to affect multiple vertebrae. Brucellosis is known to cause lytic lesions of multiple vertebrae (Ortner, 1999) however its presence has not been reported in the Pacific Islands.

Probably the most obvious disease to consider is tuberculosis. As reviewed in chapter six, tuberculosis can cause lytic lesions in vertebral bodies and endplates. The lesions of tuberculosis are characteristically discrete in type, affect a single or possibly two vertebrae, yet many of the individuals in this study had multiple vertebrae affected. The most pathognomonic change of tuberculosis is kyphosis of the spine caused by vertebral body destruction. No individuals in the Taumako or Tongan samples had these changes to the spine. However the preservation of vertebral material in these samples was poor. Therefore, it would not be expected that vertebral material with severe bone resorption would be well enough preserved to recognise these changes.

Ten of the individuals with lesions of the vertebrae also had lytic lesions of appendicular joints. These lesions may support a diagnosis of tuberculosis, such as destruction of the wrist joint (Ortner, 2003). Yet, some of the individuals with both vertebral and appendicular involvement were affected in multiple joints which is not characteristic of tuberculous destruction outside the vertebral column (Ortner, 2003).

A recent study has proposed that abnormal porosity in vertebrae, strikingly similar to that observed in these Pacific samples, is indicative of early tuberculosis lesions in several skeletal samples from diverse areas (Baker, 1999). In support of her argument Baker (1999) found individuals with classic Pott's disease in association with these less severe lytic foci. While it would be tempting to diagnose these lesions of the spine as tuberculosis based on the similarity in lesion type to those in Baker's (1999) study, without the presence of unequivocal tuberculosis spinal lesions this is not proposed. Baker (1999) found only 3 per cent of the individuals in her study had these lesions, while the frequency was considerably higher at Taumako and Tonga. If these lesions indicated early tuberculosis, then it would be expected that more severe changes would also be observed. The lesions of the endplates that were observed may be due to tuberculosis. However, based on present evidence it is difficult to determine whether these lesions are due to an infectious origin or degeneration of the endplate from a traumatic origin.

Another possibility is an abnormality of haemopoiesis causing the vascular channels of the anterior bodies to resorb. As shown in chapter six, 85 per cent of individuals with the abnormal vertebral changes had either cribra orbitalia or osteoblastic lesions of the postcranial skeleton and/or crania. The physiological response to anaemia is an overproduction of haemopoietic marrow. In dry bone, marrow hyperplasia is expressed as marked porosity of cortical bone indicating a response to the extra demands for marrow space. Trabeculae will also resorb to accommodate hyperplastic marrow (Aufderheide and Rodriguez-Martin, 1998). It is possible that the changes observed in the vertebrae of these adults is an artefact of marrow hyperplasia during childhood. Similarly, because one of the sites for haemopoiesis in adults is the vertebral bodies, these changes may represent chronic anaemia in adults as a result of continued exposure to parasitic diseases which cause haemolysis, such as hookworm and malaria. A further cause for the changes in the vertebral bodies in individuals with osteoblastic lesions may represent a response to the general hypervascularity of systemic infection.

From the brief outline above it would seem that a diagnosis of the cause of these vertebral lesions cannot be proposed with any confidence. The more likely causes are an abnormal haemopoiesis due to marrow hyperplasia and/or hypervascularity of infection. Clearly, a more specific consideration of these lytic lesions needs to be performed in the future.

Possible causes for the resorptive lesions of appendicular joints

As reviewed in chapter two, several diseases of infectious and non-infectious aetiology can cause erosive lesions of the bone in limb joints.

Firstly, brucellosis is an infectious disease that can cause bony changes of the joints. This disease is endemic in nomadic pastoral peoples of the Middle East (Ortner, 1999). It may have been present in the prehistoric Pacific Islands, but no reference to it as a cause of disease has been found for this study. Therefore, brucellosis is not considered further as a cause of these lesions.

Septic or pyogenic arthritis is a term used for a non-specific infection of a joint. These lesions are usually asymmetrical, affecting a single joint, and associated with precocious new bone production (Rogers and Waldron, 1995). Some Taumako individuals with joint lesions were affected unilaterally, however none of these lesions were associated with a significant production of new bone. Also, septic arthritis would not be expected in such a high frequency as observed in the Taumako sample. Based on these two factors, septic arthritis can be ruled out as a cause of the lytic lesions in these individuals.

Leprosy cannot be ruled out as a possible contributor to the lytic changes, especially where a single joint was involved. Two individuals from Taumako with lytic lesions showed resorption of the tufts of the distal phalanges which could be interpreted as early neurotrophic lesions of leprosy. However, in the absence of any pathognomonic *facies leprosa* lesions of leprosy a diagnosis could not be made with confidence. It is possible that some of the individuals with resorption of the nasal region interpreted as due to treponemal disease may have had concurrent leprosy. The association of both diseases has been reported elsewhere in the Pacific Islands (Stodder and Martin, 1992) however, further analysis will need to be carried out before this can be proposed with any certainty.

As discussed in chapter two, joint destruction in yaws has been suggested as a characteristic specific to this treponemal syndrome. It is possible that the lytic lesions in some of these individuals are associated with tertiary treponemal infection. However, lytic lesions would be expected to be associated with concurrent proliferative new bone characteristic of treponemal infection as shown in Figure 6.7c. This is not the case in most of the affected individuals. Six of the individuals with bilateral lesions have concurrent new bone, however the new bone is remote from the site of resorption. While the occurrence of discrete gumma causing joint destruction in some of these individuals cannot be ruled out, a diagnosis of treponemal disease as the sole cause of this pattern of lytic lesions cannot be made with any degree of confidence.

The form of the lesions observed in individuals with bilateral and unilateral involvement are very similar to those illustrated in Rogers and Waldron (1995) as examples of RA in prehistoric skeletal material. The range of lesions are also similar to illustrations of RA in Resnick and Niwayama (1995). Although it is very tempting to propose a diagnosis of RA for the individuals with symmetrical lesions, it must be noted that the manner in which these data were collected was not ideal for making a strong case for any erosive arthropathy in these samples. Rogers and Waldron (1995) state that all joints must be recorded for presence or absence in order to calculate a prevalence rate of arthritis in a population. However, the focus of this research was not to assess joint disease, therefore the presence or absence of joint elements were not systematically recorded in this study. Also, the presence of some individuals with single joints affected or multiple joints with associated proliferative margins suggests that a seronegative spondylarthropathy and/or gout must also be considered. Therefore, any subsequent discussion of the aetiology of the lesions recorded in this population must be understood as purely anecdotal. The recording of these lesions in a more systematic manner will be a focus of future research.

It can be concluded that a type of erosive arthropathy affected both of these populations in prehistory; rheumatoid arthritis may be the more likely possibility, but the seronegative spondylarthropathies and gout cannot be ruled out. The possibility of one or all of the conditions affecting these individuals to varying degrees must also be considered. Without the means to propose a more definitive diagnosis at present the effect of this arthrosis on the health of the population cannot be discussed with any degree of confidence. However, regardless of whether the aetiology of these lesions is infectious or as a result of some arthritis they probably represent significant disability in affected individuals and therefore warrant further research in the future.

7.8 Summary

Firstly, the summary results showed higher prevalence of stress and disease indicators at Taumako, except in subadult postcranial lesions. Mortality rates were similar in all ages except that males lived longer than females at Taumako, while the opposite was found at Tonga. The similarities in mortality are anomalous to the overall pattern of health and disease between the populations. It is possible that the differences are related to sampling issues and subadult age estimation, as discussed in chapter three. Similar stature results between Taumako and Tonga indicate that higher levels of growth disturbance indicators at Taumako did not adversely affect the growth potential in survivors. However, the range of heights was greater in Taumako adults which suggested that more were affected by periods of disruption in linear growth than at Tonga.

Subadults from Tonga had more postcranial lesions with a similar prevalence of cranial lesions, although younger subadults at Taumako were affected, with more endocranial lesions in young Taumako infants and children.

The population level interpretation of these results showed that the lower mortality of Taumako infants was correlated with lower levels of circular caries, cribra orbitalia and postcranial lesions. The highest levels of cranial lesions in this age-group suggests an association between mortality and this type of lesion. In the young children from Taumako the highest mortality rate was correlated with higher prevalences of all indicators. This pattern may be correlated to the cumulative effects of infant ill-health and the added pressure of the weaning period. The lower levels of mortality in older children was correlated with lower levels of postcranial and cranial lesions, although high levels of cribra orbitalia and LEH suggest a cumulative effect of early childhood illness, finally causing death in the more frail children of this age-group. Similarly, a lower mortality rate in adolescents was correlated with low levels of cribra orbitalia suggesting this was not a particularly vulnerable age at Taumako. However, higher levels of LEH and skeletal lesions suggests that early childhood stress, coupled with co-existent infectious disease, may have contributed to them succumbing to death at this age.

In adults, high mortality and higher levels of all indicators in Young adults may suggest that premature death was associated with childhood stress. Younger adults who suffered stress which caused LEH before three years of age were more likely to die than adults who suffered stress older than this. This association between early defect occurrence and premature death was stronger in males than females.

In the Tongan subadults, lower infant mortality was correlated with low or absent levels of cribra orbitalia and defects of the deciduous teeth. Higher levels of all indicators and higher mortality in young children suggest a more age-specific pattern of stress compared to Taumako. This may be related to the age of weaning at Tonga. The indicators of childhood stress were higher in older adults at Tonga which suggests these were not severe enough to predispose these individuals to premature death as at Taumako. A comparison between Taumako and Tonga at the population level showed similar mortality rates in subadults which may indicate that the threat to life was most severe in young children from both sites. However, a comparison of the differing levels of stress and disease indicators suggested that the disease experience at Taumako was more severe and prolonged than at Tonga.

At the individual level an association between all indicators of stress and disease was found in affected subadults from Taumako. This was stronger in children of two years old and younger. The overall pattern of co-existent indicators in these children suggest a highly virulent disease environment where exposure to infection may have occurred soon after birth or even in the womb. In the adults, more Young adults had co-existent LEH and postcranial lesions than older adults. This may indicate a predisposition to later, prehaps more fatal, infection in individuals suffering from non-specific stress as children.

In the Tongan sample, the individual level data showed a lower association between the indicators in subadults than at Taumako. This lower association between postcranial lesions and defects of the deciduous dentition suggests that exposure to the causal pathogen or development of the causal condition did not occur until sometime after birth. A similar pattern was observed in the Tongan adults at this level, which suggests the causal factors of postcranial lesions may have been less related to childhood stress than at Taumako.

Therefore, the individual level data was consistent with the population level interpretation. These interpretations suggest that the disease environment was more severe at Taumako and may have contributed to early mortality in some affected individuals.

The subsequent discussion on the role of nutrition in patterns of growth disturbance and anaemia between the samples argued that social inequality and periodic food shortages may have contributed to the patterns of LEH and cribra orbitalia in the Tonga population; while the possibility of food proscriptions at Taumako are suggested as contributing to the higher levels of dental growth disturbances at Taumako. These food proscriptions and the inhibitory properties of the substances in coconuts in this population may also have contributed to the higher levels of cribra orbitalia. Similarly, the interaction between chronic under-nutrition and endemic disease at Taumako may have further contributed to the more extreme expression of stress and disease in this population.

The differential diagnosis carried out on each population argued for holoendemic yaws at Taumako with a more non-specific disease pattern at Tonga. The expression of infectious disease was more severe in Taumako adults which may have been affected by the more chronic nature of non-specific stress in this population as argued above. Similarly, the more nutritious weanling and maternal diet at Tonga may have lessened the impact of infectious disease in this population. It was proposed that the anomalous pattern of subadult pathology between the samples could be explained by a different aetiology of less clinical severity at Tonga, leading to the age-specific pattern of disease expression observed.

7.9 Conclusions

This thesis has provided a synthesis of parameters of health and disease in two Pacific Island populations, in the context of examining the role of malaria in the success of human settlement of this region. This synthesis required the recording and analysis of data concerning the demographic profiles and evidence of growth disturbance, infection, and anaemia in each of the samples. The results from these analyses show that all parameters of stress and disease were higher at Taumako compared to Tonga, except premature mortality. The interpretation of these data at the population and individual levels offered explanations for the different patterns of stress and disease and attest to a more virulent disease environment at Taumako in prehistory. The discussion of the different levels of infectious disease at Taumako and Tonga, as expressed by skeletal lesions, further support the conclusions of a more virulent disease environment at Taumako. The remaining task is to directly address the aims of this research in light of the discussion presented above.

7.9.1 Research Aim 1: Were the mortality rates different between Taumako and Tonga?

The first objective of this research was to compile a demographic profile of each sample in order to test the hypothesis that malaria adversely affected population growth in prehistoric Near Oceania. Within this first research aim was the expectation that the demographic profile of Taumako would reflect a population under continual assault from disease. Based on the epidemiology of mortality associated with malaria, the death rates among young children were expected to be higher at Taumako compared to Tonga. Mortality among young females was also expected to be higher at Taumako. The skeletal data did not support the expectations of this first aim. It was found that mortality among all age groups was similar between the two populations.

Therefore, based on mortality data alone, it was not possible to assess whether the presence of malaria at Taumako might have affected the growth of the population. This failure of the data to conform to these expectations show that these samples do not support the hypothesis of Groube (1993) that malaria was responsible for the lack of population growth and technological development in prehistoric Near Oceania.

However, as reviewed in chapter five, in areas of stable endemic malaria, particularly where *P. vivax* is the predominant species, mortality associated with malaria is not high. While the correlation between death and malaria may not be very high, repeated childhood episodes of the parasite do predispose individuals to other illness. Therefore, it could be argued that in order to test the effects of malaria on population growth in prehistory, mortality rates are too crude an indicator to be truly representative.

The similar mortality patterns of these samples may also be a reflection of sample composition and preservation, as discussed in chapter two. Also because of the myriad of sampling issues associated with skeletal material, the mortality patterns of a skeletal sample cannot directly reflect the mortality patterns of a whole population (Jackes, 1992). Therefore, palaeodemography may not be the most sensitive indicator of the role of infectious disease in the success of human settlement of the Pacific Islands.

7.9.2 Research Aim 2: Could the skeletal and dental growth of the Taumako people have been adversely affected by malaria?

The second aim was to test whether there were differences in individual growth between the sites. The relative statures of the populations were assessed to test whether malaria may have adversely affected the genetic potential for growth in Taumako adults. It was expected that repeated periods of stress during childhood would be reflected in shorter stature among Taumako adults compared to Tonga. However, the results showed that Taumako adults were not particularly shorter than the Tongans.

As discussed above, the data on dental defects at Taumako attest to chronic and repeated periods of growth disturbance from prenatal life through to five years of age. These data showed statistically higher prevalences compared to Tonga and more evidence of repeated episodes of stress at Taumako. This pattern of stress during childhood may be consistent with chronic malarial infection during early childhood, while the less severe pattern at Tonga may have been associated with the stresses of weaning alone. The data of age at occurrence of dental defects also suggested that if stress occurred prior to three years of age the risk of death was increased in children and young adults at Taumako.

These data were also consistent with chronic infectious disease in this population. Other infectious and nutrition related causes for growth disturbance were also discussed. The possible differences in maternal and childhood nutrition between Taumako and Tonga are also strongly implicated as causes for the higher prevalence of growth disturbance at Taumako.

These data on dental defects support the first expectation of this research aim, while the data concerning adult stature do not. However, as discussed above, adult stature is a reflection of individuals who survived the ravages of childhood stress and disease. It could be argued therefore, that those individuals who survived to adulthood were strong enough to achieve their genetic potential for height through periods of catch-up growth. This is consistent with the age when partial immunity to malaria is reached in modern populations.

In conclusion, the skeletal and dental evidence of growth disturbance in the Taumako population suggests that periods of stress were more likely to result in early death of some individuals than shorter stature in survivors. Based on this interpretation of the dental and skeletal evidence, the expectations of the second research aim of the thesis would seem to be supported. While the data is consistent with malaria as a major contributing factor to this pattern, based on present evidence it is not possible to state this unequivocally.

7.9.3 Research Aim 3: Was the skeletal expression of infectious disease and anaemia exacerbated by the presence of malaria at Taumako?

The third aim of this research was to assess whether the relative levels of disease prevalence between populations was consistent with the presence of malaria at Taumako. It was expected that a higher prevalence of skeletal lesions, indicative of infectious disease, would be found at Taumako because of the lowered immune response of individuals suffering from malaria. As discussed above, the skeletal data support this expectation. The high prevalence of endemic treponematosis at Taumako and the more florid and severe expression of the disease compared to Tonga is consistent with the exacerbating effects of concurrent infection. This concurrent infection may have been malaria. This interpretation would seem to be supported by the higher prevalence of growth disturbance in this population compared to Tonga. The higher prevalence of postcranial lesions in older infants and young children from Tonga was interpreted as reflecting a less virulent disease environment, probably of a different aetiology to the pathology at Taumako. Based on these data the first aspect of the third research aim would appear to be supported.

The second expectation of this aim was that skeletal evidence of anaemia, due to the haemolysis associated with malaria, would be higher at Taumako. The higher prevalence of cribra orbitalia and anaemia in young infants supports this expectation. As discussed above, cribra orbitalia is probably best interpreted as a more non-specific indicator of childhood stress because of the interaction between infection and nutrition in the development of iron-deficiency anaemia. Therefore, other factors such as hookworm infection and an iron-deficient diet may have been equally responsible for this difference. In conclusion, the results of the analyses carried out in chapters five and six would seem to support both aspects of the third research aim of this thesis.

However, it is impossible to prove this assertion that malaria was the cause of more severe expression of disease at Taumako. As discussed above, endemic yaws and chronic under-nutrition were also likely contributors to this pattern of pathology. Whether or not it is possible to prove that malaria contributed strongly to this pattern of pathology at Taumako, intuitively, the differences in disease expression between these two populations may be explained by the presence or absence of this micro-predator.

This study has provided a synthesis of several aspects of health and disease from two prehistoric Pacific Island populations from different disease environments. The skeletal and dental evidence has shown that the people living within a malarious zone seemed to be more prone to periods of ill-health and consequent growth disturbance as children. The skeletal evidence of infectious disease revealed a population ravaged by holoendemic yaws and suffering from chronic iron-deficiency anaemia from infancy. The Tongans lived relatively free from virulent disease but were particularly vulnerable to infection and metabolic disease during the weaning period. The conclusions of this study represent a significant contribution to the study of health and disease in the prehistoric Pacific Islands.

However, this study has also highlighted several new questions which need addressing in future research. Firstly, it was found that the demography of the two samples were similar which made the first aim difficult to address. The small samples from the 'Atele burial mounds of Tongatapu further compounded the difficulties with comparing the measures of relative health and disease. Therefore, larger, more complete samples from Polynesia are needed for future research concerning differences in health between Near and Remote Oceania, and the role of malaria in these differences. Similarly, more complete samples of subadult remains may aid in assessing differences in skeletal growth of children between these two regions of the Pacific Islands.

Secondly, the interpretation of complex patterns of pathology was further compounded by the fact that malaria leaves no identifiable marker in dental or skeletal material. A matter for future research in this field may be seeking to identify the DNA of the parasite in skeletal material.

The enigmatic evidence of inflammatory joint disease found in both of these populations requires attention in future research. A diagnosis of the lesions observed in these sample may aid in further assessing the levels of relative health between the two regions.

This study would seem to have supported the arguments

of Groube (1993) and Kirch (2000) that the Polynesian environment was relatively free of infectious disease and therefore, favoured rapid population growth and the development of stratified societies. Kirch (2000a) has suggested that the role of infectious disease needs to be considered in the success of prehistoric Pacific Island settlement. This study has provided an initial step in addressing this question.

As outlined above, further comparative research on health and disease in skeletal samples from different regions of the Pacific Islands is required before this question can be answered with any degree of confidence. However, the people of Taumako are a testament to the impact of infectious disease on the quality of life in a population from a malarial island of the Pacific.

Appendix: Stature Equations

The stature equations are derived from Houghten et al. (1975, 333).

Single Long Bones Right Side	
$1H = 2.137F - 5.184S + 830.7$	15.7
$2H = 2.210T - 5.247S + 978.6$	4.7
$3H = 2.103f - 6.056S + 1045.0$	14.4
$4H = 1.782h - 7.339S + 1226.4$	28.4
$5H = 2.475R - 5.642S + 1160.7$	20.7
$6H = 2.257U - 6.738S + 1182.8$	24.5
Single Long Bones Left Side	
$7H = 2.176F - 4.528S + 796.8$	20.0
$8H = 2.077T - 5.602S + 1029.6$	7.1
$9H = 2.164f - 5.721S + 1023.6$	18.0
$10H = 2.520h - 4.440S + 963.1$	21.5
$11H = 2.500R - 5.415S + 1154.8$	20.5
$12H = 2.009U - 6.911S + 1257.6$	26.5

Units are millimeters H = stature (Height) F = femur T = tibia f = fibula h = humerus R =radius U = ulna S = sex (10 = male; 20 = female)

Bibliography

Abbott, I. (1991). Polynesian uses of seaweed. In P. Cox and S. Banack (Eds.), *Islands, Plants and Polynesians*, pp. 135–145. Portland, OR: Dioscorides Press.

Acsádi, G. and J. Nemeskeri (1970). *History of Human Life Span and Mortality*. Budapest, Hungary: Akademiai Kiado.

Adebajo, A. (1991). Is rheumatoid arthritis an infectious disease? *British Journal of Medicine 303*, 786.

Adebajo, A. (1996). Musculoskeletal diseases. In G. Cook (Ed.), *Manson's Tropical Disease* (20th ed.)., pp. 350–362. London, UK: W.B. Saunders.

Alemaena, O. (1986). Yaws situation in the Solomon Islands. *Southeast Asian Journal of Tropical Medicine and Public Health 17*(Supplement 4), 14–28.

Andersen, J., K. Manchester, and C. Roberts (1994). Septic bone changes in leprosy: a clinical, radiological and palaeopathological review. *International Journal of Osteoarchaeology 4*, 24–30.

Anderson, A. and G. Clark (1999). The age of Lapita settlement in Fiji. *Archaeology in Oceania 34*, 31–39.

Angel, J., J. Suchey, M. Işcan, and M. Zimmerman (1986). Age at death estimated from the skeleton and viscera. In J. Angel (Ed.), *Dating and Age Determination of Biological Materials*, pp. 179–221. Worcester, UK: Croom Helm.

Angel, J. L. (1964). Osteoporosis: Thalassemia? *American Journal of Physical Anthropology 22*(3), 369–373.

Angel, J. L. (1966). Porotic hyperostosis, anemias, malarias, and marshes in the prehistoric Eastern Mediterranean. *Science 153*(3737), 760–763.

Angel, J. L. (1984). Health as a crucial factor in the changes from hunting to developed farming in the Eastern Mediterranean. In M. N. Cohen and G. J. Armelagos (Eds.), *Paleopathology at the Origins of Agriculture*, pp. 51–73. Orlando, FL: Academic Press.

Armelagos, G., P. Brown, and B. Turner (2005). Evolutionary, historical and political economic perspectives on health and disease. *Social Science & Medicine 61*(4), 755–765.

Arya, O. (1996). Endemic treponematoses. In G. Cook (Ed.), *Manson's Tropical Diseases* (20th ed.)., pp. 940–950. London, UK: W.B. Saunders.

Aufderheide, A. C. and C. Rodriguez-Martin (1998). *The Cambridge encyclopedia of human paleopathology*. Cambridge, UK: Cambridge University Press.

Avery, J. (1974). A review of the malaria eradication programme in the British Solomon Islands 1970–1972. *PNG Medical Journal 17*(1), 50–60.

Backhouse, J., B. Hudson, P. Hamilton, and S. Nesteroff (1998). Failure of penicillin treatment of yaws on Karkar Island, Papua New Guinea. *The American Journal of Tropical Medicine and Hygiene 59*(3), 388–392. yes.

Bailey, K. (1966). Some aspects of anaemia, haemoglobin levels and iron metabolism in the New Guinea highlands. *The Medical Journal of Australia March 5*, 356–393.

Baker, B. and G. Armelagos (1988). The origin and antiquity of syphilis. *Current Anthropology 29*(5), 703–737.

Baker, B. J. (1999). Early manifestations of tuberculosis of the skeleton. In G. Pálfi, O. Dutour, J. Deák, and I. Hutás (Eds.), *Tuberculosis Past and Present*, pp. 299–307. Szeged, Hungary: Golden Book Publisher.

Baker, P., J. Hanna, and T. Baker (1986). *The Changing Samoans: Behaviour and Health in Transition*. New York, NY: Oxford University Press.

Barmes, D. (1967). Dental and nutritional surveys of primitive peoples in the Pacific Islands. *Australian Dental Journal 12*, 442–454.

Barrau, J. (1958). *Subsistence Agriculture in Melanesia*. Bulletin 219. Honolulu, HI: Bernice P. Bishop Museum.

Barrau, J. (1961). *Subsistence Agriculture in Polynesia and Micronesia*. Honolulu, HI: Bernice P. Bishop Museum.

Barrau, J. (1969). *The Oceanians and their Food Plants*, pp. 87–117. Tuscaloosa, AL: The University of Alabama Press.

Baume, L. and J. Meyer (1966). Dental dysplasia related to malnutrition, with special reference to melanodontia and odontoclasia. *Journal of Dental Research 45*, 726–741.

Bayard, D. T. (1976). *The cultural relationships of the Polynesian outliers*. Otago University studies in prehistoric anthropology. Dunedin: University of Otago, Department of Anthropology.

Bedford, S., H. Buckley, F. Valentin, N. Tayles, and N. Longga (2011). Lapita burials, a new Lapita cemetery and post-Lapita burials from Malakula, Northern Vanuatu. Southwest Pacific. *Journal of Pacific Archaeology 2*(2), 26–48.

Bedford, S. and C. Sand (2007). Lapita and Western Pacific settlement: pregress, prospects, and persistent problems. In S. Bedford, C. Sand, and S. P. Connaughton (Eds.), *Oceania explorations: Lapita and Western Pacific settlement*, Volume 26 of *Terra Australis*, Chapter 1, pp. 1–16. Canberra, Australia: ANU.

Bedford, S., C. Sand, and S. Connaughton (2007). *Oceanic Explorations: Lapita and Western Pacific Settlement*, Volume 26 of *terra australis*. Canberra, Australia: ANU E Press.

Bedford, S. and M. Spriggs (2008). Northern Vanuatu as a Pacific crossroads: the archaeology of discovery, interaction, and the emergence of the "ethnographic present". *Asian Perspectives 47*(1), 95–120.

Bedford, S., M. Spriggs, H. Buckley, F. Valentin, and R. Regenvanu (2009). The Teouma Lapita site, South Efate, Vanuatu: A summary of three field seasons (2004-2006). In P. Sheppard, T. Thomas, and G. R. Summerhayes (Eds.), *Lapita: Ancestors and descendants*, New Zealand Archaeological Association Monograph, pp. 215–234. Auckland: New Zealand Archaeological Association.

Bedford, S., M. Spriggs, H. Buckley, F. Valentin, R. Regenvanu, and M. Abong (2010). A cemetery of first settlement: the site of Teouma, South Efate, Vanuatu. In C. Sand and S. Bedford (Eds.), *Lapita: Oceanic Ancestors*, pp. 140–161. Paris, France: Musée du Quai Branly and Somogy Éditions d'Art.

Belkin, J. (1962). *The Mosquitoes of the South Pacific; (Diptera. Culcidae)*. Berkeley, CA: University of California Press.

Bellwood, P. (1978). *Man's Conquest of the Pacific: The Prehistory of Southeast Asia and Oceania*. Auckland, New Zealand: Collins.

Bellwood, P. (1989). The colonization of the Pacific: Some current hypotheses. In S. Serjeantson (Ed.), *The Colonisation of the Pacific: A genetic trail*, pp. 1–159. Oxford, UK: Oxford University Press.

Bentley, R. A., H. R. Buckley, M. Spriggs, S. Bedford, C. J. Ottley, G. M. Nowell, C. G. Macpherson, and D. G. Pearson (2007). Lapita migrants in the Pacific's oldest cemetery: Isotopic analysis at Teouma, Vanuatu. *American Antiquity 72*(4), 645–656.

Bindon, J. and S. Zansky (1986). Growth and body composition. In P. T. Baker, J. M. Henna, and T. S. Baker (Eds.), *The Changing Samoans: Behaviour and Health in Transition*, Research Monographs on Human Population Biology, pp. 222–253. New York, NY: Oxford University Press.

Blakey, M. L., T. E. Leslie, and J. P. Reidy (1994). Frequency and chronological distribution of dental enamel hypoplasia in enslaved African Americans: A test of the weaning hypothesis. *American Journal of Physical Anthropology 95*, 371–383.

Blondiaux, J., A. Cotten, C. Fontaine, C. Hanni, A. Bera, and R. Flipo (1997). Two Roman and Medieval cases of symmetrical erosive polyarthropathy from Normandy: Anatomico-pathological and radiological evidence for rheumatoid arthritis. *International Journal of Osteoarchaeology 7*, 451–466.

Bogan, K. and R. Crittenden (1987). Environment, economic development and the nutritional status of children in the southern highlands of Papua New Guinea. *Ecology of Food and Nutrition 20*, 29–49.

Bogdan, G. and D. Weaver (1992). Pre-Columbian treponematosis in coastal North Carolina. In D. Ubelaker (Ed.), *Disease and Demography in the Americas*, pp. 155–164. Washington D.C.: Smithsonian Institution Press.

Bogin, B. (1999). *Patterns of Human Growth* (2nd ed.). Cambridge, UK: Cambridge University Press.

Bogin, B. (2001). *The Growth of Humanity*. Foundations of Human Biology. New York, NY: Wiley-Liss.

Bogin, B. and J. Loucky (1997). Plasticity, political economy, and physical growth status of Guatemala Maya children living in the United States. *American Journal of Physical Anthropology 102*, 17–32.

Bouwman, A. S. and T. A. Brown (2005). The limits of biomolecular palaeopathology: ancient DNA cannot be used to study venereal syphilis. *Journal of Archaeological Science 32*(5), 703–713.

Bowden, D., A. Hill, D. Higgs, and D. Weatherall (1985). Relative roles of genetic factors, dietary deficiency, and infection in anaemia in Vanuatu, south-west Pacific. *The Lancet 2*, 1025–1028.

Boyce, B. F., M. J. Zuscik, and L. Xing (2013). Biology of bone and cartilage. In R. V. Thakker, J. A. Eisman, T. Igarashi, and M. P. Whyte (Eds.), *Genetics of Bone Biology and Skeletal Disease*, Chapter 1, pp. 3–24. San Diego, California: Academic Press.

Brickley, M. and R. Ives (2008). *The Bioarchaeology of Metabolic Bone Disease*. San Diego, CA: Academic Press.

Brothwell, D. (1976). Further evidence of Treponematosis in a pre-European population from Oceania. *Bulletin of History of Medicine 50*, 435–442.

Brothwell, D. R. and A. T. Sandison (1967). *Diseases in antiquity; a survey of the diseases, injuries, and surgery of early populations*. Springfield, IL: C. C. Thomas.

Bruce-Chwatt, L. (1965). Paleogenesis and paleo-epidemiology of primate malaria. *WHO Bulletin 32*, 363–387.

Bruce-Chwatt, L. and J. de Zulueta (1980). *The Rise and Fall of Malaria in Europe*. New York, NY: Oxford University Press.

Buck, P. (1910). *Medicine Amongst the Maoris, in Ancient and Modern Times*. Doctor of medicine, University of Otago, Dunedin, New Zealand.

Buckley, H. (2000a). A possible fatal wounding in the prehistoric Pacific Islands. *International Journal of Osteoarchaeology 10*, 135–141.

Buckley, H. (2006). 'The predators within': investigating the relationship between malaria and health in the prehistoric Pacific Islands. In M. F. Oxenham and N. Tayles (Eds.), *Bioarchaeology of Southeast Asia*, pp. 309–332. Cambridge, MA: Cambridge University Press.

Buckley, H. (2011). Epidemiology of gout: perspectives from the past. *Current Rheumatology Reviews 7*(2), 106–113.

Buckley, H. and N. Tayles (2003a). The functional cost of tertiary yaws (*Treponema pertenue*) in a prehistoric Pacific Island skeletal sample. *Journal of Archaeological Science 30*(10), 1301–1314.

Buckley, H. R. (2000b). Subadult health and disease in prehistoric Tonga, Polynesia. *American Journal of Physical Anthropology 113*(4), 481–505.

Buckley, H. R. (2007). Possible gouty arthritis in Lapita-associated skeletons from Teouma, Efate Island, Central Vanuatu. *Current Anthropology 48*(5), 741–749.

Buckley, H. R. and G. J. Dias (2002). The distribution of skeletal lesions in treponemal disease: is the lymphatic system responsible? *International Journal of Osteoarchaeology 12*(3), 178–188.

Buckley, H. R., R. Kinaston, S. E. Halcrow, A. Foster, M. Spriggs, and S. Bedford (2014). Scurvy in a tropical paradise? evaluating the possibility of infant and adult vitamin C deficiency in the Lapita skeletal sample of Teouma, Vanuatu, Pacific islands. *International Journal of Paleopathology 5*(0), 72–85.

Buckley, H. R. and M. F. Oxenham (2016). Bioarchaeology in the Pacific Islands: A temporal and geographical examination of nutritional and infectious disease. In H. R. Buckley and M. F. Oxenham (Eds.), *The Routledge Handbook of Bioarchaeology in Southeast Asia and the Pacific Islands*, Chapter 17, pp. 363–389. Oxon, UK: Routledge.

Buckley, H. R. and N. Tayles (2003b). Skeletal pathology in a prehistoric Pacific Island sample: Issues in lesion recording, quantification, and interpretation. *American Journal of Physical Anthropology 122*(4), 303–324.

Buckley, H. R., N. Tayles, S. E. Halcrow, K. Robb, and R. Fyfe (2010). The people of Wairau Bar: a re-examination. *Journal of Pacific Archaeology 1*(1), 1–20.

Buckley, H. R., N. G. Tayles, M. J. T. Spriggs, and S. Bedford (2008). A preliminary report on health and disease in early Lapita skeletons, Vanuatu: Possible biological costs of island colonization. *The Journal of Island and Coastal Archaeology 3*(1), 87–114.

Buikstra, J. (1976). Status and social complexity. In *Hopewell in the lower Illinois valley: a regional approach to the study of human biological variability and prehistoric behavior* (2nd ed.)., Scientific Papers. Evanston, IL: Northwestern University.

Buikstra, J. E. and L. Beck (2006). *Bioarchaeology: The Contextual Analysis of Human Remains.* Burlington, MA: Elsevier Inc.

Buikstra, J. E. and D. C. Cook (1980). Paleopathology: An American account. *Annual Review of Anthropology 9*, 433–470.

Buikstra, J. E. and J. H. Mielke (1985). Demography, diet, and health. In R. I. Gilbert and J. H. Mielke (Eds.), *The Analysis of Prehistoric Diets*, pp. 360–422. Orlando, FL: Academic Press.

Buikstra, J. E. and D. H. Ubelaker (1994). *Standards for Data Collection from Human Skeletal Remains.* Arkansas Archaeological Survey Research Series No. 44. Fayetteville, AK: Arkansas Archaeological Survey.

Bullough, P. (1992). *Atlas of Orthopedic Pathology: with clinical and radiological correlations* (2nd ed.). New York, NY: Gower Medical Publishing.

Burley, D., K. Edinborough, M. Weisler, and J.-x. Zhao (2015). Bayesian modeling and chronological precision for Polynesian settlement of Tonga. *PloS one 10*(3).

Burley, D. V. (1998). Tongan archaeology and the Tongan past, 2850–150 B.P. *Journal of World Prehistory 12*(3), 337–392.

Burley, D. V., W. R. Dickinson, A. Barton, and R. S. Jr (2001). Lapita on the periphery. new data on old problems in the Kingdom of Tonga. *Archaeology in Oceania 36*(2), 89–104.

Buxton, A. (1928). *Researches in Polynesia and Melanesia: An account of investigations in Samoa, Tonga, the Ellice Group , and the New Hebrides, in 1924,1925*, Volume 2 of *Memoir Series of the London School of Hygiene and Tropical Medicine.* London, UK: The London School of Hygiene and Tropical Medicine.

Caffey, J. (1957). Cooley's anemia: A review of the roentological findings in the skeleton. *American journal of roentgenology 78*(3), 381–391.

Carpenter, R. (1999). The prevalence and distribution of bruising in babies. *Archives of Disease in Childhood 80*, 363–366.

Chandrapanond, A., D. Suttapryasri, S. Tansuphasiri, and P. Harinasata (1973). Nutrition, growth and development of Thai village children. *Journal of the Medical Association of Thailand 56*, 723–731.

Clark, G., D. Burley, and T. Murray (2008). Monumentality and the development of the Tongan maritime chiefdom. *Antiquity 82*(318), 994–1008.

Clark, G., E. Grono, E. Ussher, and C. Reepmeyer (2015). Early settlement and subsistence on Tongatapu, Kingdom of Tonga: Insights from a 2700–2650calBP midden deposit. *Journal of Archaeological Science: Reports 3*, 513–524.

Clark, G., F. Leach, and S. O'Conner (2008). *Islands of Inquiry: Colonisation, Seafaring and the Archaeology of Maritime Landscapes.* terra australis 29. Canberra, Australia: ANU E Press.

Clark, J. and K. Kelly (1993). Human genetics, paleoenvironments, and malaria: Relationships and implications for the settlement of Oceania. *American Anthropologist 95*(3), 612–630.

Cockburn, A. (1963). *The Evolution and Eradication of Infectious Diseases.* Baltimore, MA: The Johns Hopkins Press.

Cohen, M. and G. Armelagos (Eds.) (1984). *Paleopathology at the Origins of Agriculture.* Orlando, FL: Academic Press, Inc.

Cohen, M. and G. Crane-Kramer (2007). *Ancient Health: Skeletal indicators of agricultural and economic intensification.* Gainsville, FL: University Press of Florida.

Collins, A. (1944). Pacific Islands: Western Pacific (Tonga to the Solomon Islands). Geographical handbook series, Naval Intelligence Division.

Cook, D. and J. Buikstra (1979). Health and differential survival in prehistoric populations: Prenatal dental defects. *American Journal of Physical Anthropology 51*, 649–664.

Cook, D. C. (1976). *Pathologic states and disease process in Illinois Woodland populations: an Epidemiologic approach.* Ph.d. thesis, Department of Anthropology, University of Chicago, Chicago, Illinois.

Cook, G. (1996). *Manson's Tropical Diseases* (20th ed.). Norfolk, UK: WB Saunders.

Cooney, J. and E. Crosby (1944). Absorptive bone changes in leprosy. *Radiology 42*(1), 14–19.

Corruccini, R. S., J. S. Handler, and K. P. Jacobi (1985). Chronological distribution of enamel hypoplasias and weaning in a Carribean slave population. *Human Biology 57*(4), 699–711.

Costa, R. (1986). Determination of age at death: Dentition analysis. In J. Angel (Ed.), *Dating and Age Determination of Biological Materials*, pp. 248–270. Worcester, UK: Croom Helm.

Cotran, R., V. Kumar, and S. Robbins (1994). *Robbins' Pathological Basis of Disease* (5th ed.). Philadelpha, PA: W. B. Saunders.

Cox, P. A. and S. A. Banack (1991). *Islands, plants, and Polynesians: an introduction to Polynesian ethnobotany*. Portland, OR: Dioscorides Press.

Davenport, W. (1968). Social organization notes on the northern Santa Cruz islands: the Duff Islands (Taumako). *Baessler-Archiv, Neue Folge 16*, 137–205.

Davidson, J. (1969). Archaeological excavations in two burial mounds at 'Atele, Tongatapu. *Records of the Auckland Institute and Museum 6*, 251–286.

Davidson, J. (1974). Cultural replacement on small islands: New evidence from Polynesian Outliers. *Mankind 9*, 273–277.

Davidson, J. and B. Leach (1991). Bird-man amulets and *Tridacna* shell discs from Taumako, Solomon Islands. In A. Pawley (Ed.), *Man and a Half: Essays in Pacific Anthropology and Ethnobiology in Honour of Ralph Bulmer*, Memoir. Auckland, New Zealand: The Polynesian Society.

Davies, G. (1958). A comparative epidemiological study of diet and dental caries in three isolated communities. *Alabama Dental Review 6*(1), 19–44.

De Zulueta, J. (1994). Malaria and ecosystems: From prehistory to posteradication. *Parassitologia 36*, 7–15.

DeFant, D. G. (2008). Early human burials from the Naton Beach site, Tumon Bay, Island of Guam, Mariana Islands. *Journal of Island and Coastal Archaeology 3*, 149–153.

Denham, T. (2011). Early agriculture and plant domestication in New Guinea and Island SEAsia. *Current Anthropology 52*(S4), S379–S395.

Dignan, C. (1994). *The Pacific Islands Food Composition Tables*. Suva, Fiji: South Pacific Commission.

Domett, K. (2001). *Health in Late Prehistoric Thailand*. Oxford, UK: BAR Publishing.

Domett, K. M. (1999). *Health in late prehistoric Thailand*. Ph. D. thesis, Department of Anatomy, University of Otago.

Dooley, J. and C. Binford (1976). Treponematoses. In D. Connor (Ed.), *Pathology of Tropical and Extraordinary Diseases: An Atlas*, Volume 1. Washington, D.C: Armed Forces Institute of Pathology.

Dudley, N. (2010). *Field Anthropology at Namu: The Mortuary Practices of a Polynesian Outlier*. unpublished dissertation (ba hons), University of Otago, Dunedin, New Zealand.

Duray, S. (1996). Dental indicators of stress and reduced age at death in prehistoric Native Americans. *American Journal of Physical Anthropology 99*, 275–286.

El-Najjar, M. Y., D. J. Ryan, C. G. Turner, and B. Lozoff (1976). The etiology of porotic hyperostosis among the prehistoric and historic Anasazi Indians of Southwestern United States. *American Journal of Physical Anthropology 44*(3), 477–487.

Evans, A. (1998). Epidemiological concepts. In P. Brachman (Ed.), *Bacterial Infections of Humans: Epidemiology and Control* (3rd ed.)., pp. 13–64. New York, NY: Plenum Medical Book Company.

Evans, J. A. (1987). *Comparative dental conditions in several prehistoric Oceanic groups*. Ph. D. thesis, Department of Anatomy, University of Otago.

Eveleth, P. and J. Tanner (1990). *Worldwide Variation in Human Growth* (2nd ed.). Cambridge, UK: Cambridge University Press.

Ewald, P. (1994). *Evolution of Infectious Disease*. Oxford, UK: Oxford University Press.

Exarchou, E., C. Politou, E. Vretou, D. Pasparakis, G. Medessis, and A. Caramerou (1984). Fractures and epiphyseal deformites in beta-thalassemia. *Clinical Orthopaedics 89*, 229–233.

Faget, G. and A. Mayoral (1944). Bone changes in leprosy: A clinical and roentological study of 505 cases. *Radiology 42*(1), 1–13.

Faine, S. and C. E. Hercus (1951, 12). The nutritional status of Cook Islanders. *British Journal of Nutrition 5*, 327–343.

Fegan, D., M. Glennon, G. Macbride-Stewart, and T. Moore (1990). Yaws in the Solomon Islands. *The American Journal of Tropical Medicine and Hygiene 93*, 52–57.

Feinberg, R. and R. Scaglion (2012). Introduction: the Polynesian Outliers. In R. Feinberg and R. Scaglion (Eds.), *Polynesian Outliers: the states of the art*, Chapter 1, pp. 1–16. Pittsburgh, PA: University of Pittsburgh.

Flannery, K. and J. Marcus (2012). *The Creation of Inequality: How our Prehistoric Ancestors Set the Stage for Monarchy, Slavery and Empire*. Cambridge, MA: Harvard University Press.

Fleming, A. (1996). Haematological diseases in the tropics. In G. Cook (Ed.), *Manson's Tropical Diseases* (20th ed.)., pp. 101–173. London, UK: W.B. Saunders Company Ltd.

Flint, J., A. Hill, D. Bowden, S. Oppenheimer, P. Sill, S. Serjeantson, J. Bana-Koiri, K. Bhatia, M. Alpers, A. Boyce, D. Weatherall, and J. Clegg (1986). High frequencies of a-thalassaemia are the result of natural selection by malaria. *Nature 321*, 744–750.

Foster, A. (2007). *Pottery Production and Markers of Occupational Stress: Musculoskeletal Stress Markers and Osteoarthritis in Two Pacific Island Skeletal Samples*. unpublished dissertation (ba hons), University of Otago, Dunedin, New Zealand.

Foster, A. V. (2011). *Gendered Divisions of Labour in Southeast Asian and Pacific Island Prehistory*. Ph. D. thesis, University of Otago.

Foster, A. V., H. R. Buckley, N. Tayles, M. Spriggs, and S. Bedford (2013). Gender, labour division and the skeleton: a case study from the Teouma Lapita cemetery. In G. Summerhayes and H. Buckley (Eds.), *Pacific Archaeology: Documenting the past 50,000 years*, Chapter 2, pp. 76–90. Dunedin: University of Otago Press.

Fyfe, D., N. Chandler, and N. H. F. Wilson (1993). Alveolar bone status of some pre-seventeenth century inhabitants of Taumako, Solomon Islands. *International Journal of Osteoarchaeology 3*, 29–35.

Ganczakowski, M., D. Bowden, K. Maitland, T. Williams, D. O'Shaughnessy, J. Viji, A. Lucassen, J. Clegg, and D. Weatherall (1995). Thalassaemia in Vanuatu, South-West Pacific: Frequency and haematological phenotypes of young children. *British Journal of Haematology 89*, 485–495.

Garn, S. (1992). *The iron-deficiency anaemias and their skeletal manifestations*, pp. 33–61. New York, NY: Aldine De Gruyter.

Gason, S. (1879). *The Manners and Customs of the Dieyerie Tribe of Australian Aborigines*. Adelaide, Australia: ES Wigg and Son.

Geizer, I. (1986). Yaws in the Western Pacific region: An overview. *Southeast Asian Journal of Tropical Medicine and Public Health 17*(Supplement 4), 8–13.

Genton, B., F. Al-Yaman, H. Beck, J. Hii, S. Mellor, A. Narara, N. Gibson, T. Smith, and M. Alpers (1995). The epidemiology of malaria in the Wosera area, East Sepik Province, Papua New Guinea, in preparation for vaccine trials, I. Malariometric indices and immunity. *Annals of Tropical Medicine and Parasitology 89*(4), 359–376.

Genton, B., F. Al-Yaman, H. Beck, J. Hii, S. Mellor, L. Rare, M. Ginny, T. Smith, and M. Alpers (1995). The epidemiology of malaria in the Wosera area, East Sepik Province, Papua New Guinea, in preparation for vaccine trials. II. Mortality and morbidity. *Annals of Tropical Medicine & Parasitology 89*(4), 377–390.

Gershman, K., R. Rolfs, S. Larsen, A. Zaidi, and N. Palafox (1992). Seroepidemiological characterization of a syphilis epidemic in the Republic of the Marshall Islands, formerly a yaws endemic area. *International Journal of Epidemiology 21*(3), 599–606.

Gilbert, B. M. and T. W. McKern (1973). A method for aging the female os pubis. *American Journal of Physical Anthropology 38*, 31–38.

Gilles, H. (1996). Soil-transmitted helminths (geohelminths). In G. Cook (Ed.), *Manson's Tropical Diseases* (20th ed.)., pp. 1369–1412. London, UK: WB Saunders.

Gilles, H. and D. Warrell (1993). *Bruce-Chwatt's Essential Malariology* (3rd ed.). London, UK: Edward Arnold; a division of Hodder and Stoughton.

Glasse, R. (1965). Leprosy at Karamui. *PNG Medical Journal 8*(3), 95–98.

Golson, J. (1977). No room at the top: Agricultural intensification in the New Guinea Highlands. In J. Allen, J. Golson, and R. Jones (Eds.), *Sunda and Sahul: Prehistoric Studies in Southeast Asia, Melanesia and Australia*, pp. 601–637. London, UK: Academic Press.

Goodman, A., L. Allen, G. Hernandez, A. Amador, L. Arriola, A. Chavez, and G. Pelto (1987). Prevalence and age at development of enamel hypoplasias in Mexican children. *American Journal of Physical Anthropology 72*, 7–19.

Goodman, A. and R. Song (1999). Sources of variation in estimated ages at formation of linear enamel hypoplasias. In C. Fitzgerald (Ed.), *Human Growth in the Past: Studies from Bones and Teeth*, Cambridge Studies in Biological and Evolutionary Anthropology, pp. 210–240. Cambridge, UK: Cambridge University Press.

Goodman, A. H. (1993). On the interpretation of health from skeletal remains. *Current Anthropology 34*(3), 281–288.

Goodman, A. H. (1994). Cartesian reductionism and vulgar adaptationism: issues in the interpretation of nutritional status in prehistory. In K. Sobolik (Ed.), *Paleonutrition: The Diet and Health of the Prehistoric Americans, Occasional Papers No.22, Center for Archaeological Investigations*, pp. 163–177. Carbondale, IL: Southern Illinois University.

Goodman, A. H. and G. J. Armelagos (1985). Factors affecting the distribution of enamel hypoplasias within the human permanent dentition. *American Journal of Physical Anthropology 68*, 479–493.

Goodman, A. H. and G. J. Armelagos (1988). Childhood stress and decreased longevity in a prehistoric population. *American Anthropologist 90*, 936–944.

Goodman, A. H., G. J. Armelagos, and J. C. Rose (1980). Enamel hypoplasias as indicators of stress in three prehistoric populations from Illinois. *Human Biology 52*, 515–528.

Goodman, A. H., D. L. Martin, G. J. Armelagos, and G. Clark (1984). Indications of stress from bones and teeth. In M. N. Cohen and G. J. Armelagos (Eds.), *Paleopathology at the Origins of Agriculture*, pp. 13–49. Orlando, FL: Academic Press.

Goodman, A. H. and J. C. Rose (1990). Assessment of systemic physiological perturbations from dental enamel hypoplasias and associated histological structures. *Yearbook of Physical Anthropology 33*, 59–110.

Goodman, A. H., R. B. Thomas, A. C. Swedlund, and G. J. Armelagos (1988). Biocultural perspectives on stress in prehistoric, historical, and contemporary population research. *Yearbook of Physical Anthropology 31*, 169–202.

Gordon, J., I. Chitkara, and J. Wyon (1963). Weanling diarrhoea. *The American Journal of the Medical Sciences 245*, 345–377.

Gosling, A. L., H. Buckley, E. Matisoo-Smith, and T. Merriman (2015). Pacific populations, metabolic disease and 'just-so stories': A critique of the 'thrifty genotype' hypothesis in Oceania. *Annals of Human Genetics 79*(6), 470–480.

Gosling, A. L., E. Matisoo-Smith, and T. R. Merriman (2014). Gout in Māori. *Rheumatology 53*(5), 773–774.

Gowland, R. (2015). Entangled lives: Implications of the developmental origins fo health and disease hypothesis for bioarchaeology and the life course. *American Journal of Physical Anthropology Early view*, 00–00.

Gowland, R. and A. Western (2012). Morbidity in the marshes: using spatial epidemiology to investigate skeletal evidence for malaria in Anglo-Saxon England (AD 410-1050). *American Journal of Physical Anthropology 147*, 301–311.

Gray, B. (1982). Enga birth, maturation and survival: Physiological characteristics of the life cycle in the New Guinea Highlands. In C. McCormack (Ed.), *Ethnography of Fertility*, pp. 75–113. London, UK: Academic Press.

Green, A. (1999). *Iron Deficiency Anaemia in New Zealand Prehistory*. Ph. D. thesis, University of Otago.

Green, R. C. (1991). Near and Remote Oceania- disestablishing "Melanesia" in culture history. In A. Pawley (Ed.), *Man and a Half: Essays in Pacific Anthropology and Ethnobiology in Honour of Ralph Bulmer*, pp. 491–502. Auckland, New Zealand: The Polynesian Society.

Green, R. C. and P. V. Kirch (1997). Lapita exchange systems and their Polynesian transformations: Seeking explanatory models. In M. Weisler (Ed.), *Prehistoric Long-Distance Interaction in Oceania: An interdisciplinary Approach*, Volume 21, Chapter 21, pp. 19–37. Auckland, New Zealand: New Zealand Archaeological Association.

Greenwood, B. (1997). The epidemiology of malaria. *Annals of Tropical Medicine and Parasitology 91*(7), 763–769.

Grolleau-Raoux, J., E. Crubezy, D. Rouge, J. Brugne, and S. Saunders (1997). Harris lines: a study of age-associated bias in counting and interpretation. *American Journal of Physical Antrhopology 103*(2), 209–217.

Groube, L. (1993). Contradictions and malaria in Melanesian and Australian prehistory. In M. Spriggs, D. E. Yen, W. Ambrose, R. Jones, A. Thorne, and A. Andrews (Eds.), *A Community of Culture: A People and Prehistory of the Pacific*, Occasional Papers in Prehistory No 21, pp. 164–186. Canberra, Australia: The Australian National University.

Gustafson, G. and G. Koch (1974). Age estimation up to 16 years of age based on dental development. *Odontologisk revy 25*(3), 297.

Hackett, C. J. (1936a). Boomerang legs and yaws in Australian Aborigines. *Transactions of the Royal Society of Tropical Medicine and Hygiene 30*(2), 139–143.

Hackett, C. J. (1936b). A critical survey of some references to syphilis and yaws among the Australian Aborigines. *The Medical Journal of Australia 1*(22), 733–745.

Hackett, C. J. (1946). The clinical course of yaws in Lango, Uganda. *Transactions of the Royal Society of Tropical Medicine and Hygiene 40*, 206–227.

Hackett, C. J. (1951). *Bone Lesions of Yaws in Uganda*. Oxford, UK: Blackwell Scientific Publications.

Hackett, C. J. (1976). *Diagnostic criteria of syphilis, yaws, and treponarid (treponematoses) and of some other diseases in dry bones (for use in Osteo-archaeology)*. Berlin, Germany: Springer-Verlag.

Halcrow, S. E., N. Tayles, and C. L. King (2016). To follow in their footsteps: An examination of the burial identity of the elderly from Non Nok Tha. In M. F. Oxenham and H. R. Buckley (Eds.), *The Routledge Handbook of Bioarchaeology in Southeast Asia and the Pacific*, Chapter 10, pp. 187–219. Oxen, UK: Routledge.

Halcrow, S. E., N. Tayles, and V. Livingstone (2008). Infant death in prehistoric mainland Southeast Asia. *Asian Perspectives 47*, 371–404.

Hanson, D. (1990). Paleopathological observations on human skeletal remains from Rota, Mariana Islands: Epidemiological implications. *Micronesia Supplement 2*, 349–362.

Harper, K. and G. Armelagos (2013). Genomics, the origins of agriculture, and our changing microbe-scape: Time to revisit some old tales and tell some new ones. *American Journal of Physical Anthroplogy 57*, 135–152.

Harrison, G. (1993a). *Human Biology: An Introduction to Human Evolution, Variability, Growth and Adaptability*. Oxford, UK: Oxford University Press.

Harrison, G. (1993b). Physiological adaptation. In G. Harrison (Ed.), *Human Adaptation*, pp. 55–71. Oxford, UK: Oxford Science Publications.

Heathcote, G. M., A. L. W. Stodder, H. R. Buckley, D. B. Hanson, M. T. Douglas, J. H. Underwood, T. F. Taisipic, and V. P. Diego (1998). On treponemal disease in the Western Pacific: corrections and critique. *Current Anthropology 39*(3), 359–368.

Hendrickse, R. (1987). Malaria and child health. *Annals of Tropical Medicine and Parasitology 81*(5), 499–509.

Hendrickse, R. and B. Brabin (1996). Paediatrics in the tropics. In G. C. Cook (Ed.), *Manson's Tropical Diseases* (20th ed.)., pp. 363–377. London, UK: W.B. Saunders.

Hengen, O. P. (1971). Cribra orbitalia: Pathogenesis and probable etiology. *HOMO 22*, 57–75.

Hershkovitz, A., B. Ring, M. Speirs, E. Galili, M. Kislev, G. Edelson, and A. Hershkovitz (1991). Possible congenital hemolytic anemia in prehistoric coastal inhabitants of israel. *American Journal of Physical Anthropology 85*, 7–13.

Hill, A., D. O'Shaughnessy, and J. Clegg (1989). Haemoglobin and globin gene variants in the pacific. In S. Serjeantson (Ed.), *The Colonization of the Pacific: A Genetic Trail*, Research Monographs on Human Population Biology, pp. 246–285. Oxford, UK: Oxford University Press.

Hillson, S. (1986). *Teeth*. Cambridge Manuals in Archeaology. Cambridge, UK: Cambridge University Press.

Hillson, S. (1996). *Dental Anthropology*. Cambridge, UK: Cambridge University Press.

Hillson, S. (2006). *Teeth* (2nd ed.). Manuals in Archaeology. Cambridge, UK: Cambridge University Press.

Hillson, S. (2014). *Tooth Development in Human Evolution and Bioarchaeology*. Cambridge, UK: Cambridge University Press.

Hillson, S., C. Grigson, and S. Bond (1998). Dental defects of congenital syphilis. *American Journal of Physical Anthropology 107*(1), 25–40.

Hoffman, J. (1979). Age estimation from diaphyseal lengths: Two months to twelve years. *Journal of Forensic Science 24*, 461–469.

Holland, T. D. and M. J. O'Brien (1997). Parasites, porotic hyperostosis, and the implications of changing perspectives. *American Antiquity 62*(2), 183–193.

Horwood, M. (1988). *Trace elements and prehistoric diet in the Pacific*. Ma thesis, Department of Anatomy.

Houghton, P. (1974). The relationship of the pre-auricular groove of the ilium to pregnancy. *American Journal of Physical Anthropology 41*, 381–389.

Houghton, P. (1996). *People of the Great Ocean; Aspects of Human Biology of the Early Pacific*. Oxon, UK: Cambridge University Press.

Houghton, P. (2008). The people of Namu. In F. Leach and J. Davidson (Eds.), *Archaeology on Taumako: A Polynesian Outlier in the Eastern Solomon Islands*, New Zealand Journal of Archaeology Special Publication, pp. 325–352. Dunedin, New Zealand: New Zealand Journal of Archaeology.

Houghton, P. (n.d.). The people of Namu. Osteology, Unpublished report. University of Otago, Dunedin, New Zealand.

Houghton, P., B. Leach, and D. Sutton (1975). The estimation of stature of prehistoric Polynesians in New Zealand. *Journal of the Polynesian Society 84*(3), 325–336.

Huntsman, R. and H. Lehman (1984). The haemoglobinopathies. In A. Woodruff (Ed.), *Medicine in the Tropics* (2nd ed.). New York, NY: Churchill Livingstone.

Infante, P. and G. Gillespie (1974). An epidemiological study of linear enamel hypoplasia of deciduous anterior teeth in Guatemalan children. *Archives of Oral Biology 19*, 1055–1061.

Jackes, M. (1992). Paleodemography: Problems and techniques. In S. R. Saunders and M. A. Katzenberg (Eds.), *Skeletal Biology of Past Peoples: Research Methods*, pp. 189–224. New York, NY: Wiley-Liss.

Jacobs, A. (1982). *Clinics in Haematology*. Disorders of Iron Metabolism. London, UK: W.B. Saunders Company Ltd.

Jaffe, H. (1972). *Metabolic, Degenerative, and Inflammatory Diseases of Bones and Joints*. Philadelphia, PA: Lea and Febiger.

Jansen, R. (2002). *Deciduous Dental Enamel Defects as Indicators of Growth Disturbance in a Prehistoric Skeletal Sample*. Ba (hons) dissertation, University of Otago, Dunedin, New Zealand.

Jelliffe, D. B. (1968). *Infant Nutrition in the Subtropics and Tropics* (2nd ed.), Volume 29 of *World Health Organization Monograph Series*. Geneva, Switzerland: World Health Organization.

Jelliffe, D. B. (1970). *Diseases of Children in the Subtropics and Tropics* (2nd ed.). London, UK: Edward Arnold.

Johnston, F. (1962). Growth of the long bones of infants and young children at Indian Knoll. *American Journal of Physical Anthropology 20*, 249–254.

Johnston, F. E. and L. O. Zimmer (1989). Assessment of growth and age in the immature skeleton. In M. Y. Işcan and K. A. R. Kennedy (Eds.), *Reconstruction of Life from the Skeleton*, pp. 11–21. New York, NY: Wiley-Liss.

Jones, B. (1972). Doigt en lorgnette and concentric bone atrophy associated with healed yaws osteitis: report of two cases. *Journal of Bone and Joint Surgery 54b*(No.2), 341–345.

Jones, M. R., N. P. Larson, and G. P. Prichard (1930). Odontoclasia: a clinically unrecognized form of tooth decay in the pre-school child of Honolulu. *Dental Cosmos 72*, 439–450.

Kampmeier, R. (1954). *Essentials in Syphilology*. Oxford, UK: Blackwell.

Kariks, J. and D. Woodfield (1972). Anaemia in Papua New Guinea-a review. *PNG Medical Journal 15*(1), 15–24.

Katz, D. and J. Suchey (1989). Race differences in pubic symphyseal aging patterns in the male. *American Journal of Physical Antrhopology 80*, 167–172.

Katzenberg, M. A., D. A. Herring, and S. R. Saunders (1996). Weaning and infant mortality: Evaluating the skeletal evidence. *American Journal of Physical Anthropology 101*, 177–199.

Kelley, M. A. and M. S. Micozzi (1984). Rib lesions in chronic pulmonary tuberculosis. *American Journal of Physical Anthropology 65*, 381–386.

Kelly, C. (1966). *La Austrialia Del Espiritu Santo: The Journal of Fray Martin de Munilla O.F.M. and other documents relating to the Voyage of Pedro Fernandez de Quiros to the South Sea (1605-1606) and the Franciscan Missionary Plan (1617-1627)*, Volume I and II of *The Hakluyt Society Series*. Cambridge, UK: Cambridge University Press.

Kent, S. (1992). Anemia through the ages: Changing perspectives and their implications. In P. Stuart-Macadam and S. Kent (Eds.), *Diet, Demography and Disease*, pp. 1–30. New York, NY: Aldine de Gruyter.

Keyes, E. (1908). *Syphilis: A treatise for practitioners*. London, UK: Sidney Appleton.

Kieser, J. (2012). Biomechanics of bone and bony trauma. In J. Kieser, M. Taylor, and D. Carr (Eds.), *Forensic Biomechanics*, Chapter 3, pp. 35–69. Chichester, UK: Wiley.

Kieser, J., C. Preston, and W. Evans (1983). Skeletal age at death: An evaluation of the Miles method of ageing. *Journal of Archaeological Science 10*, 9–12.

Kilgore, L. (1989). Possible case of rheumatoid arthritis from Sudanese Nubia. *American Journal of Physical Anthropology 79*, 177–183.

Kinaston, R. (2010). *Prehistoric Diet and Health in the Western Pacific Islands*. Ph. D. thesis, University of Otago.

Kinaston, R., S. Bedford, M. Richards, S. Hawkins, A. Gray, K. Jaouen, F. Valentin, and H. Buckley (2014). Diet and human mobility from the Lapita to the Early Historic Period on Uripiv Island, Northeast Malakula, Vanuatu. *PLoS ONE 9*(8), e104071.

Kinaston, R., H. Buckley, F. Valentin, S. Bedford, M. Spriggs, S. Hawkins, and E. Herrscher (2014). Lapita diet in Remote Oceania: New stable isotope evidence from the 3000-year-old Teouma site, Efate Island, Vanuatu. *PLoS ONE 9*(3), e90376.

Kinaston, R. L., D. Anson, P. Petchey, R. Walter, K. Robb, and H. Buckley (2015). Lapita diet and subsistence strategies on Watom Island, Papua New Guinea: New stable isotope evidence from humans and animals. *American Journal of Physical Anthropology 157*(1), 30–41.

Kinaston, R. L., S. B. Bedford, M. Spriggs, D. Anson, and H. R. Buckley (2016). Is there a 'Lapita diet'? a comparison of Lapita and post-Lapita skeletal samples from four Pacific island archaeological sites. In M. F. Oxenham and H. R. Buckley (Eds.), *The Routledge Handbook of Bioarchaeology in Southeast Asia and the Pacific*, Chapter 20, pp. 427–461. Oxen, UK: Routledge.

Kinaston, R. L., H. R. Buckley, and A. Gray (2013). Diet and social status on Taumako, a Polynesian outlier in the Southeastern Solomon Islands. *American Journal of Physical Anthropology 151*(4), 589–603.

Kinaston, R. L., H. R. Buckley, S. E. Halcrow, M. J. T. Spriggs, S. Bedford, K. Neal, and A. Gray (2009). Investigating foetal and perinatal mortality in prehistoric skeletal samples: a case study from a 3000-year-old Pacific Island cemetery site. *Journal of Archaeological Science 36*(12), 2780–2787.

Kirch, P. V. (1984). *Evolution of the Polynesian Chiefdoms*. New Studies in Archaeology. Cambridge, England: Cambridge University Press.

Kirch, P. V. (1997). *The Lapita peoples: Ancestors of the Oceanic world*. The Peoples of South-East Asia and the Pacific. Cambridge, MA: Blackwell Publishers.

Kirch, P. V. (2000a). *On the Road of the Winds: An Archaeological History of the Pacific Islands before European Contact*. Berkeley, CA: University of California Press.

Kirch, P. V. (2000b). Pigs, humans, and trophic competition on small Oceanic islands. In A. Anderson and T. Murray (Eds.), *Australian Archaeologist: Collected papers in honour of Jim Allen*, Chapter 30, pp. 412–426. Canberra, Austalia: Coombs Academic Publishing.

Kirch, P. V. (2001). Polynesian feasting in ethnohistoric, ethnographic, and archaeological context: A comparison of three societies. In M. Dietler and B. Hayden (Eds.), *Feasts: Archaeological and Ethnographic Perspectives on Food, Politics, and Power*, Chapter 4, pp. 168–184. Washington, D.C.: Smithsonian Insitution Press.

Kirch, P. V. (2010). Peopling of the Pacific: a holistic anthropological perspective. *Annual Review of Anthropology 39*, 131–148.

Kirch, P. V. and R. C. Green (2001). *Hawaiki, Ancestral Polynesia: An Essay in Historical Anthropology*. Cambridge, UK: Cambridge University Press.

Kirch, P. V. and T. L. Hunt (1997). *Historical ecology in the Pacific Islands: prehistoric environmental and landscape change*. New Haven, CT: Yale University Press.

Kirch, P. V. and J. G. Kahn (2007). Advances in Polynesian prehistory: A review and assessment of the past decade. *Journal of Archaeological Research 15*(3), 191–238.

Kirkpatrick, R. (1935). Dental caries and odontoclasia in New Guinea. *Dental Journal of Australia 7*(1), 707–714.

Klemp, P., S. Stansfield, B. Castle, and M. Robertson (1997). Gout is on the increase in New Zealand. *Annals of the Rheumatic Diseases 56*, 22–26.

Krogman, W. and M. Y. İşcan (1986). *The Human Skeleton in Forensic Medicine* (2nd ed.). Springfield, IL: Charles C. Thomas.

Krugman, S. and S. Katz (1981). *Infectious Diseases in Children* (7th ed.). St. Louis, MO: Mosby.

Laird, M. (1956). *Studies of Mosquito Fresh Water Ecology in the South Pacific*. Auckland, New Zealand: Royal Society of New Zealand Bulletin No. 6.

Lallo, J., G. Armelagos, and J. Rose (1977). The role of diet, disease, and physiology in the origin of porotic hyperostosis. *Human Biology 49*, 471–483.

Lambert, S. (1941). *A Doctor in Paradise*. Melbourne, Australia: J.M Dent and Sons.

Lanphear, K. (1990). Frequency and distribution of enamel hypoplasias in a Historic skeletal sample. *American Journal of Physical Anthropology 81*, 35–43.

Laor, E., A. Garfunkel, and E. Koyoumdjisky-Kaye (1982). Skeletal and dental retardation in B-thalassemia major. *Human Biology 54*(1), 85–92.

Larsen, C. (2015). *Bioarchaeology: Interpreting behaviour from the human skeleton* (2nd ed.). Cambridge, MA: Cambridge University Press.

Leach, B. F. and J. Davidson (1977-1978). Taumako: A Polynesian outlier in the outer eastern Solomon Islands. Unpublished report.

Leach, B. F. and J. Davidson (2008). *Archaeology on Taumako: A Polynesian Outlier in the Eastern Solomon Islands*. New Zealand Journal of Archaeology.

Leach, B. F., C. Quinn, J. Morrison, and G. Lyon (2003). The use of multiple isotope signatures in reconstructing prehistoric human diet from archaeological bone from the Pacific and New Zealand. *New Zealand Journal of Archaeology 23*, 31–98.

Leach, H. (2003). Did East Polynesians have a concept of luxury foods? *World Archaeology 34*(3), 442–457.

Lee, R. V. (1988). Parasites and pregnancy: the problems of malaria and toxoplasmosis. *Clinics in perinatology 15*(2), 351–363.

Lepowsky, M. (1987). Food taboos and child survival: A case study from the coral sea. In N. Scheper-Hughes (Ed.), *Child Survival: Anthropological Perspectives on the Treatment and Maltreatment of Children*, pp. 71–92. Dordrecht, Germany: Reidel Publishing Company.

Lewis, M. and C. Roberts (1997). Growing pains: the interpretation of stress indicators. *International Journal of Oseoarchaeology 7*, 581–586.

Lewis, M. E. (2007). *The bioarchaeology of children*. Cambridge Studies in Biological and Evolutionary Anthropology. Cambridge, UK: Cambridge University Press.

Lewis, M. E. (2010). Life and death in a civitas capital: Metabolic disease and trauma in the children from late Roman Dorchester, Dorset. *American Journal of Physical Anthropology 142*(3), 405–416.

Lewis, M. E. and R. Gowland (2007). Brief and precarious lives: Infant mortality in contrasting sites from medieval and post-medieval England (AD 850—1859). *American Journal of Physical Anthropology 134*(1), 117–129.

Lovejoy, C. O. (1985). Dental wear in the Libben population: Its functional pattern and role in the determination of adult skeletal age at death. *American Journal of Physical Anthropology 68*(1), 47–56.

Lovejoy, C. O., R. S. Meindl, R. Mensforth, and T. J. Barton (1985). Multifactorial determination of skeletal age at death: A method and blind test of its accuracy. *American Journal of Physical Anthropology 68*, 1–14.

Lovell, N. (2000). *Paleopathological description and diagnosis*, pp. 217–248. New York, NY: Wiley-Liss.

MacKay, C. (1938). Some pathological changes in Australian Aboriginal bones. *Medical Journal of Australia 2*(14), 537–555.

Maitland, K., T. Williams, S. Bennett, C. Newbold, T. Peto, J. Viji, R. Timothy, J. Clegg, D. Weatherall, and D. Bowden (1996). The interaction of *Plasmodium falciparum* and *P vivax* in children on Espirito Santo island, Vanuatu. *Transactions of the Royal Society of Tropical Medicine and Hygiene 90*, 614–620.

Maitland, K., T. Williams, T. Peto, K. Day, J. Clegg, D. Weatherall, and D. Bowden (1997). Absence of malaria-specific mortality in children in an area of hyperendemic malaria. *Transactions of the Royal Society of Tropical Medicine and Hygiene 91*, 562–566.

Malcolm, L. A. (1969). Growth and development of the Kaiapat of the Markham Valley, New Guinea. *American Journal of Physical Anthropology 31*, 39–52.

Malcolm, L. A. (1970a). *Growth and Development in New Guinea- A Study of the Bundi People of the Madang District*. Institute of Human Biology Papua - New Guinea: Monograph series No 1. Madang, PNG: Institute of Human Biology.

Malcolm, L. A. (1970b). Growth and development of the Bundi child of the New Guinea highlands. *Human Biology 42*(2), 293–328.

Manchester, K. (1991). Tuberculosis and leprosy: Evidence for interaction of disease. In D. J. Ortner and A. C. Aufderheide (Eds.), *Human Paleopathology: Current Syntheses and Future Options*, pp. 23–35. Washington, D.C.: Smithsonian Institution Press.

Mann, S. (2007). *The People of Taumako: The Health and Wealth of a Polynesian Outlier*. unpublished dissertation (ba hons), University of Otago.

Mariner, W. and J. Martin (1827). *An account of the natives of the Tonga Islands in the South Pacific Ocean: with an original grammar and vocabulary of their language* (3rd ed.). Edinburgh, Scotland: Constable and Hurst Chance.

Marples, M. and D. Bacon (1953). Observation on yaws and certain skin diseases in Manono, Western Samoa. *Transactions of the Royal Society of Tropical Medicine and Hygiene 47*(2), 141–147.

Martorell, R. and T. Ho (1983). Malnutrition, morbidity and mortality. In L. Chen (Ed.), *Child Survival: Strategies for Research*, pp. 49–68. London, UK: Cambridge University Press.

Masawe, A., J. Mundi, and G. Swai (1974). Infections in iron deficiency and other types of anaemia in the tropics. *The Lancet August 10*, 314–317.

Massler, M., I. Sehoun, and H. Pauchen (1941). Developmental patters of the child as reflected in the teeth. *American Journal of Diseases of Children 62*, 33–67.

May, J. (1958). *The Ecology of Human Disease*. New York, NY: MD Publications.

Mays, S. and G. Taylor (2002). Osteological and biomolecular study of two possible cases of hypertrophic osteoarthropathy from Mediaeval England. *Journal of Archaeological Science 29*, 1267–1276.

McKern, T. (1970). Estimation of skeletal age: From puberty to about 30 years of age. In T. Stewart (Ed.), *Personal Identification in Mass Disasters*. Washington D.C.: Smithsonian Institution.

McMahon, J. (1974). Malaria endemicity amongst the semi-nomadic people of the Karimui area of Papua New Guinea. *PNG. Medical Journal 17*(1), 99–107.

Meindl, R. S. and C. O. Lovejoy (1985). Ectocranial suture closure: A revised method for the determination of skeletal age at death based on the lateral-anterior sutures. *American Journal of Physical Anthropology 68*, 57–66.

Meindl, R. S., C. O. Lovejoy, and R. P. Mensforth (1985). Accuracy and direction of error in the sexing of the skeleton: Implications for paleodemography. *American Journal of Physical Anthropology 68*, 79–85.

Menendez, C., J. Todd, P. Alonso, N. Francis, S. Lulat, S. Ceesay, B. M'Boge, and B. Greenwood (1994). The effects of iron supplementation during pregnancy, given by traditional birth attendants, on the prevalence of anaemia and malaria. *Transactions of the Royal Society of Tropical Medicine and Hygiene 88*, 590–593.

Mensforth, R., C. Lovejoy, J. Lallo, and G. Armelagos (1978). The role of constitutional factors, diet, and infectious disease in the etiology of porotic hyperostosis and periosteal reactions in prehistoric infants and children. *Medical Anthropology 2*(1), 1–59.

Mensforth, R. P. (1985). Relative tibia long bone growth in the Libben and BT-5 prehistoric skeletal samples. *American Journal of Physical Anthropology 68*, 247–262.

Merchant, V. and D. Ubelaker (1977). Skeletal growth in protohistoric Arikara. *American Journal Physical Anthropology 46*, 61–72.

Middlemiss, J. and B. Raper (1966). Skeletal changes in the haemoglobinopathies. *Journal of Bone and Joint Surgery 48B*(4), 693–702.

Miles, A. (1963). Dentition in the estimation of age. *Journal of Dental Research 42*(1), 255–263.

Miles, A. (1978). Teeth as an indicator of age in man. In K. Joysey (Ed.), *Development, Function and Evolution of Teeth*, pp. 454–464. London, UK: Academic Press.

Miles, J. (1997). *Infectious Diseases: Colonising the Pacific?* Dunedin, New Zealand: Otago University Press.

Mills, A. (1955). The incidence of yaws in the New Hebrides. *Transactions of the Royal Society of Tropical Medicine and Hygiene 49*(1), 58–61.

Mims, C. (1993). *Medical Microbiology*. St. Louis, MO: Mosby.

Mittler, D. and D. Van Gerven (1994). Developmental, diachronic, and demographic analysis of cribra orbitalia in the medieval Christian populations of Kulubnarti. *American Journal of Physical Anthropology 93*(3), 287–297.

Möller-Christensen, V. (1967). Evidence of leprosy in earlier peoples. In A. Sandison (Ed.), *Diseases in Antiquity: A survey of the Diseases, Injuries and Surgery of Early Populations*, pp. 295–305. Chicago, IL: Charles C Thomas.

Molnar, S. (1971). Human tooth wear, tooth function and cultural variability. *American Journal of Physical Anthropology 34*(2), 175–189.

Moorees, C., E. Fanning, and E. Hunt (1963a). Age variation of formation stages for ten permanent teeth. *Journal of Dental Research 42*, 1490–1502.

Moorees, C., E. Fanning, and E. Hunt (1963b). Formation and resorption of three deciduous teeth in children. *American Journal of Physical Anthropology 21*, 205–213.

Murray, J., A. Murray, M. Murray, and C. Murray (1978a). The biological suppression of malaria: an ecological and nutritional interrelationship of a host and two parasites. *American Journal of Clinical Nutrition 31*(8), 1363–1366.

Murray, M., A. Murray, M. Murray, and C. Murray (1978b). The adverse effect of iron repletion on the course of certain infections. *BMJ 2*, 1113–1115.

Nelson, G. C., J. H. Stone, and S. M. Fitzpatrick (2016). Adapting to Palau. In M. F. Oxenham and H. R. Buckley (Eds.), *The Routledge Handbook of Bioarchaeology in Southeast Asia and the Pacific Islands*, Chapter 23, pp. 502–526. Oxon, UK: Routledge.

Noordeen, S. and V. Pannikar (1996). Leprosy. In G. Cook (Ed.), *Manson's Tropical Diseases* (20th ed.)., pp. 1016–1044. Norfolk, UK: WB Saunders Company Ltd.

Nowell, G. (1978). An evaluation of the miles method of ageing using the Tepe Hissar dental sample. *American Journal of Physical Anthropology 49*, 271–276.

Nurse, G. (1979). Iron, the thalassaemias, and malaria. *The Lancet 2*, 938–940.

Oppenheimer, S., F. Gibson, S. Macfarlane, J. Moody, C. Harrison, A. Spencer, and O. Bunari (1986). Iron supplementation increases prevalence and effects of malaria: report on clinical studies in Papua New Guinea. *Transactions of the Royal Society of Tropical Medicine and Hygiene 80*, 603–612.

Oppenheimer, S., S. Macfarlane, J. Moody, O. Bunari, and H. R. (1986). Effect of iron prophylaxis on morbidity due to infectious disease: report on clinical studies in Papua New Guinea. *Transactions of the Royal Society of Tropical Medicine and Hygiene 80*, 596–602.

Ortner, D. (1979). Disease and mortality in the Early Bronze Age people of Bab edh-Dhra, Jordon. *American Journal of Physical Anthropology 51*, 589–598.

Ortner, D. (1996). Case report no 19: Early inflammatory change in leprosy affecting the rhinomaxillary region. *Paleopathology Newsletter Dec. 1996*, 8–11.

Ortner, D. (1999). Paleopathology: Implications for the history and evolution of tuberculosis. In G. Pálfi, O. Dutour, J. Deák, and I. Hutás (Eds.), *Tuberculosis Past and Present*, pp. 255–262. Golden Book Publisher.

Ortner, D., N. Tuross, and A. Stix (1992). New approaches to the study of disease in archaeological New World populations. *Human Biology 64*(3), 337–360.

Ortner, D. J. (1991). Theoretical and methodological issues in paleopathology. In D. J. Ortner and A. C. Aufderheide (Eds.), *Human Paleopathology: Current Syntheses and Future Options*, Chapter 1, pp. 5–11. Washington, DC: Smithsonian Institution Press.

Ortner, D. J. (1992). Skeletal paleopathology: Probabilities, possibilities, and impossibilities. In J. W. Verano and D. H. Ubelaker (Eds.), *Disease and Demography in the America*, pp. 5–14. Washington, D.C.: Smithsonian Institution Press.

Ortner, D. J. (1998). Male-female immune reactivity and its implications for interpreting evidence in human skeletal paleopathology. In A. L. Grauer and P. Stuart-Macadam (Eds.), *Sex and Gender in Paleopathological Perspective*, pp. 79–92. Cambridge, UK: Cambridge University Press.

Ortner, D. J. (2003). *Identification of Pathological Conditions in Human Skeletal Remains* (2nd ed.). Washington, D.C.: Smithsonian Institution Press.

Ortner, D. J. and M. F. Ericksen (1997). Bone changes in the human skull probably resulting from scurvy in infancy and childhood. *International Journal of Osteoarchaeology 7*, 212–220.

Ortner, D. J., E. H. Kimmerle, and M. Diez (1999). Probable evidence of scurvy in subadults from archaeological sites in Peru. *American Journal of Physical Anthropology 108*, 321–331.

Oxenham, M. F. and H. R. Buckley (2016). The population history of mainland and island Southeast Asia. In M. F. Oxenham and H. R. Buckley (Eds.), *The Routledge Handbook of Bioarchaeology in Southeast Asia and the Pacific Islands*, Chapter 2, pp. 9–23. Oxon, UK: Routledge.

Oxenham, M. F. and I. Cavill (2010). Porotic hyperostosis and cribra orbitalia: the erythropoietic response to iron-deficiency anaemia. *Anthropological Science 118*(3), 199–200.

Palkovich, A. M. (1987). Endemic disease patterns in paleopathology: Porotic hyperostosis. *American Journal of Physical Anthropology 74*(4), 527–537.

Parkinson, A. (1974). Malaria in Papua New Guinea 1973. *PNG Medical Journal 17*(1), 8–16.

Parramore, T. (1970). Non-venereal treponematosis in colonial North America. *Bulletin of the History of Medicine 44*, 571–581.

Parsonson, G. (1966). Artificial islands in Melanesia: The role of malaria in the settlement of the Southwest Pacific. *New Zealand Geographer 22*(1), 1–21.

Pasvol, G. and D. Weatherall (1976). Fetal haemoglobin and malaria. *The Lancet June 12*, 1268–1272.

Peters, W. and S. Christian (1960). Studies on the epidemiology of malaria in New Guinea: Part IV. unstable Highland malaria- the clinical picture. *Transactions of the Royal Society of Tropical Medicine and Hygiene 54*(6), 529–536.

Phenice, T. W. (1969). A newly developed visual method of sexing the os pubis. *American Journal of Physical Anthropology 30*, 297–302.

Pietrusewsky, M. (1969). An osteological study of cranial and infracranial remains from Tonga. *Records of the Auckland Institute and Museum 6*, 287–402.

Pietrusewsky, M. (1976). *Prehistoric Skeletal Remains from Papua New Guinea and the Marquesas*. Honolulu, HI: University of Hawai'i.

Pietrusewsky, M. (1994). Pacific-Asian relationships: A physical anthropological perspective. *Oceanic Linguistics 33*(2), 407–429.

Pirie, P. (1972). The effects of treponematosis and gonorrhea on the populations of the Pacific Islands. *Human Biology in Oceania 1*, 187–206.

Pollock, N. J. (1992). *These Roots Remain: Food Habits in Islands of Central and Eastern Pacific Since Western Contact*. Honolulu, Hawai'i: Institute for Polynesian Studies.

Poolsuwan, S. (1995). Malaria in prehistoric southeastern Asia. *Southeast Asian Journal of Tropical Medicine and Public Health 26*(1), 3–22.

Poulsen, J. (1987). *Early Tongan Prehistory: The Lapita period on Tongatapu and its relationships.* Terra Australis 12. Canberra, Australia: Department of Prehistory, Australian National University.

Poulson, J. (1987). *Early Tongan prehistory: the Lapita period on Tongatapu and its relationships*, Volume 12 of *Terra Australis.* Canberra, Australia: Dept. of Prehistory, Research School of Pacific Studies, Australian National University.

Powell, M. (1991). Endemic treponematosis and tuberculosis in the prehistoric Southeastern United States: biological costs of chronic endemic disease. In A. Aufderhelde (Ed.), *Human Paleopathology: Current Synthesis and Future Options.*, pp. 173–181. Washington, D.C.: Smithsonian Institution Press.

Powell, M. L. (1988). *Status and health in prehistory: a case study of the Moundville Chiefdom.* Smithsonian series in archaeological inquiry. Washington, D.C.: Smithsonian Institution Press.

Powell, M. L. (1992). Health and disease in the late prehistoric southeast. In J. W. Verano and D. H. Ubelaker (Eds.), *Disease and Demography in the America*, pp. 41–54. Washington, D.C.: Smithsonian Institution Press.

Powell, M. L. and D. C. Cook (Eds.) (2005). *The Myth of Syphilis: the Natural History of Treponematosis in North America.* Gainesville, FL: University Press of Florida.

Prestney, M. (1997). *Skeletal evidence of environmental effects on human body form.* Master of science, Department of Anatomy. Dept library.

Prior, I. (1976). Nutritional problems in Pacific Islanders. Technical report, Wellington Hospital.

Quinn, C. J. (1990). *Stable Isotopes and Diet: Indications of the Marine and Terrestrial Component in the Diets of Prehistoric Populations from New Zealand and the Pacific.* MA, Department of Anthropology.

Ragsdale, B. (1996). The irrelevance of contemporary orthopedic pathology to specimens from antiquity. *Paleopathology News 95*, 6–10.

Redvers-Newton, N. (1995). Radiocarbon dating human bone from Namu, Taumako, Solomon Islands; a comparison of two methods of bone pretreatment. *Archaeology in New Zealand 38*(2), 107–116.

Resnick, D. (1995a). *Diagnosis of Bone and Joint Disorders.* Philadelphia, PA: W.B. Saunders Company.

Resnick, D. (1995b). Hemoglobinopathies and other anemias. In D. Resnick (Ed.), *Diagnosis of Bone and Joint Disorders* (3rd ed.), Volume 4, pp. 2105–2295. Philadelphia, PA: W.B. Saunders.

Resnick, D. and G. Niwayama (1981). *Diagnosis of Bone and Joint Disorders: With Particular Emphasis on Articular Abnormalities*, Volume 1. Philadelphia, PA: W. B. Saunders.

Resnick, D. and G. Niwayama (1995). Rheumatoid arthritis and related disease. In D. Resnick (Ed.), *Diagnosis of Bone and Joint Disorders*, Volume 2, pp. 805–1150. Philadelphia, PA: W.B. Saumders.

Robb, K. (2005). *Degenerative Joint Disease in a Prehistoric Pacific Island Sample: Activity Induced Pathology at Taumako, Solomon Islands.* unpublished dissertation (BSc Hons), University of Otago.

Robb, K. (2009). *Adaptation to the Pacific Island Environment: An Investigation of Non-specific Indicators of Stress in Prehistoric Human Skeletal Remains.* PhD thesis, University of Otago.

Robb, K. F., H. R. Buckley, M. Spriggs, and S. Bedford (2012). Cortical index of three prehistoric human Pacific Island samples. *International Journal of Osteoarchaeology 22*(3), 284–293.

Roberts, C. A. and K. Manchester (1995a). *The archaeology of disease* (3rd ed.). Ithaca, N.Y.: Cornell University Press.

Roberts, C. A. and K. Manchester (1995b). *The archaeology of disease* (2nd ed.). Ithaca, N.Y.: Cornell University Press.

Rodda, C., E. Reid, S. Johnson, J. Doery, R. Matthews, and D. Bowden (1995). Short stature in homozygous b-thalassaemia is due to disproportionate truncal shortening. *Clinical Endocrinology 42*, 587–592.

Rogers, J. and T. Waldron (1995). *A Field Guide to Joint Disease in Archaeology.* Chichester,UK: John Wiley and Sons.

Rosen, L., D. Gublek, and P. Bennett (1981). Epidemic polyarthritis (ross river) virus infection in the Cook Islands. *The American Journal of Tropical Medicine and Hygiene 30*(6), 1294–1302.

Rothschild, B. and G. Heathcote (1993). Characterization of the skeletal manifestations of the treponemal disease yaws as a population phenomenon. *Clinical Infectious Diseases 17*, 198–203.

Rothschild, B. and G. Heathcote (1995). Characterization of gout in a skeletal population sample: Presumptive diagnosis in a Micronesian population. *American Journal of Physical Anthropology 98*(4), 519–525.

Rothschild, B. and C. Rothschild (1995). Treponemal disease revisited: Skeletal discriminators for yaws, bejel, and venereal syphilis. *Clinical Infectious Diseases 20*, 1402–1408.

Rothschild, B. and C. Rothschild (1996). Treponemal disease in the New World. *Current Anthropology 37*(3), 555–561.

Roy, K. (1989). *A Study of Early Marianas Islanders- The Skeletons Under Their Skins*. Unpublished master of arts thesis, Department of Anthropology, University of Otago. yaws.

Russell, D. (1960). Leprosy in Papua and New Guinea. *PNG Medical Journal 4*(2), 49–54.

Ryan, A. F. (1997). Iron-deficiency anemia in infant development: Implications for growth cognitive development, resistance to infection and iron supplementation. *Yearbook of Physical Anthropology 40*, 25–62.

Sand, C. and S. Bedford (2011). Lapita, archaeological signature of the first Austronesian settlement of the Southwest Pacific. In C. Sand and S. Bedford (Eds.), *Lapita: Oceanic Ancestors*. Paris, France: Somogy Editions d'Art.

Santos, R. and C. Coimbra (1999). Hardships of contact: Enamel hypoplasias in Tupi-Monde Amerindians from the Brazilian Amazonia. *American Journal of Physical Anthropology 109*, 111–127.

Saunders, S., C. DeVito, A. Herring, R. Southern, and R. Hoppa (1993). Accuracy tests of tooth formation age estimations for human skeletal remains. *American Journal of Physical Anthropology 92*(2), 173–188.

Saunders, S. and R. Hoppa (1993). Growth deficit in survivors and non-survivors: biological mortality bias in subadult skeletal samples. *Yearbook of Physical Anthropology 36*, 127–151.

Saunders, S. R. (1992). Subadult skeletons and growth related studies. In S. R. Saunders and A. Katzenberg (Eds.), *Skeletal Biology of Past Peoples: Research Methods*, pp. 1–20. New York, NY: Wiley-Liss, Inc.

Saunders, S. R. and A. M. Katzenberg (Eds.) (1992). *Skeletal Biology of Past Peoples: Research Methods*. New York, NY: Wiley-Liss, Inc.

Say, L., D. Chou, A. Gemmill, O. Tunçalp, A.-B. Moller, J. Daniels, A. M. Gülmezoglu, M. Temmerman, and L. Alkema (2014). Global causes of maternal death: a WHO systematic analysis. *The Lancet Global Health 2*(6), e323–e333.

Sayers, E. (1943). *Malaria in the South Pacific*. Wellington, NZ: E. V. Paul.

Schour, I. and M. Massler (1941). The development of the human dentition. *The Journal of the American Dental Association 20*, 379–427.

Schultz, M. (2001). Paleohistopathology of bone: A new approach to the study of ancient diseases. *American Journal of Physical Anthropology 116*(S33), 106–147.

Sciulli, P. (1992). Estimating age of occurence of enamel defects in deciduous teeth. *Journal of Paleopathology 2*, 31–39.

Scott, E. C. (1979). Dental wear scoring technique. *American Journal of Physical Anthropology 51*(2), 213–217.

Scott, R. and H. Buckley (2010). Biocultural interpretations of trauma in two prehistoric Pacific Island populations from Papua New Guinea and the Solomon Islands. *American Journal of Physical Anthropology 142*(4), 509–518.

Scott, R. M. (2009). *Skeletal Trauma in Prehistoric Oceania*. MSc, University of Otago, Dunedin, New Zealand.

Scott, R. M. and H. R. Buckley (2014). Exploring prehistoric violence in Tonga. *Current Anthropology 55*(3), 335–347.

Scrimshaw, N. (1981). Significance of the interactions of nutrition and infection in children. In R. Suskind (Ed.), *Textbook of Pediatric Nutrition*, pp. 229–240. New York, NY: Raven Press.

Scrimshaw, N., C. Taylor, and J. Gordon (1968). *Interactions of Nutrition and Infection*. Monograph series No. 57. Geneva, Switzerland: World Health Organization.

Sengupta, S. (1983). Musculoskeletal lesions in yaws. *Clinical Orthopedics 192*, 193–198.

Shaw, B. J., H. R. Buckley, G. Summerhayes, C. Stirling, and M. Reid (2011). Prehistoric migration at Nebira, south coast of Papua New Guinea: New insights into interaction using isotope and trace element concentration analyses. *Journal of Anthropological Archaeology 30*(3), 344–358.

Shaw, B. J., G. R. Summerhayes, H. R. Buckley, and J. A. Baker (2009). The use of strontium isotopes as an indicator of migration in human and pig Lapita populations in the Bismarck Archipelago, Papua New Guinea. *Journal of Archaeological Science 36*(4), 1079–1091.

Sheppard, P. (2011). Lapita colonization across the Near/Remote Oceania boundary. *Current Anthropology 52*(6), 799–840.

Sheppard, P. and R. Walter (2006). A revised model of Solomon Islands culture history. *Journal of the Polynesian Society 115*, 47–76.

Silman, A. (1991). Is rheumatoid arthritis an infectious disease? *British Journal of Medicine 303*, 200–201.

Silverman, F. (1985). *Caffey's Pediatric X-Ray Diagnosis: An Integrated Imaging Approach* (8th ed.), Volume 1. Chicago, IL: Year Book Medical Publishers, Inc.

Simpson, A. (1979). *An assessment of health in the prehistoric inhabitants of New Zealand and the Chatham Islands*. Unpublished bmedl. sci. thesis, University of Otago. Dept library.

Skinner, M. and A. Goodman (1992). Anthropological uses of developmental defects of enamel. In M. Katzenberg (Ed.), *Skeletal Biology of Past Peoples: Research Methods*, pp. 153–174. New York, NY: Wiley-Liss.

Smillie, A., J. Rodda, and K. Kawasaki (1986). Some aspects of hereditary defects of dental enamel, including some observations on pigmented Polynesian enamel. *New Zealand Dental Journal 82*, 122–125.

Snow, C. E. (1974). *Early Hawaiians: An initial study of skeletal remains from Mokapu, Oahu*. Lexington, KY: The University Press of Kentucky.

Spencer, M. (1971). Bionomics of vector *Anophelines* in Papuan Islands. *PNG. Med. J. 14*(1), 14–23. yes.

Spencer, T., M. Spencer, and D. Venters (1974). Malaria vectors in Papua New Guinea. *PNG Medical Journal 17*, 22–30.

Spennemann, D. H. R. (1987). Availability of shellfish resources on prehistoric Tongatapu, Tonga: Effects of human predation and changing environment. *Archaeology in Oceania 22*(3), 81–96.

Spriggs, M. (1984). The Lapita Cultural Complex: Origins, distribution, contemporaries and successors. *Journal of Pacific History 19*, 202.

Spriggs, M. (1997). *The Island Melanesians*. The Peoples of South-East Asia and the Pacific. Cambridge, MA: Blackwell Publishers.

Stahl, A. (1989). Plant-food processing: implications for dietary quality. In D. R. Harris and G. Hillman (Eds.), *Foraging and Farming: The Evolution of Plant Exploitation*, pp. 171–194. London,UK: Unwin Hyman.

Stantis, C. (2015). *Diet and Migration in Prehistoric Remote Oceania*. Phd thesis, University of Otago, Dunedin, New Zealand.

Stantis, C., R. L. Kinaston, M. P. Richards, J. M. Davidson, and H. R. Buckley (2015). Assessing human diet and movement in the Tongan maritime chiefdom using isotopic analyses. *PLoS ONE 10*(3), e0123156.

Stantis, C., N. Tayles, R. L. Kinaston, C. Cameron, P. D. Nunn, M. P. Richards, and H. R. Buckley (2016). Diet and subsistence in Remote Oceania: an analysis using oral indicators of diet. In M. F. Oxenham and H. R. Buckley (Eds.), *The Routledge Handbook of Bioarchaeology in Southeast Asia and the Pacific Islands*, Chapter 26, pp. 569–598. Oxon, UK: Routledge.

Steckel, R., J. Rose, C. S. Larsen, and P. L. Walker (2002). Skeletal health in the Western Hemisphere from 4000 B.C. to the present. *Evolutionary Anthropology 11*, 142–155.

Steinbock, R. T. (1976). *Paleopathological diagnosis and interpretation: Bone diseases in ancient human populations*. Springfield, IL: Thomas.

Stewart, T. and A. Spoehr (1952). Evidence on the paleopathology of yaws. *Bulletin of the History of Medicine 26*, 538–553.

Stiles, D. (2006). *Vertebral Joint Disease in a Prehistoric Sample*. MSc dissertation, University of Otago, Dunedin, New Zealand.

Stini, W. A. (1969). Nutritional stress and growth: Sex differences in adaptive response. *American Journal of Physical Anthropology 31*, 123–147.

Stini, W. A. (1985). Growth rates and sexual dimorphism in evolutionary perspective. In R. Gilbert and J. Mielke (Eds.), *The Analysis of Prehistoric Diets*, pp. 191–226. Orlando, FL: Academic Press.

Stinson, S. (1985). Sex differences in environmental sensitivity during growth and development. *Yearbook of Physical Anthropology 28*, 123–147.

Stirland, A. (1994). The origin of syphilis in Europe: Before or after 1493? *International Journal of Osteoarchaeology 4*(1), 53–54.

Stodder, A. L. and D. L. Martin (1992). Health and disease in the Southwest before and after Spanish Contact. In J. W. Verano and D. H. Ubelaker (Eds.), *Disease and Demography in the America*, pp. 55–74. Washington, D.C.: Smithsonian Institution Press.

Stodder, A. L. W. (1997). Subadult stress, morbidity, and longevity in latte period populations on Guam, Mariana Islands. *American Journal of Physical Anthropology 104*(3), 363–380.

Stuart-Macadam, P. (1985). Porotic hyperostosis: Representative of a childhood condition. *American Journal of Physical Anthropology 66*(4), 391–398.

Stuart-Macadam, P. (1987). Porotic hyperostosis: New evidence to support the anemia theory. *American Journal of Physical Anthropology 74*(4), 521–526.

Stuart-Macadam, P. (1989a). Nutritional deficiency diseases: A survey of scurvy, rickets, and iron-deficiency anemia. In M. Y. Iscan and K. A. Kennedy (Eds.), *Reconstruction of Life from the Skeleton*. New York, NY: Alan R. Liss.

Stuart-Macadam, P. (1989b). Porotic hyperostosis: Relationship between orbital and vault lesions. *American Journal of Physical Anthropology 80*, 187–193.

Stuart-Macadam, P. (1992). Anemias in past human populations. In P. Stuart-Macadam and S. Kent (Eds.), *Diet, Demography, and Disease*, pp. 151–173. New York: Aldine de Gruyter.

Stuart-Macadam, P. and S. Kent (1992). *Diet, Demography and Disease: Changing Perspectives on Anemia*. New York City: Aldine de Gruyter.

Sturt, C. (1833). *Two Expeditions into the Interior of Southern Australia during the years 1828, 1829, 1830 and 1831*. London, UK: Smith, Elder and Co.

Summerhayes, G. (2001a). Defining the chronology of Lapita in the Bismarck Archipelago. In A. G. Clark, A. Anderson, and T. Vunidilo (Eds.), *The archaeology of Lapita dispersal in Oceania*, Chapter 3, pp. 25–38. Canberra, Australia: Pandanus Books.

Summerhayes, G. (2001b). Lapita in the Far West: recent developments. *Archaeology in Oceania 36*, 53–63.

Sundick, R. (1978). Human skeletal growth and age determination. *Homo 29*, 228–249.

Swärdstedt, T. (1966). *Odontological aspects of a medieval population in the province of Jüntland, mid-Sweden*. Ph. D. thesis, Stockholm University.

Sweeney, E., J. Cabrera, J. Urrutia, and L. Mata (1969). Factors associated with linear hypoplasia of human deciduous incisors. *Journal of Dental Research 48*, 1275–1279.

Sweeney, E., A. Saffir, and R. De Leon (1971). Linear hypoplasia of diciduous incisor teeth in malnourished children. *American Journal of Clinical Nutrition 24*, 29–31.

Tayles, N. (1996). Anemia, genetic diseases, and malaria in prehistoric mainland Southeast Asia. *American Journal of Physical Anthropology 101*, 11–27.

Tayles, N. (1999). *The excavation of Khok Phanom Di: A prehistoric site in Central Thailand*, Volume V of *Reports of the Research Committee of the Society of Antiquaries of London No. LXI*. Dunedin, New Zealand: The Society of Antiquaries of London.

Temple, D. H. and A. H. Goodman (2014). Bioarcheology has a "health" problem: Conceptualizing "stress" and "health" in bioarcheological research. *American Journal of Physical Anthropology 155*, 186–191.

Terrell, J. (1986). *Prehistory in the Pacific Islands: A study of variation in language, customs, and human biology*. Oxon, UK: Cambridge University Press.

Tesh, R., R. NcClean, D. Shroyer, C. Calisher, and L. Rosen (1981). Ross river virus (Togaviridae: Alphavirus) infection (epidemic polyarthritis) in American Samoa. *Transactions of the Royal Society of Tropical Medicine and Hygiene 75*(3), 426–431.

Than, T., T. Toe, and A. Batu (1975). Inhibition of iron by coconut milk. *American Journal of Clinical Nutrition 28*, 1348.

Thomason, J., C. Jenkins, and P. Heywood (1986). Child feeding patterns amongst the Au of the West Sepik, Papua New Guinea. *Journal of Tropical Pediatrics 32*, 90–92.

Thurnham, D. (1986). Nutrient deficiencies and malaria: a curse or a blessing? In N. Jenkins (Ed.), *Proceedings of the XIII International Congress of Nutrition 1985*, pp. 129–131. London, UK: John Libbey.

Tilley, L. and M. F. Oxenham (2011). Survival against the odds: Modeling the social implications of care provision to seriously disabled individuals. *International Journal of Paleopathology 1*(1), 35–42.

Todd, T. W. (1920). Age changes in the pubic bone. I. the male white pubis. *American Journal of Physical Anthropology 3*(3), 285–334.

Tonkin, S. (1966). Growth survey of Maori infants and the effect of iron fortified milk powder on their anaemia. *New Zealand Medical Journal 65*, 942–946.

Trembly, D. (1995). On the antiquity of leprosy in Western Micronesia. *International Journal of Osteoarchaeology 5*, 377–384.

Trembly, D. (1996). Treponematosis in pre-Spanish western Micronesia. *International Journal of Osteoarchaeology 6*, 397–402.

Trembly, D. (1997). A Germ's journey to isolated islands. *International Journal of Osteoarchaeology 7*, 621–624.

Trent, R. (1991). a-thalassaemia in the Polynesians. *Today's Life Science March*, 24–28.

Trotter, M. (1970). Estimation of stature from intact long limb bones. In T. Stewart (Ed.), *Personal Identification in Mass Disasters*, pp. 71–83. Washington, D.C.: National Museum of Natural History, Smithsonian Institution.

Ubelaker, D. H. (1978). *Human Skeletal Remains: Excavation, Analysis and Interpretation* (1st ed.). Washington, D.C.: Smithsonian Institute Press.

Ubelaker, D. H. (1979). Skeletal evidence for kneeling in prehistoric Ecuador. *American Journal of Physical Anthropology 51*, 679–686.

Ubelaker, D. H. and L. G. Grant (1989). Human skeletal remains: Preservation or reburial? *Yearbook of Physical Anthropology 32*, 249–287.

van der Hoeven, J. (1958). Taboos for pregnant women, lactating mothers and infants on the North coast of Netherlands New Guinea. *Tropical and Geographical Medicine 10*, 71–76.

van Der Sluis, I. (1969). *The Treponematosis of Tahiti: Its origin and evolution, a study of sources.* Amsterdam, The Netherlands: Peco.

van Dijk, W. and A. Parkinson (1974). Epidemiology of malaria in New Guinea. *PNG Med. J. 17*(1), 17–21.

Vathakanon, R. and B. Chavalittamrong (1978). Height and weight of Bangkok children; 2. height and weight of Bangkok children in comparison to the previous studies. *Journal of the Medical Association of Thailand 61*(Supplement 2), 30–41.

Wadsworth, G. (1992). Physiological, pathological and dietary influences on the hemoglobin level. In P. Stuart-Macadam and S. Kent (Eds.), *Diet, Demography, and Disease: Changing Perspectives on Anemia*, pp. 63–104. New York, NY: Aldine De Gruyter.

Waldron, T. (1994). *Counting the dead: the epidemiology of skeletal populations.* Chichester, England: John Wiley & Sons.

Waldron, T. (2007). *Palaeoepidemiology: The Measure of Disease in the Human Past.* Walnut Creek, CA: Left Coast Press, Inc.

Waldron, T. (2009). *Palaeopathology.* Manuals in Archaeology. Cambridge, MA: Cambridge University Press.

Walker, P. (1986). Porotic hyperostosis in a marine-dependent California Indian population. *American Journal of Physical Anthropology 69*(3), 345–354.

Walker, P. L., R. R. Bathurst, R. Richman, T. Gjerdrum, and V. A. Andrushko (2009). The causes of porotic hyperostosis and cribra orbitalia: A reappraisal of the iron-deficiency-anemia hypothesis. *American Journal of Physical Anthropology 139*(2), 109–125.

Wall, C. E. (1991). Evidence of weaning stress and catch-up growth in the long bones of a Central California Amerindian sample. *Annals of Human Biology 18*(1), 9–22.

Walter, R., C. Jacomb, and S. Bowron-Muth (2010). Colonisation, mobility and exchange in New Zealand prehistory. *Antiquity 84*, 497–513.

Wark, L. and L. A. Malcolm (1969). Growth and development of the Lumi child in the Sepik district of New Guinea. *The Medical Journal of Australia 2*, 129–136.

Washburn, S. L. (1948). Sex differences in the pubic bone. *American Journal of Physical Anthropology 6*, 199–208.

Waters, A., D. Higgins, and T. McCutchan (1991). *Plasmodium falciparum* appears to have arisen as a result of lateral transfer between avian and human hosts. *Proceedings of the National Academy of Science 88*(2), 3140–3144.

Weatherall, D. (1997). Thalassaemia and malaria, revisited. *Annals of Tropical Medicine and Parasitology 91*(7), 885–890.

Weatherall, D. and J. Clegg (1981). *The Thalassaemia Syndromes* (3rd ed.). Oxford, UK: Blackwell Scientific Publications.

Weinberg, E. (1974). Iron and susceptibility to infectious disease. *Science 184*, 952–956.

Weinberg, E. (1992). Iron withholding in prevention of disease. In P. Stuart-Macadam and S. Kent (Eds.), *Diet, Demography, and Disease: Changing Perspectives on Anemia*, pp. 105–150. New York, NY: Aldine De Gruyter.

Weinberg, E. D. (1984). Iron withholding: a defense against infection and neoplasia. *Physiological Review 64*(1), 65–102.

Weiss, K. M. (1973). Demographic models for anthropology. In *Society for American Archaeologists*, Volume 38 American Antiquity Vol. 2 Part 2, Washington, DC. SAA.

White, N. (1996). Malaria. In G. Cook (Ed.), *Manson's Tropical Diseases* (20th ed.)., pp. 1087–1164. London, UK: WB Saunders Company Ltd.

White, T. D. and P. A. Folkens (1991). *Human Osteology*. San Diego, CA: Academic Press.

Whitehead, N., S. Devine, and B. F. Leach (1986). Electron spin resonance dating of human teeth from the Namu burial ground, Taumako, Solomon Islands. *New Zealand Journal of Geology and Geophysics 29*, 359–361.

Wickes, I. (1953). A history of infant feeding: Part 1. *Archives of Diseases in Childhood 28*, 151–158.

Wigley, R. (1987). Rheumatic complaints in Tokelau. *Rheumatology International 7*(2), 53–59.

Williams, T., K. Maitland, S. Bennett, M. Ganczakowski, T. Peto, C. Newbold, D. Bowden, D. Weatherall, and J. Clegg (1996). High incidence of malaria in a-thalassaemic children. *Nature 383*(6600), 522–525.

Williams, T., K. Maitland, L. Phelps, S. Bennett, T. Peto, J. Viji, R. Timothy, J. Clegg, D. Weatherall, and D. Bowden (1997). *Plasmodium vivax*: a cause of malnutrition in young children. *Quarterly Journal of Medicine 90*, 751–757.

Willmott, J. (1969a). *Nutrition in the British Solomon Islands Protectorate*. Suva, Fiji: South Pacific Health Service.

Willmott, J. (1969b). Nutrition in the British Solomon Islands protectorate. Technical report, South Pacific Health Service.

Wolff, R. (1965). Meanings of food. *Tropical and Geographical Medicine 1*, 45–51.

Wood, J. W., G. R. Milner, H. C. Harpending, and K. M. Weiss (1992). The osteological paradox: Problems of inferring prehistoric health from skeletal samples. *Current Anthropology 33*(4), 343–370.

Wood, L. (1996). Frequency and chronological distribution of linear enamel hypoplasia in a North Amercian colonial skeletal sample. *American Journal of Physical Anthropology 100*, 247–259.

Woodruff, A. and S. Wright (1987). *A Synopsis of Infectious and Tropical Diseases* (3rd ed.). Bristol, UK: Wright.

Woods, R. and B. Rothschild (1988). Population analysis of symmetrical erosive arthritis in Ohio Woodland Indians (1200 years ago). *The Journal of Rheumatology 15*(2), 1258–1263.

Worboys, M. (1994). From miasmas to germs: Malaria 1850–1879. *Parassitologia 36*, 61–68.

Wright, L. E. (1997). Intertooth patterns of hypoplasia expression: implications for childhood health in the Classic Maya collapse. *American Journal of Physical Anthropology 102*(2), 233–247.

Wyler, D. (1982). Malaria: host-pathogen pathology. *Reviews of Infectious Diseases 4*(4), 785–797.

Yenchitsomanus, P., K. Summers, K. Bhatia, J. Cattani, and P. Board (1985). Extremely high frequencies of a-globin gene deletion in Madang and on Kar Kar Island, Papua New Guinea. *American Journal of Human Genetics 37*, 778–784.

Yenchitsomanus, P., K. Summers, C. Chockkalingam, and P. Board (1986). Characterization of G6PD deficiency and thalassaemia in Papua New Guinea. *PNG Medical Journal 29*, 53–58.

Zigas, V. and V. Morea (1974). Pre-operational malariometric survey Kairuku subdistrict. *PNG. Medical Journal 17*(1), 93–98.

Milton Keynes UK
Ingram Content Group UK Ltd.
UKHW052308201123
432927UK00008B/101

9 781407 314433